A New Kind of Computational Biology

Parimal Pal Chaudhuri · Soumyabrata Ghosh
Adip Dutta · Somshubhro Pal Choudhury

A New Kind
of Computational Biology

Cellular Automata Based Models
for Genomics and Proteomics

 Springer

Parimal Pal Chaudhuri
CARLBio Pvt. Ltd.
Kolkata, West Bengal, India

Adip Dutta
CARLBio Pvt. Ltd.
Kolkata, West Bengal, India

Soumyabrata Ghosh
CARLBio Pvt. Ltd.
Kolkata, West Bengal, India

Somshubhro Pal Choudhury
CARLBio Pvt. Ltd.
Kolkata, West Bengal, India

ISBN 978-981-13-4658-3   ISBN 978-981-13-1639-5   (eBook)
https://doi.org/10.1007/978-981-13-1639-5

This Springer imprint is published by the registered company Springer Nature Singapore Pte Ltd.
The registered company address is: 152 Beach Road, #21-01/04 Gateway East, Singapore 189721, Singapore

Prologue

Throughout my professional career, since the early 1960s, I had been associated with computers and computer science. However, by the end of the last century, I had the realization that I will be able to derive higher level of academic satisfaction from the study of life sciences. I also developed the conviction that Mother Nature has provided us with a voluminous encyclopedia from which we can learn new things every day. The knowledge gathered from the study of nature can be utilized for solving the problems in different fields of science and technology. Let me illustrate the point with a few examples. If we study the invention process of airplane by the Wright brothers, we can notice that during the period of 1899–1902, they spent considerable time to study birds in flight. From this study, they formulated the basic technology of airplane design. The next example comes from the recent history of the development of the currently popular gene-editing technology of CRISPR/Cas9. Its origin dates back to mid-1980s from the study of virus–host interaction observed by the scientists. For the last illustration, I fall back on the discipline of 'computer science and technology'. A computer system employs a class of memory referred to as 'content addressable memory' that retrieves data from a part of its content. Real-life application of this class of memory can be observed in human behaviour as illustrated next. An ex-student whom I have taught long back asked me 'how are you sir?' I forgot his name, but his facial features are stored in my memory along with his name. So my memory is searched by such facial features to retrieve his name. A host of different processing techniques employed in a computer system and computing models have been developed from the study of behavioural pattern of human being (e.g. neural network model mimicking the human nervous system and brain).

Scientific development of the last few centuries concentrated heavily on the study of inanimate objects around us. Human civilization has flourished from the advances of basic sciences—physics, chemistry, mathematics and the technology developed out of such scientific advances. The study of life sciences, no doubt, attracted considerable attention. However, its advancement got restricted since the technology was not mature enough to study the 'living cells' of an organism.

Initiation of the 'human genome project' in the last decade of the last century along with various other projects for the study of living organisms has revolutionized the study of life sciences. The technology for in vivo study of living cells has improved significantly. Latest in this series is the cryo-electron microscopy. The 2017 Noble Prize in chemistry was awarded to three inventors for the development of this technology for biomolecule imaging. All such advances lead to the development of different voluminous databases storing wet lab experimental results of structure and behaviour of biomolecules—DNA, RNA and protein. The motivation behind such an intensive study of biomolecules with a major focus on drug development can be ascertained from the following fact. Scientific and research establishments have become aware of the alarming problem posed by life-threatening diseases like cancer, diabetes, infection due to antibiotic-resistant bacteria affecting human society around the globe. So drug development got a new impetus to fight against such deadly diseases for the survival of human society. The commercial aspect of the discovery of new drugs can also be found to be a strong motivator for the study of life sciences.

From the analysis of all such developments, my conviction got deep-rooted on two issues—life sciences will be a melting pot of different disciplines—physics, chemistry, biology, mathematics, computer sciences, etc., and the twenty-first century will be the century of life sciences. All the other disciplines of science and technology will derive immense benefit from the knowledge generated from the study of life sciences. In this background of my personal conviction, the motivation to write this book is a logical culmination of the journey of my academic/research career I started at the Indian Institute of Technology (IIT) in the mid-1970s.

Subsequent to launching the undergraduate programme in computer science and engineering at IIT Kharagpur in the late 1970s, the emphasis was shifted to postgraduate and research programmes. Building a computational model for physical systems was identified as one of the major research thrust of my research group. John von Neumann, one of the father figures in the field of computers, proposed the discrete dynamical system of Cellular Automata (CA) in the 1950s as a modelling tool for physical systems. Interaction of a few gifted minds, John v Neumann, Stanislaw Ulam, and Arther Burks, enriched the field of CA for designing an abstract model of self-reproduction in biology. By 1970, John Conway proposed 'the Game of Life' that exhibited a complex behaviour out of simple CA rules. Another major contributor in the field of CA was S. Wolfram whose comprehensive study resulted in the foundation of simple three-neighbourhood, two-state per cell CA rules exhibiting complex evolution patterns. His fundamental research in the field culminated in the publication of his book entitled 'A New Kind of Science' in 2002. During the period from the 1960s to 2018, research of a large number of authors has enriched the field of CA.

In general, it can be observed that a set of simple subsystems constitute a physical system that poses a mystery when considered as a whole. The interaction among simple processes of each subsystem exhibits a behaviour that is too complex to be visualized by considering a subset in isolation. For example, each neuron cell is simple and limited by its capability. But their interactions make the neural system exhibit a complex behaviour that is in no way comparable to the capabilities of an individual neuron. A wide variety of illustrations of interacting simple homogeneous subsystems exhibiting a complex behaviour can be observed in nature. This feature of a physical system can be observed in the evolution of CA. The homogeneous structure of CA consists of a set of simple cells which interact with neighbouring cells employing simple CA rules. The CA evolution over time can display complex patterns that cannot be ascertained from the initial state of the cells unless the CA rules are restricted to fourteen rules which employ only XOR/XNOR logical functions. This class of CA is referred to as 'additive CA', while the other class is noted as nonlinear CA that employs nonlinear CA rules implementing AND/OR logic functions.

From the early 1980s, my research group started the study of CA evolution patterns employing additive CA rules. The matrix algebraic tools provided the theoretical framework to analyse wide varieties of CA behaviour. Around the mid-1980s, a sponsored research project was funded by the Department of Electronics, Government of India, to this research group. The project's goal was to develop very large-scale integrated (VLSI) systems with a specific emphasis on testable VLSI chip design. The research group concentrated on the theory and applications of CA for testable VLSI chip design. By the mid-1990s, nine research students of this research group submitted their Ph.D. thesis on this area. All these research works were compiled in the book entitled 'Additive Cellular Automata—Theory and Applications' and published in 1997 by IEEE Computer Society Press, NJ, USA (ISBN 0-8186-7717-1). By the end of 1996, Intel Research Labs, Oregon, offered me a visiting research professor assignment to explore the possibility of applying CA technology for testable asynchronous circuit design. On completion of this assignment, since the end of 1997, I shifted to Bengal Engineering and Science University, my alma mater. I continued to build the CA research group in this university with sponsored research projects funded by three US multinationals—Intel, Nortel and Fujitsu. All these projects were focussed on the application of CA models in diverse application domains including VLSI design. The focus of this new CA research group was shifted to the study of nonlinear CA dynamics since it was observed that additive CA has certain limitations for modelling real-life physical systems. A group of five research students concentrated on this area and submitted their thesis to receive Ph.D. degree. By the beginning of the current century, I had the following realizations from my experience of CA research for more than two decades.

1. (a) Wide varieties of publications were reported in the field of CA to model different physical systems. CA-based model has been explored for the study of complex nonlinear systems—from the study of ant colonies to atmospheric research, aerodynamics to biological models and ecology, etc. CA evolutions displayed interesting patterns that are identified to represent the complex behaviour of different physical systems.

(b) Many research groups designed multi-dimensional CA with complex multiple states per cell. However, from our study, we observed that higher complexity of CA rules makes the analysis of CA evolution pattern extremely difficult. Mapping of the patterns/parameters extracted from CA evolution to the features of the physical system being modelled is an upheaval task. For a specific initial state, it might be possible to model some features of the physical system. However, with a small perturbation of the initial state, the evolved CA patterns may differ significantly, providing no clue to CA rule design and modelling the physical environment prevailing for the system. Such a frustrating experience might have forced many researchers to give up the CA model building methodology for applications of their interest.

(c) Based on such analysis, we concluded that in order to achieve desired success, a new approach is a necessity for design of CA rules for modelling a physical system.

2. Another major shift has dawned in my thinking by the end of the last century. Rather than building computational models for physical systems studied under different branches of science and engineering, it will be highly rewarding to build successful CA model in the field of life sciences. In this background, I concluded that the major focus of CA research should be directed towards building CA model to study life sciences. The underlying reason for such conclusion stems from the following observation. One-dimensional structure of simple CA with a limited set of rules suits ideally to model linear strand/chain of two nucleic acids (DNA and RNA) and protein—the biomolecules having limited number of building blocks.

3. By the end of the last century, different international projects were initiated to address various problems of life sciences including the urgent need to design new drugs to fight wide varieties of diseases threatening the human community worldwide.

4. Molecular dynamics simulation is the de facto standard computational biology tool that demands increasing computing power to handle biomolecules of larger size. Due consideration to prevailing environmental factors add further complexity to this computational model. However, biomolecules, in general, are built out of a limited set of building blocks. Rather than building computational biology tools based on the molecular structure of a biomolecule, it is worthwhile to explore building alternative computational biology tools based on the information content of the constituent building blocks (sugar-phosphate molecules, nucleotide bases and amino acid residues) of the three basic biomolecules —DNA, RNA and protein.

In the background of my realization summarized above, I started my next journey on CA research from the first decade of the current century with the establishment of a research laboratory named as Cellular Automata Research Lab (CARL). Two research scholars, Soumyabrata Ghosh and Nirmalya S. Maiti, joined me in this journey. Soumya is a student of biology, while Nirmalya worked with me for several years to develop CA model for innovative applications. From 2008 onwards, this research effort to develop CA model for life sciences was supported by Alumnus Software Ltd. a Kolkata-based software company. Arindam Mukherjee (an ex-student of IIT Kharagpur and CEO of Alumnus Software Ltd.) provided all the necessary support to the CARL team from 2008 onwards. As a team, we were new to this discipline. However, we set our ambitious goal to develop a technology that should have industrial applications. Our emphasis was to invent a new methodology to develop CA model for biomolecules. This technology should be able to add value to drug design methodology to reduce the time and cost for design of new drugs. The energy and drive of two young researchers yielded some encouraging results for modelling protein structure and function. Further down the lane as we developed some more results in the field of proteomics study, we started approaching biotech industries in India and abroad with our model along with the results of real-life applications.

At this stage, a company named CARLBio Pvt. Ltd. was formed around 2012 to explore the commercial application of CA model for biomolecules. Since 2014, my two sons, Somshubhro (co-author of this book) and Santashil, provided the financial support to continue this research effort with the goal of developing CA–based in silico platform for drug design. Somshubhro took the responsibility of exploring the commercial success of the new technology we had been developing. From 2015 onwards, we were also planning to write a book on our CA research in the domain of life sciences. The publisher Springer India office approached me sometime at the end of 2016 to write a book for them based on our research. I accepted the offer from Springer to write the book entitled *A New Kind of Computational Biology*. It is a totally a new approach for design of an in silico platform based on some compatible relationships or properties of the entities that build up the system in question. Rather than building a generic approach, we focussed our attention on three biomolecules of major interest. We planned the layout of the book covering the following areas projecting the new kind of computational biology.

1. We started from the first principle of CA rule design based on the analysis of three basic molecules which play a dominant role for all living organisms—RNA, DNA and protein. The constituent building blocks of these three macromolecules are sugar-phosphate molecules, nucleotide bases and amino acids. CA rules are designed to develop an alternative representation of the information contained in the base triplet of codon and also the atomic structure of these micromolecules. We designed the simplest three-neighbourhood, two-state per cell CA rules for codons. The five-neighbourhood CA rules are designed to represent sugar-phosphate molecules and five nucleotide bases—adenine (A), guanine (G), cytosine (C), uracil (U) and its methylated version, thiamine (T). Further, three-neighbourhood CA rules are deigned for amino acid backbone and side chain. The design of CA rules is based on symmetry, presence/absence of atoms in its molecular structure and type of dominant bonds they form in a micromolecule. An individual bit in the CA rule structure does not interpret any property of the atoms themselves or the forces they are

associated with, but is a representative value of their presence or absence in associated sub-structures.

2. The bottom-up design of CA rules from the first principle leads to the design of one-dimensional CA for the biological strings of nucleic acids (DNA and RNA) and amino acid residue chain of protein. Next, we extracted the relevant parameters/patterns out of CA evolution modelling each class of biomolecules. Signal graphs are designed out of the CA model parameters. A signal graph refers to the graphical representation of CA parameter values derived out of a CA cell modelling a nucleotide base in a DNA/RNA strand or a codon in a mRNA chain. Chapter 2 reports the details of CA rule design.

3. For top-down validation of CA model, we developed signal graph analytics based on machine learning (ML) and artificial intelligence (AI) methodologies to map the CA model parameters/patterns to the physical domain features of known results published out of wet lab experiments. Fortunately, voluminous biological databases are currently available due to enormous emphasis directed to study life sciences. We had to manually extract the relevant information from these databases.

4. The in silico experimental set-up is designed based on consistent mapping of the physical domain features onto the CA model parameters/patterns. We have reported the underlying design approaches of the in silico experimental set-up for RNA, DNA and protein in Chaps. 3, 4 and 5, respectively.

5. On successful implementation of the in silico experimental set-up, algorithms are designed followed by program development in Python language. Subsequent experimentation for the prediction of results out of CA model has been reported in the case studies presented in the respective chapters.

6. Chapter 6 of the book reports a summary of all the other chapters in addition to reporting the problems we encountered in respect of the in silico experimental set-up we designed for cLife (CA-based Life) in different chapters. The chapter also covers our plan for addressing such problems in the next phase of development of cLife and its future extension.

In addition, we planned for an exhaustive study of different classes of RNA molecules followed by an extension of the cLife to model the primitive life form which, as hypothesized by some authors, gave way to the current life form with DNA–RNA–protein. We would like to explore whether a consistent cLife model is viable to support the 'RNA world' hypothesis. We strongly feel that such a model with a detailed study of RNA molecules will enable better understanding of the structure and function of different classes of RNA molecules including small-size RNAs having 20–24 bases. These small-size RNA molecules are identified to be associated with different biological functions/malfunctions. We hope the upcoming technology of cryo-electron microscopy will get matured in the coming years for wet lab analysis/imaging of the structure of such small-size molecules including their interaction with other biomolecules—RNA transcript, protein and DNA.

   As I close my presentation of the journey I undertook for preparing this book, I should acknowledge the all-out support I received from my co-authors, CARLBio team members, my students and fellow researchers in the field of cellular automata, and the help I received from my family members, in addition to the support I received from Alumnus Software Ltd. to set up CARL in 2008.

   I would like to dedicate this book to the memory of my parents who taught me the art of good living through sacrifice and contribution to society.

Parimal Pal Chaudhuri

Acknowledgements

During my cellular automata research since the mid-1980s, I have received an excellent support from a large number of colleagues and research students. Notable among them are Dr. Dipanwita Roy Chowdhury, Professor at the Indian Institute of Technology Kharagpur, and Dr. Biplab K. Sikdar, Professor at the Indian Institute of Engineering Science and Technology, Shibpur. Both the institutions—Indian Institute of Technology Kharagpur and Indian Institute of Engineering Science and Technology, Shibpur—provided me the platform to carry on my research for several decades. I wish to thank all the corporate and government agencies who have funded my cellular automata research at different points during the last three decades. I sincerely acknowledge and thank Arindam Mukherjee and Puspendu K. Samanta of Alumnus Software Ltd. of Kolkata, India, for their wholehearted support for the development of CA applications in the field of biotechnology.

I also acknowledge with sincere thanks the names of two biotechnology professionals for their insights and regular enlightening discussions during our development of CA applications in the field of biotechnology—Dr. Samik Ghosh of Systems Biology Institute, Japan, and Dr. Dinesh Palanivelu, currently with Amgen, Singapore and previously with Biocon Ltd., India.

Adip Datta, one of my co-authors, developed his keen interest in cellular automata since his high school days in the 1980s. He came forward enthusiastically as we proposed to write a book related to our CA research. Along with my other co-authors, I sincerely acknowledge his valuable contribution to this book since its planning stage. He shouldered the responsibility of organizing the layout of different chapters. He also coined the terminology 'cLife'—an alternative representation of current life form as we know of. All my co-authors join me to appreciate the support Adip and CARLBio team received from Sandhya (Adip's wife) and Aditya (Adip's son) to complete this book.

My another co-author, Dr. Soumyabrata Ghosh, is Founding Member of CARLBio Pvt. Ltd., a start-up engaged in the commercial application of CA model for biomolecules. Being a student of biology, Soumyabrata provided the critical link between the CA research team and biotechnology. Single-handedly, he steered through the application of CA model for proteomics. I also appreciate the moral support provided by Dipannita (Soumya's wife) to CARLBio team. My third co-author, Somshubhro Pal Choudhury, has been an investor and adviser to CARLBio since its inception. His valuable contributions to editing and refining the text of different chapters are highly appreciated by all other authors. Souvik Das, the youngest member of CARLBio team, developed the programs to predict the results reported in Chaps. 2, 3, and 4. All the co-authors of the book join me to acknowledge the contributions of Souvik for preparing the final manuscript. To sum it up, this book is the result of a coordinated effort of a dedicated team with a focus on the development of a new technology based on the potential of cellular automata framework.

In addition to my co-authors and members of CARLBio team, I sincerely acknowledge the contributions of Jayasree (my wife) to support my research, building the CARLBio team and finalizing this book. Being a medical doctor, she had to forego her professional assignments to support my research throughout my academic/research career. In addition to the support I got from my sons, Somshubhro and Santashil, I do appreciate the moral support I received from my two daughters-in-law, Arundhati and Yashodhara, for building CARLBio and completing this book. Finally, let me also admit the fact that the interaction with my three grandchildren—Shreyaj, Oishik and Ritoja—provided an exit route for me during the frustrating moments of my research career.

Before I close, I convey my sincere thanks to Springer India officials for inviting me to write this book on our CA research in the field of life sciences.

About the Authors

Prof. Parimal Pal Chaudhuri is chief scientific officer and founder of CARLBio. He is a renowned academician and researcher in Cellular Automata (CA) for over 30 years. He started his career at IBM in 1963 and shifted to academia in 1975 where he founded the Computer Science Department of the Indian Institute of Technology (IIT) Kharagpur. He was the head of department for several years where he pioneered industry–academia research in India and initiated the research in cellular automata. He was a visiting research professor at Intel Research Labs, Oregon, USA, where his CA research was applied in testing of asynchronous circuits of Intel chips. On his return to India, he joined his alma mater, Indian Institute of Engineering Science and Technology, where he continued his CA research focussing on the end applications in partnership with several MNCs. He held several assignments on sabbaticals—as a consultant to Cadence Design Systems and as visiting professor at the University of Illinois, Urbana–Champaign, USA. He has supervised the research of 20 Ph.D. students and authored 2 books and over 200 research publications. He is a fellow of the National Academy of Engineers (India) and was awarded the Thomas Ward memorial gold medal by the Institution of Engineers (India).

Soumyabrata Ghosh is a founder and director of research at CARLBio. He received his Ph.D. degree from the Indian Institute of Engineering Science and Technology, Shibpur, formerly Bengal Engineering and Science University (IIEST). He did his M.Sc. in biophysics and bioinformatics from the University of Calcutta. He has more than 10 years of industrial experience in the field of data analytics, machine learning and computational biology. His research interest includes predictive analytics, cognitive computing and AI.

Adip Dutta is a Senior researcher at CARLBio. For more than 33 years, he has worked in diverse domains—developing algorithms, smart devices, sensor networks, power conversion, automation, motion control and instrumentation. He was a Defence Research and Development Organization (DRDO) scientist for several years working on high-power laser systems, thermal imaging and optimal control algorithms. He worked with Raman Research Institute and Indian Space Research Organization (ISRO). He was the chief of R&D division in National Instruments and was senior architect in the advanced technology division of PortalPlayer, the designer of the first SOC used in Apple's iPOD. He was chief knowledge officer of Hamilton Research and was research consultant to NComputing Inc. USA. Since 2012, he is associated with the development of CA applications in biology pioneered by CARLBio. He completed his graduation in physics in 1982 and B.Tech. (1985) in electronics.

Somshubhro Pal Choudhury is an investor, advisor and director at CARLBio. He is a partner at Bharat Innovation Fund, a $100M fund focussed on building disruptive and innovation-led startups in India. He is a thought leader in Internet of Things (IoT) space, formerly board and executive council member of India Electronics and Semiconductor Association (IESA), co-founder and co-chair of India's premier IoT event, IoTNext, and ecosystem building initiative, IoTForum. He was previously managing director of Analog Devices India, a wholly owned subsidiary of a US-based $3.5B semiconductor MNC. He was based for almost 2 decades in silicon valley in various roles: global director of product management and marketing at NETGEAR, first employee of an early M2M/IoT company in the energy management and smart grid space, American Grid, and held several management and engineering roles at Cadence Design Systems. He holds a Wharton-University of Pennsylvania MBA and engineering degrees from the Indian Institute of Technology (IIT) Kharagpur and North Carolina State University.

Contents

Chapter 1
Computational Biology—First Principle Informatics Modelling

"I think that modern physics has definitely decided in favour of Plato. In fact the smallest units of matter are not physical objects in the ordinary sense; they are forms, ideas which can be expressed unambiguously only in mathematical language".
—Werner Heisenberg

1.1 Introduction

Human civilization has progressed from the Stone Age to Agricultural and then to the Industrial Age. We are now in the present 'information' age. Humans are basically makers of tools. From the stone age, we have strengthened our capability to survive by enhancing the quality of our tools, mostly weaponry. Once we started to convey ideas in abstract forms through drawings and language and as we improved methods of storing information, the capability to make sophisticated tools increased near exponentially by the industrial age. Humans no longer remain stuck in their regular and mundane survival activities like hunting for food. They started developing methods and means to understand the nature surrounding them. The very first step in this direction was to understand the material world.

If we read the mythologies of the ancient world, we find 'gods' making life from inanimate objects through their 'godly' methods. Life on earth as we know is too complicated to be described through material objects and hence science has progressed through building mathematical models. This book explores the search for a mathematical framework to model the basic building blocks of life. It follows from the study of underlying principles employed to model inanimate objects presented in Sect. 1.2. We have tried to understand the underlying founding principles through a brief journey of scientific developments in basic sciences in physics and chemistry. We have started from physics fundamentals to show that theoretical physics is dependent on representative method—mathematics. Using mathematical tools, we try to create a mathematical universe which subsequently is verified through experiments. This method is bottom-up rules building of a model which is

© Springer Nature Singapore Pte Ltd. 2018
P. P. Chaudhuri et al., *A New Kind of Computational Biology*,
https://doi.org/10.1007/978-981-13-1639-5_1

verified by top-down experiments, through a number of iterations we try to build up a consistent model.

Sections 1.3.1 and 1.4 present the basic building blocks of life and their constituent parts. The principles learned from the earlier sections are next employed in Sect. 1.4 to build the foundation for developing the model for biomolecules observed in life sciences. All the remaining chapters of this book elaborate this model building process and its real-life applications to address the problems of genomics and proteomics. In the process, we have proposed 'A New Kind of Computational Biology'.

1.2 Modelling the Inanimate

The underlying methodology followed throughout this book has been derived from the study of information embedded in the elements observed in nature. In Sects. 1.2, 1.2.1 and 1.2.2 of this chapter, we traverse down the history of scientific developments of the basic model of an element—its atomic and subatomic structures.

The word 'element' refers to a component of a whole. Philosophers of earlier centuries believed and tried to experiment in alchemy to analyse elements observed in nature. Even Sir Isaac Newton practised it; much of Newton's writing on alchemy has been lost in a fire in his laboratory. Perhaps this is the only accident we may not mind, so far as the loss of works of a genius is concerned. The concept of alchemy went against the term 'elements'; if an element is viewed as a building block, then it should not be able to create another building block. Mercury cannot be an element if its chemical combination with other 'elements' can produce the element gold. At the best, they can be used to make alloys. Prior to twentieth century, we had no means to break open complex objects into smaller parts. Today, we employ particle accelerators to study elementary particles, set up chemical reactions to analyse the constituents of an object, or use an ultracentrifuge to separate building blocks of a macromolecule. The motivation is to visualize ourselves and our surroundings as stream of discrete objects, each built out of its constituent parts.

Seventeenth-century English scientist, John Dalton proposed his atomic theory [1]. The philosophy of the atom, the least size constituent, existed back in the ancient times and mentioned in Indian and Greek discourses, but the scientific basis was missing. Dalton had the reputation of proposing multiple groundbreaking work in varied fields of science. He made his contemporaries take a serious view of atoms. His law of multiple proportions expressed the idea that chemical combination is a discrete rather than continuous chemical phenomenon. Any amount of oxygen does not 'mix' with any amount of carbon to make carbon dioxide; there are weight ratios of each element—O_2 and C to make CO_2. He even put forward, although incorrectly, that atoms combine in only one ratio, Dalton said '… it must be presumed to be a binary one, unless some cause appear to the contrary'. His error was in thinking that elements could be represented in multiples of a reference

Reihen	Gruppo I. — R'O	Gruppo II. — RO	Gruppo III. — R'O³	Gruppo IV. RH⁴ RO'	Gruppo V. RH' R'O⁵	Gruppo VI. RH' RO'	Gruppo VII. RH R'O'	Gruppo VIII. — RO⁴
1	H=1							
2	Li=7	Bo=9,4	B=11	C=12	N=14	O=16	F=19	
3	Na=23	Mg=24	Al=27,3	Si=28	P=31	S=32	Cl=35,5	
4	K=39	Ca=40	—=44	Ti=48	V=51	Cr=52	Mn=55	Fo=56, Co=59, Ni=59, Cu=63.
5	(Cu=63)	Zn=65	—=68	—=72	As=75	So=78	Br=80	
6	Rb=85	Sr=87	?Yt=88	Zr=90	Nb=94	Mo=96	—=100	Ru=104, Rh=104, Pd=106, Ag=108.
7	(Ag=108)	Cd=112	In=113	Sn=118	Sb=122	To=125	J=127	
8	Cs=133	Ba=137	?Di=138	?Co=140	—	—	—	— — — —
9	(—)		—	—	—	—	—	
10	—	—	?Er=178	?La=180	Ta=182	W=184	—	Os=195, Ir=197, Pt=198, Au=199.
11	(Au=199)	Hg=200	Tl=204	Pb=207	Bi=208	—	—	
12	—	—	—	Th=231	—	U=240	—	— — — —

Mendeleev's periodic table

Fig. 1.1 Mendeleev's periodic table

weight. Nearly a century after Dalton's work, another chemist put the same weight ratios to form a link between the elements.

Dmitri Ivanovich Mendeleev [2] was a chemist and inventor born in Czarist Russia on 8 February 1834. He constructed his famous table of elements (Fig. 1.1) considering all the elements discovered, till that time.

Tables were made to tabulate the number of elements in various ways, but Mendeleev in his time took the proportional weights of elements and had the genius to interpret the observation. Mendeleev's systematic method of filling data in his table enabled him to predict numbers of elements to be discovered in future. He himself predicted the existence of eight elements. For example, between aluminium and eka-aluminium, there are no elements in Group III (column 4 in Fig. 1.1); his genius lies in putting the dash; the dash was filled up with Gallium later on. In Mendeleev's original version of periodic table, we could see many such dashes filled with elements to be discovered later. One of the important contributions of this table is that we could think of the elements in relation to numbers. Though still inaccurate, it proved elemental discreetness.

1.2.1 Below the Elements

The elemental discreteness, as visualized in the earlier section, was not enough for the concept of atom proposed by Dalton in early 1800. The classical atomic philosophy was—there is something indivisible making up all matters. So the elements are assumed to be made of atoms. The question arises—if a common 'something' is the building block, how different elements can have different properties, both physical and chemical?

It took more than 150 years after Dalton's theory to get down to a model where we could break down all elements into two parts: electrons and protons. There are positively charged particles named protons in a small central position of any atom (called the nucleus). The number of protons dictates the difference between elements. Thus, the characteristic of elements is encoded by the number of protons in its nucleus—a profound knowledge. It is the number of protons that gives the elements their properties. Consequently, the behaviour of elements could now be based on an intrinsically simpler thing—a number. Once the elements could be defined, so could be the molecules which are combination of atoms. This was the model developed by Ernest Rutherford around 1911.

The model of an element now became a positively charged nucleus made of protons along with a negatively charged electron cloud around the protons. All the weight of an atom is concentrated in the positively charged protons that dictated its atomic number. The number of protons decides how many electrons there will be so that an atom remains electrically neutral. Depending on the number of electrons, the negatively charged electron clouds follow definite rules specifying how many of them will occupy which orbit. These rules were understood later with the advent of quantum mechanics. This set of rules enabled us to predict the number of electrons furthest from the nucleus, which in turn decided the chemical behaviour of elements and how they will interact among themselves and with other elements. It was a beautiful picture but incomplete. Experiments did not match with the relationship:

Atomic number = Number of Protons = > Decides the Element

There are many elements that have the same chemical characteristics but with varying atomic weights. The simplest example is hydrogen which has one proton in its nucleus, but there are instances where hydrogen that was two times heavier was identified. Two protons cannot stay together in a nucleus as both have the same positive charge and hence would repel each other. This disparity of relation between the atomic weight and atomic number began to show up in early 1914. Around that time, scientists developed the technology to bombard atoms with subatomic particles. At the same time, they were generating electromagnetic radiations of different frequencies. It was observed that frequencies of light (a very small part of electromagnetic radiation) could be generated or absorbed in discrete quantities. Not only was matter composed of discrete quantities but so was energy. By 1920, we were assured that all interactions are taking place, (i) between matter, (ii) between matter and energy in discrete quantities. Further, now matter and energy were equivalent with Einstein's famous $E = mc^2$ equation postulated in 1905. Let us now go back to the mysterious missing weight mentioned at the beginning of this paragraph. The next step in this direction is to study the behaviour of subatomic particles.

1.2.2 Below the Atoms

Theoretically, there was no known mechanism in Rutherford's model for more than one proton to be together in the central nucleus. Like charges repel each other, hence protons would repel each other due to coulombic force. Rutherford himself in 1920 suggested the presence of a neutral particle within the nucleus. This was the time for great discoveries; just to name two of the rules and tools for further exploration of nature—special theory of relativity and quantum theory. Quantum theory provided us with the rules to predict the general behaviour and properties of atomic and subatomic particles. The element now turned out to be not the basic building block. Using quantum theory, scientists could calculate and predict the properties inside the nucleus. The predicted neutral particle inside the nucleus was finally discovered in 1932 by James Chadwick and was named neutron. We understood that except for the element hydrogen all nucleus contains neutrons. The heavier hydrogen mentioned in the last section had one neutron in its nucleus along with one proton. The elements with the same number of protons but different numbers of neutrons were named isotopes of the element. This single discovery now interpreted 'alchemy' in a scientific manner. Elements could be unstable, depending on the ratio of neutron to proton. We could theoretically change any element to another, some we could do practically. Without understanding the basic building blocks, we could not have understood natural radioactivity (where unstable atomic nucleus losing its energy by emitting radiation), never imagine artificial radioactivity and never would have asked how the elements could be created in the first place. We now had the basic building blocks, proton, neutron and electron.

The number of electrons and the rules specifying how they would be placed in orbits and sub-orbits (orbital) came through in the earlier discoveries. The reactivity of the elements depends on these rules. We could now very accurately predict which elements would willingly combine with another element and make a molecule generating energy; the energy would come out because the combination had energy lesser than that of their individual free states. For example, two hydrogen atoms combine with one oxygen atom to make a water molecule while releasing energy in the form of heat.

So far, the basic model of an atom and its constituents holds good until we discovered subatomic particles. Smashing complex system, in general, brings out debris. For example, if we break a mechanical clock, we get pinions and springs. To make sense of the debris, we must know their relevancies in building up the complex system and the rules to discard or account for the existence of other debris. Breaking the atom gave the model of the electron, proton and neutron complex. Breaking atoms at higher and higher energies (as we do in all the particle accelerators like the Large Hadron Collider) gave out debris which was very difficult to explain. Elements could be explained with proton, neutron and electron [3] but doubt arose when at higher energies, we found many short-lived subatomic particles; short life means they decayed into other particles in fractions of a second. We had the periodic table of elements that was based on atomic number. We knew that the subatomic particles—protons, electrons and neutrons—were the basic constituents of an atom. As we discovered more subatomic particles, it was now necessary to build a table characterizing their interrelationship. At the present state, we know that proton and neutron (members of group of particles called hadrons) are made of further and final constituents called quarks. Three types of quarks, the electron and a highly illusive particle named neutrino could model most of the matter in which life sustains. In 1964, Murray Gell-Mann and Carl Zwig build up the 'table' of the fundamental subatomic particles now known as the Standard Model (Fig. 1.2) which we are tempted to produce hereafter Fig. 1.1. Proton is made of three quarks (UUD) and also neutron. So the nucleons (proton and neutron) are made of quark triplets. The electron has no substructure. In a very strange way, we would write down the word 'triplets' in a different context in Sect. 1.4.3. At this point of presentation, we can state that the extremely complex system of molecules, made up of basic elements, ultimately get reduced to a simple representation of numbers called quantum numbers. Each member of Fig. 1.2, as discussed in Annexure 1.1, has a set of quantum number associated with it. These numbers represent almost all the information we have of the known matter and forces

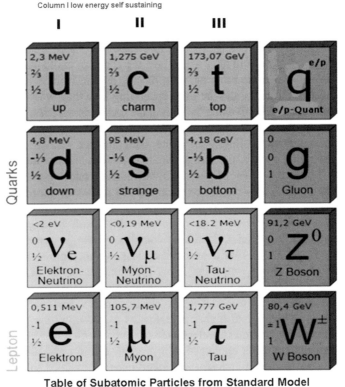

Table of Subatomic Particles from Standard Model

Fig. 1.2 Subatomic particles

working between them.[1] Through the earlier discussion, our goal was to establish a mindset that the matter–energy in the known universe could be described in numbers and a set of rules. In our endeavour to create a top level model of life, we would be placing multiple tables and rules for modelling the biological building blocks of life in Chap. 2 of this book.

Mendeleev's original table as shown in Fig. 1.1 had gaps which motivated us to find the missing elements. Table shown in Fig. 1.2 described systematically the position and properties of subatomic particles that appeared in high energy experiments. It predicted the existence of unfounded subatomic particles.

The Standard Model is still incomplete and scientists are working on it. We end this section with the idea that we came so far in describing the inanimate universe by using top-down (breaking atoms) and bottom-up methods with mathematical techniques employing rules operating on numbers (quantum numbers).

In the remaining sections of this chapter, we will explore the elements of life and traverse both top-down behavioural models and bottom-up mathematical techniques with a set of rules operating on binary numbers to study the animated universe, i.e. life.

1.3   The Elements for Life

The principal elements required for life are hydrogen, oxygen, nitrogen, carbon, phosphorus, sulphur and iron. At the present state of our knowledge, the universe came into being around 13.7 billion years ago, in an event named as 'big bang'. Hydrogen formed within 10 seconds after the creation of the universe. However, the rest of the elements were formed inside the stars.

The first group of stars came into being about 200 million years after the big bang. The oldest star died 10 billion years back. As most stars die, in their incredible death spasm, they explode. This is called Nova; this process spews out the heavier elements required for life. We mentioned earlier that theoretically, we could change from one element to another if any amount of energy is available. Nova-like events creates such energies. Once the required elements come together, organic compounds form—they are found throughout the universe. The required elements for life formed 10 billion years back, i.e. after the death of the first star. Earth-like planets formed at the same time when their corresponding star was formed. Earth as per current estimation was formed about 4.54 billion years ago.

About 400 million years back, the current form of carbon-based life emerged on earth. It is based on the elements mentioned earlier with carbon as the major constituent. Out of all the elements in the periodic table only one element, carbon is small enough (only six protons in the nucleus and two electrons in inner orbit and

---

[1]It is only when a quantum mechanical system is in a so-called eigenstate that some of its properties can be expressed with quantized value.

Table 1.1 Bond energy

Bond	Bond energy (kcal/mol approximately)
C–C	83.0
C–H	97.0
C–O	C–O
C–N	73.0

four in outer, called the valence electrons). The size and number of valence electrons enable carbon to combine with itself and other elements in an incredible number of ways. These combinations gave rise to more than ten million organic compounds and this number continues to increase further. The carbon-based chemistry is called organic because all the molecules that make up any known life form are organic, but the reverse is not true. The small size of carbon enables it to combine with another carbon and the C–C bond energy is very close to its bonding capability with other atoms, as elaborated in Table 1.1. We do not know yet how the sheer complexity and interconnection of different organic compounds ultimately creates an externally complex behaviour—life, which contemplates, writes and reads about itself, and is *aware* of itself [4].

But, we do know

- How it stores the information to replicate itself, molecule by molecule (some details still to be worked out).
- How it uses some random changes, and how such changes increase or decrease its chance of survival in a changing environment.
- How this information can be changed using external agents like chemical reactions and radiation.
- That every complex living thing coexists with other smaller living things in a mutually beneficial way.

We are still learning

- How this information gets destroyed or modified causing the onset of diseases. Thereby, we are trying to learn how to restore the original information with recent gene editing techniques like CRISPER/Cas9.
- The capability to modify (mutate) the information and change life forms to behave in a different way.

Most of the above knowledge is being enhanced by breaking down complex things into smaller parts, and then modelling the parts in different ways.

Can we model life from Fig. 1.2 which depicts subatomic particles? We hardly could arrive at Fig. 1.1, the periodic table of elements, from Fig. 1.2. So, none of the tables above help us in modelling life. We have reduced matter and energy to numbers and a set of rules for their interactions. But this information is too fundamental to build up a model of life. We need more information regarding what happens when they combine together, we have to find the rules behind their emergent behaviour.

Let us examine a line credited to Danish philosopher Søren Kierkegaard:

"Life must be lived forwards, but can only be understood backwards".

The alphabets, atoms of a language, convey nothing in itself. Each alphabet has their properties and sequence of use to form a meaningful word; a sequence of words can carry one's emotion and describe feelings. To analyse the feeling the quoted sentence creates, we cannot begin with the alphabets or words. Thus, life cannot be described by way of behaviour of quarks or protons or even organic molecular interactions. We have to find a different way to model it. We visualize the above quotation first by using English grammar to get a raw meaning, and then we process the meaning according to our acquired semantics.

The grammar of nature is the laws of forces. From these laws, we could find out how atoms form molecules. Without having this higher order semantics, we would find it extremely difficult, if not impossible, to find out how tens of thousands of molecules interact.

1.3.1 Life Viewed Top-Down

In science, we can either break things and find out what it is made of or build up our test bench from mathematical modelling to predict its behaviour. We have used both methodologies in earlier sections for modelling elements observed in nature. We have smashed atoms and used mathematics to build up the bottom-up model. The material world as we have seen have been reduced to numbers and rules, incorporated as an information processing system. In this background, we shall try to view life from top-down, from the observed behaviour to find a way explaining —how life behaves the way it does.

The organic compounds came together in an unknown process to form something akin to a primitive life form. Nature had billions of years and all the space to experiment with different combinations. We want to replicate that experiment or parts of it in a much shorter period of time, say within days. Nature's random experiments had failures, part of them fossilized and vital information went missing. All that remain are the results, information and us. For filling the missing gaps, we are taking the help of increasingly powerful information processing machines, the computers.

Scientists created the Standard Model (Fig. 1.2) and reinforced its validity by creating higher and higher energy densities. Energy densities can be visualized in a simple manner with the next illustration. An average house fly hitting your cheek will deliver 1600 Tev (Terra electron-volt unit of energy), and around the same amount of energy is spent by the fly to rise from ground to a height of 10 m. The largest particle collider at CERN is generating 13 Tev but that energy is packed into a space billion-billion-billion times smaller than a fruit fly. We have not seen the quarks or even electrons; we can only visualize their interaction. We have seen the artefacts in the high energy collisions which are consistent with the mathematical models (as elaborated in Annexure 1.1).

To model life, we have to start using different methods, but first of all, we have to choose the fundamental blocks we want to replicate and which will, at the least, ensure us that we are on the right path.

Children use LEGO blocks to either copy documented shapes representing cars and houses, or sometimes totally imaginary objects on their own. Scientists use mathematical models to explain physical phenomena and could sometimes predict a phenomenon not yet discovered, be it elements or subatomic particle. Apart from that, they try to replicate in a different form, a phenomenon that mimics a physical reality; the behaviour of these simulations is verified in the laboratories for consistencies or in the abstract world of mathematics. If we find the right LEGO blocks, we have the tools to make our own models of life. Physical experiments in biology (called the wet lab experiments) are extremely time consuming and expensive. It should be guided by some mathematical model just as we did for other branches of science. The essential tools for making models are mathematics, rules and algorithms of higher complexity executed by using the ever-increasing power of computers.

In this book, we will explore a method of model-making of life in a different way. Life, as we know it, in essence, is a carbon-based self-replicating, self-modifying system which brings molecules within itself and in its neighbourhood into the desired order. To sustain itself, it will consume energy from the external world. When we mention 'external world' , it implicates that life is taken as an isolated system (thermodynamically) different from the environment sustaining it. We will describe it essentially as an information processing system.

Around 1965, it was established that three types of large organic macromolecules are essential to sustain the present form of life—one set is called DNA and the other called protein. A third molecule called RNA is an essential run between the two. Once the basic biomolecules of life were identified, it is not so difficult to find detailed information about them. Their interaction essentially explains the incredibly vast amount of variation seen in life. The information gathered from their coordinated behaviour began to grow rapidly to one of the largest existing databases. It may not be a coincidence that it is only now we are equipped to store and retrieve such vast quantity (and ever increasing) information and managing them. The reason for such an explosion of information and associated activities probably lies in the fact that the focus on life sciences is a comparatively recent phenomenon. While other branches of science flourished in last few centuries, emphasis on study of biological systems was comparatively lesser due to various reasons; primary reason being that life in micro detail could only be studied after it was inanimate, that is dead.

With computers, we can now cope with the information storage and retrieval of behaviour and structure of life—this field of study is currently referred to as bioinformatics. We have also started to model the natural life's chemical blocks using numbers. We have initiated a move to build such a model to understand how the chemistry-based information processing system works; most importantly—how the vast amount of information already generated (and being generated continuously) out of wet lab experiments can be adequately interpreted as the cause leading to a disorder/disease.

1.4   Introduction to Basic Building Blocks of Life

The basic building blocks of life are DNA, RNA and proteins.

Deoxyribonucleic acid (DNA)—deoxy means that it originates from sugar ribose owing to the loss of an oxygen atom. It is the hereditary macromolecule found in all living organism. Nearly, every cell in the human body has the identical DNA. Most of this DNA is located in the cell nucleus (hence called nuclear DNA). A smaller quantity of DNA is also found in the mitochondria, a special organelle (hence called mitochondrial DNA or mtDNA).

Human DNA consists of approximately 3 billion bases (detailed in Sect. 1.4.2). More than 99% of DNA bases are common in all humans, but only 55% of human DNA is common with potato. The sequence of these DNA bases determines the genetic information available for building and maintaining a life form. The common analogy is the way in which letters of the alphabet appear in a certain order and sequence to form meaningful words and sentences.

An important and interesting trait of DNA is the fact that it can replicate, i.e. make copies of itself. Each strand of DNA in the double helix can serve when required as a pattern for duplicating the sequence of bases. This is the most critical ability. When cells divide, each new cell now has an exact replica of the DNA from the old cell. A brief description of cell anatomy is covered in Annexure 1.2.

RNA stands for ribonucleic acid. Similar to the DNA, RNA nucleotides also contain three major components.

- Nitrogenous base,
- Five-carbon sugar and
- Phosphate group (which acts as the backbone to the nucleotide chain).

However, there are multiple types of RNA named after their functionality. We will focus on mRNA and tRNA in this chapter.

Proteins are made from chains of Amino acids.

Next few sections introduce the three building blocks (proteins, DNA, RNA), just enough to understand the rest of this book without frequent references from other sources.

1.4.1 Proteins

Proteins are large molecules that are composed of one or more long chains of amino acids. They are the most critical and essential component of all living organisms; wherever there exists life as we know it, there are proteins. They form the essential structure of a living organism and different combinations of proteins are responsible for important properties of life. Living organisms use proteins for many functions, including repairing and building tissues, acting as enzymes, aiding the immune system, and serving as hormones. Each of these important essential functions necessary to sustain life requires different classes of proteins. In spite of their differences in structure, all proteins contain the same basic sub-components (micromolecules). As we look deeper into the mechanism of life, we notice that proteins can communicate, acquire and store energy, modify themselves and do many more things.

Protein is a single class of macromolecule consisting of amino acid residues joined by peptide bonds. The building blocks of an amino acid are the atoms—carbon, hydrogen, nitrogen, oxygen, sulphur and occasionally phosphorus—the elements that are most abundant in the universe.

Amino acids are organic compounds containing two functional groups, amine and carboxyl and a R group (basically alkyl group or hydrogen). **Amines** (derivative of ammonia) are compounds and functional groups composed of a basic (opposite to acidic) nitrogen atom with a two valence electron (lone pair). There are four classes of amines—primary, secondary, tertiary and cyclic. Morphines and heroin are tertiary amines, while antidepressants like amoxapine are secondary amines. The breakdown of amino acids releases amines. The smell of decaying fish is due to break down of amino acids; it is of trimethylamine, another tertiary amine. The colour convention used for rendering atoms is as follows:

Carbon = Black   Nitrogen = Blue

Phosphorus = Orange

Sulphur = Yellow   Oxygen = Red

The R group in amino acid proline forms complex linkages (Fig. 1.3). The amino acid molecules are not rigid like ball and stick, they are vibrating, dynamic and influenced by the medium they are in, which in the present case is water. The dynamism gives rise to redistribution of charges over the surface of the molecules.

Figure 1.4 shows the structure of glycine, the simplest amino acid, this is a white substance. We do not take it through our food, it is synthesized in our body. It is also found in outer space in comet tails. When two glycines join together (Fig. 1.4), they must get rid of one water molecule thus forming glycine residue. We can now envisage the behaviour of linkages between amino acids that result in complex properties.

Some amino acid residues are **polar**, meaning they have an electric charge difference across themselves. These polar amino acid residues happily interact with

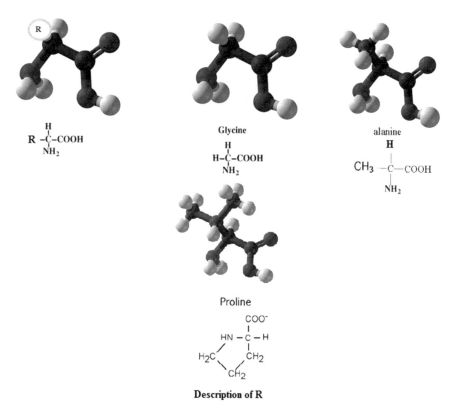

Description of R

Fig. 1.3 R group of amino acid proline

water and are called **hydrophilic**. Other amino acids are **non-polar** without a charge. They do not like water and are called **hydrophobic**.

We know about 500 amino acids—proteinogenic and non-proteinogenic. Interestingly, only 20 common ones are synthesized out of the genetic code during translation. These are referred to as proteinogenic amino acids, each having a common backbone linked to a side chain that differs for each one. One end of the backbone covers an N–H group, while its other end covers a C=O group of atoms with an alpha carbon in between. A chain of amino acids is linked by peptide bonds. Formation of a peptide bond linking a pair of amino acids releases two hydrogen and one oxygen atom; the resulting amino acid is referred to as amino acid residue. The specific sequence of residues in a polypeptide chain imparts a unique property to a protein with respect to its structure and function. One particular amino acid, cysteine, has a sulphur atom. Sulphur atoms on two separate cysteine residues can bond with each other, forming a disulphide **bridge**; **disulphide** bonds (Fig. 1.5) are very important in determining the quaternary structure of some proteins.

Fig. 1.4 Amino acid glycine
structure

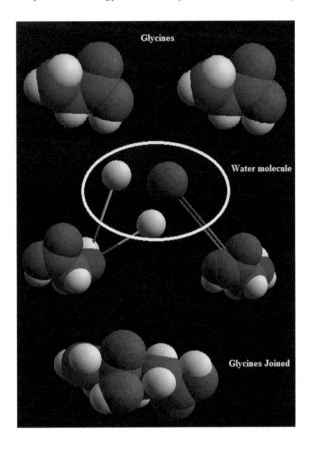

Fig. 1.5 Cysteine structure
with disulphide bond

The amino acid residue chain making up the long strand is the main determining factor that decides what shape a particular protein will fold into. The sequence of amino acids behaves like a string of instructions, and their properties depend on their physical and chemical characteristics and also on their neighbours. The 3D structure of the proteins is derived from uniquely encoded primary structure (the amino acid sequence).

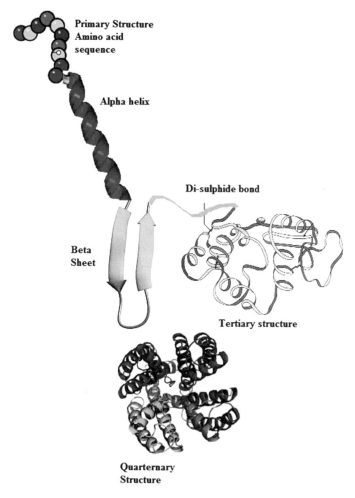

Fig. 1.6 Structures of protein

Proteins fold into unique three-dimensional structures during protein folding. They fold through the interaction of the different amino acids in the medium. Prior to settling down to its quaternary structural conformation, the intermediate conformations of a protein structure are summarized below and illustrated in Fig. 1.6.

- Primary structure of the protein is made of amino acid residue sequence.
- Secondary structure of the protein is composed of regularly repeating local structures. It is stabilized by hydrogen bonds that get formed. Some of the most common examples are the alpha helix, beta sheet and turns.
- Tertiary structure of the protein is the overall shape that the single protein molecule folds into. It essentially characterizes the spatial relationship of the

secondary structures to one another. The tertiary structure of the protein essentially dictates the basic function of the protein.

- Quaternary structure of the protein is formed by several chains of multichain protein that combines to function as a single protein complex.

Figure 1.6 shows a collage of structures of proteins. It shows the primary amino acid sequence, the alpha helix and beta sheet; a quaternary structure of hae-moglobin, the protein constituents in blood. Also, it shows P13 protein, the nuclear antigen expressed in human granulocytes (a white blood cell with secretory gran-ules in its cytoplasm) and monocytes (a type of white blood cell) in lymphoid tissues. The P13 is also present in non-lymphoid tissues that are expressed during the early phases of myeloid differentiation (the process by which stem cells develop into various types of mature blood cell).

The primary structure of the protein is the sequence of amino acid residues in the polypeptide chain. The primary structure is held together by covalent bonds, mostly the peptide bonds. The primary structure then evolves to the secondary structure. In view of different electrostatic forces, the secondary structure is formed out of the primary structure linear chain. By convention, common structural blocks are referred to as 'motif'. The alpha helix is a common motif that gets formed in the secondary structure of proteins. It is a right-handed coil like formation. Beta sheets in the secondary structures are made of beta strands connected laterally by two or more hydrogen bonds forming the backbone. Each beta strand, or chain, is made of 3 or more amino acid residues. For a multichain protein, the chains are aligned to form quaternary structure.

The proteins that work on other molecules are called enzymes. Enzymes accelerate chemical reactions. Almost all forms of metabolic activities in the living cell of any life form need enzymes to accelerate them, making them fast enough to sustain life. Their *specificity* originates from their unique 3D structure. There exists at least one kind of enzyme for every task that is performed. Figure 1.7 animates how a cell (consisting of required enzymes for its survival) needs to split a food molecule into two small particles. It is obvious that for an enzyme to break any

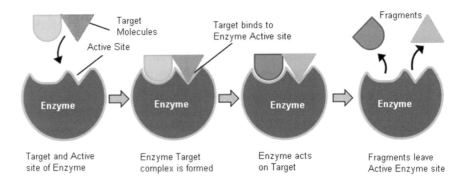

Fig. 1.7 Enzyme action on target

particular molecule, the shape and size of the 'hole' in the enzyme must match to that of the particular target molecule. As a result of the target entering, the enzyme changes its shape and then breaks the target in two. Once the target is broken, the enzyme changes its shape again and releases the broken fragments. A single type of enzyme can do only one job, but it can do the job many times over. The shape of the region in this enzyme is hence extremely important.

Modelling and predicting how a protein would fold could have been easier if they were all very structured; unfortunately, they are not. There are unstructured regions that cannot be captured by current methods. The oldest method to find the structure of a protein is by protein crystallography. However, many large regions of the protein could not be assigned in X-ray data sets. This happens because they occupy multiple positions. This suggests that regions are disordered. New micro-biological imaging methods such as Nuclear Magnetic Resonance (NMR) demonstrated the presence of large flexible linkers that terminates in many structural ensembles. In present state of art, it is accepted that proteins exist as motifs, which are ensemble of similar structure and some regions more structurally constrained than others. At this stage, we must mention that Intrinsically Unstructured Proteins (IUP)s occupy the extreme end of structural flexibility. Intrinsically unstructured proteins are highly abundant among disease-related proteins.

Proteins are highly dynamic, although they are presented as solid rigid bodies that impart an important feature in their function and regulation. When in a solution, fragments of proteins and sometimes the entire protein, do not actually have a well-defined structure. When in a specific functional state, they assume such a structure. Nevertheless, the presence of unstructured regions in a protein is a reality and that probably explains why the prediction of protein structure has not achieved the desired success even after serious effort of the last few decades.

Databases of 3D structures of proteins obtained from X-ray crystallography, NMR spectroscopy and cryo-electron microscopy are in the Protein Data Bank (PDB); this is freely available through the Internet. Predicting structure through different means mostly through computer-based modelling is a continuous endeavour through Critical Assessment of protein Structure Prediction (CASP) competition. The large-scale study of proteins is called Proteomics. It is one of the most difficult branches of study in biological systems. The proteome is the entire set of proteins that are produced or modified by an organism or system.

The flexibility and variety of protein molecules depend on how they fold and function. Proteins working as a team perform complex functions, depending on a sequence of amino acids (the information). Study of Proteomics provokes us to think that proteins as alive—without having capability or mechanism to reproduce themselves.

Proteins are regularly destroyed or broken down; therefore there must exist a mechanism that can make proteins. To achieve this goal, different amino acid sequence data must be stored in encrypted form and a physical process must exist in a biological system to construct the required protein chain out of that form.

How do living organisms make a new protein? They are made in a protein-making molecular factory occupying the inside of every living cell. This factory is called the ribosome, which itself is made mostly of proteins. Ribosome can construct proteins with sequence of amino acids as directed from another molecule called messenger RNA or mRNA transcripted out of a DNA molecule.

1.4.2   DNA

We may represent the current life form in macromolecular level as:

$$Life = Nucleic\ Acid + Protein$$

Nucleic acids are large biomolecules. The building blocks of proteins (polypeptides) and the monomers are amino acids. The nucleotides are the mono-mers of nucleic acids. Nucleotides contain either a purine or a pyrimidine—two major groups of bases. The three building blocks (Fig. 1.8) of a nucleic acid are:

- A five-carbon sugar
- A phosphate group
- Nitrogen-containing base. The nitrogen-containing bases (or just base) come in five types—Adenine (A), Uracil (U), Guanine (G), Thymine (T) and Cytosine (C). The base U in RNA gets replaced with T in DNA.

A and G are referred to as purines, while U, T and C belong to pyrimidine group. Purines and pyrimidine are the nitrogen-containing 'bases' found in the nucleotides that make up DNA and RNA. They have been synthesized artificially.

All the bases (Fig. 1.8) are nearly two-dimensional flattish rings and therefore stackable; due to these shapes, they can form hydrogen bond only on the plane of this paper. G and A are larger than T and the other base—uracil is shown in Fig. 1.9. Thymine (T) of DNA is replaced with Uracil (U) in the RNA base sequence. Uracil misses a methyl group (CH_3) present in thymine.

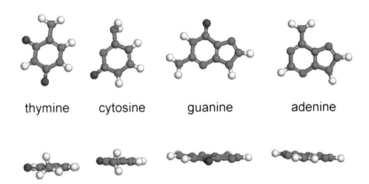

thymine   cytosine   guanine   adenine

Fig. 1.8 Structure of nucleic acids

Fig. 1.9 Structure of uracil

uracil

In a living organism, nucleotides function to create, encode and then store information. The encoded information is contained and conveyed via the nucleic acid sequence.

Strings of nucleotides are linked to form the sugar–phosphate backbone of a nucleic acid (Fig. 1.10). The backbone of DNA covers two chains forming a stable helical structure. On the other hand, the RNA backbone is formed out of a single chain and its structure displays a set of structural motifs (loop, turn, etc.). The deoxyribonucleic acid (DNA) has deoxy sugars in its molecule, meaning that it is derived from the sugar ribose by the loss of an oxygen atom. In the model shown in Fig. 1.10, we can visualize how the bases stick out, and due to this unique shape they occupy the same plane.

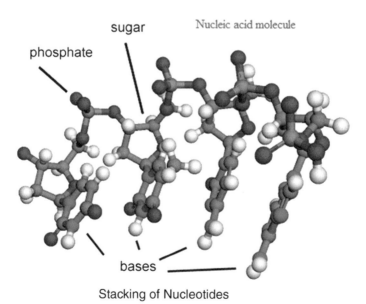

Stacking of Nucleotides

Fig. 1.10 Structure of nucleotides

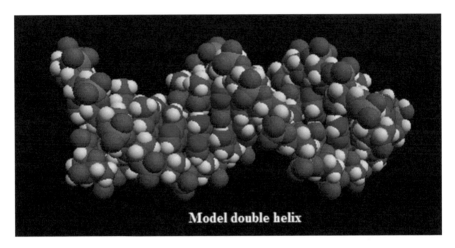

Fig. 1.11 Model of double helix structure of DNA

In DNA, the complementary base pairs (C–G and A–T) are linked in two strands with H-bonds, as illustrated below:

A-A-T-T-A-A-C-G-T-T-A-C-T-A   (one strand)

| | | | | | | | | | ||| | | (hydrogen bonds)

T-T-A-A-T-T-G-C-A-A-T-G-A-T   (another strand with complimentary bases)

If we model this tiny double strand it will look like as shown in Fig. 1.11. We could remove the van der Waal's sphere (detailed in Sect. 1.4.8) and view the hydrogen bonds linking the two strands (Fig. 1.12).

The two strands of DNA have redundant information and reading the sequence of one strand is good enough. However in genetics, they are differentiated; the segment that runs from 5′ to 3′ (explained in Sect. 1.4.3) is called sense strand or coding strand, while the other strand runs from 3′ to 5′ and is called the antisense strand. The sense strand is the strand of DNA that is taken as the template for generation of mRNA string, as depicted below. In the process, the mRNA becomes the complementary to sense strand, and so it becomes identical to antisense strand. RNA is later translated to protein.

A-A-U-U-A-A-C-G-U-U-A-C-U-A   (DNA sense strand)

U-U-A-A-U-U-G-C-A-A-U-G-A-U   (strand of mRNA)

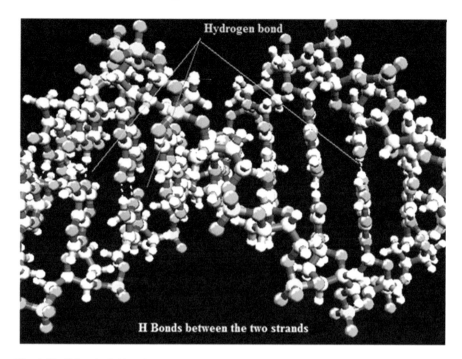

Fig. 1.12 H-bonds linking the two strands of DNA

The pairing of A–T and C–G in DNA, as noted earlier, is realized through hydrogen bonds (Figs. 1.13 and 1.14). Hydrogen bond is a weak bond between two molecules created because of an electrostatic attraction between a proton in one of the molecules and a negatively charged atom. Section 1.4.8 covers further details of H-bond.

Before proceeding further, we should mention the 'gene' which is a part of DNA. Human genes actually vary in size from a few hundred DNA bases and may go up to more than 2 million bases. A gene is a section of DNA that is responsible for coding a specific protein. Thus protein, as explained earlier, is the final product

Fig. 1.13 Two hydrogen bonds between thymine and adenine

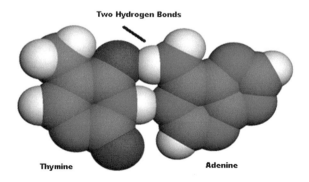

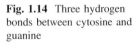

Fig. 1.14 Three hydrogen bonds between cytosine and guanine

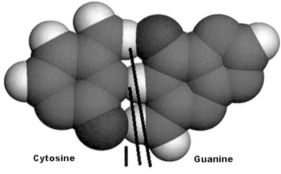

Cytosine   Guanine

Three Hydrogen Bonds

of a gene of a DNA molecule. For example in case of humans, one gene will code for the protein insulin that has the most important role to control the amount of sugar in our blood.

If we take the DNA from all the cells in human body and line it up, end to end, it would form a very thin (order of 10^{-4} mm) 5000 million miles long. In order to store such a large volume of information in a structure called chromosomes, DNA molecules are tightly packed around proteins called histones. Nature has achieved some really efficient packing mechanism to store voluminous information coded in DNA.

One can see that if an enzyme pulls the two strands of DNA apart, each could be used to make the other strand. The earlier discussions point to the following two important characteristics of DNA:

- It has the capability of induced self-replication.
- Each strand carries information. The order of the nucleotide bases along each of the strands stores the code for making proteins.

We will see later in Sect. 1.4.7 that the single-strand RNA does have the above capabilities but with some deficiencies.

DNA is an information storage molecule [5]. The DNA contains the complete instruction set a biological cell requires to sustain itself. Specific section of nucleotides is called genes. Genes carry the instruction. To execute an order, the instruction contained in the genes must be expressed. They must be copied into a form to produce proteins.

The instruction sets stored in the DNA are read and processed by a living cell in two distinct steps: transcription and translation. Each step is a complex biochemical process involving multiple molecules.

1.4.3 Transcription and RNA

During the transcription process, a portion of the cell's DNA serves as a template for the creation of an RNA molecule. The newly created RNA molecule produced out of the non-coding DNA may itself become a finished product; this is referred to as non-coding RNA that performs important functions within the cell. Typical examples are messenger RNA (mRNA), tRNA (transfer RNA), ribosomal RNA (rRNA) and a set of regulatory RNAs. While the first three types of RNAs are associated with protein synthesis, the fourth group performs different regulatory functions within a cell. If one can see that DNA is working like a code memory, the various other types of RNA are putting the code into action.

The RNA molecule, called the messenger RNA, is produced out of coding region of DNA sequence. The mRNA carries the messages from the DNA to other parts of the cell for processing. This information is used to manufacture proteins.

The first step in the transcription process is the mechanism of unzipping of the two strands of a part of DNA gene and then using one strand (sense strand) as a template to make a copy of the information which is the sequence of nucleotides. This process of transcription is done by a set of proteins called polymerases (Fig. 1.15).

The polymerases transcribe a gene (both non-coding and coding DNA) to produce an RNA (Ribo Nucleic Acid) molecule. It is called 'ribo' because the sugars in RNA are different from the sugars in DNA, the ribose means to have not lost an oxygen atom.

The copy of the coding DNA sequence is called messenger ribonucleic acid, mRNA. If we consider a mRNA strand transcripted from the DNA sequence A-A-T-T-A-A-C-G-T-T-A-C-T-A illustrated earlier, it will look as shown in Fig. 1.16.

At this stage, we reintroduce what is meant by 5′ and 3′ (called 5 prime and 3 prime). The 5′ and 3′ indicate the specific carbon numbers in the DNA's sugar backbone. A phosphate group is attached to the 5′ carbon and the 3′ carbon has a hydroxyl (–OH) group attached. This asymmetry provides the DNA strand a direction.

Transcription process itself can be divided into sub-processes and we describe through each process for clarity. First, the RNA polymerase binding on to the DNA template strand. This first step is called 'initiation'.

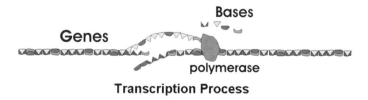

Fig. 1.15 Transcription process

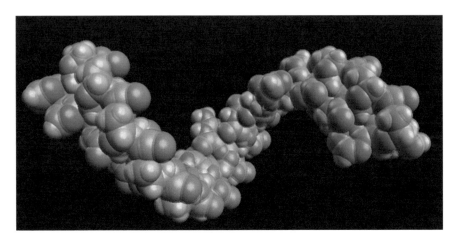

Fig. 1.16 mRNA strand transcripted from DNA

In the initiation phase, the RNA polymerase and its associated transcription factors bind to the DNA strand at a specific site that facilitates the transcription process. This area of binding is called the promoter region. Sometimes, it includes a specialized nucleotide sequence, TATAAA, referred to as the 'TATA box'. Approximately, 23% of human genes contain this 'TATA box' within the core promoter.

Once the RNA polymerase and its related transcription factors are in position, the single-stranded DNA is exposed and is ready for transcription. The RNA polymerase now begins moving down the DNA template strand in the 3' to 5' direction, stringing together complementary nucleotides as it does so. The very process of complementary base pairing creates a new strand of mRNA that is organized in the 5' to 3' direction. As the RNA polymerase continues down the DNA strand, more nucleotides are progressively added to the mRNA leading to the formation of a longer chain of nucleotides. This process is aptly named 'elongation'. So the process of elongation follows the initiation process.

RNA contains all the bases of DNA except that the base uracil replaces thiamine. Thus during the elongation process, the presence of thiamine in the DNA template strand tells RNA polymerase to attach a uracil in the corresponding area of the growing RNA strand.

In the elongation phase of transcription, a new mRNA molecule is created from a single template strand of DNA. As the mRNA elongates, it peels away (Fig. 1.15) from the template. This mRNA molecule carries information to an organelle called ribosome.

The mRNA can only perform its assigned function after the elongation ends, then it separates from the DNA template. This process is referred to as 'termination'. In some cases, termination occurs as soon as the polymerase reaches a specific series of nucleotides along the DNA template. In other cases, the presence of a special protein named termination factor is also required for termination to occur.

At the completion of termination phase, the mRNA molecule falls off the DNA template. The newly synthesized mRNA, called precursor mRNA, undergoes a process in which non-coding nucleotide sequences, called introns, are clipped out of the mRNA strand. This process is referred to as 'splicing' tidies up the molecule and removes nucleotides that are not involved in protein production. The splicing process requires a special complex molecule called spliceosome.

A typical mRNA has two Untranslated Regions (UTR)—5′ UTR and 3′ UTR on 5′ and 3′ ends. Further, prior to export of mRNA to cytoplasm for translation, a sequence of adenine nucleotides is added to the 3′ end and a 5′ cap added to the 5′ end. The UTRs and caps cover a polyspecific sequence of nucleotide bases. The polyadenylation process or addition of the poly-A tail sequence signals to the cell that the mRNA molecule is now ready to leave the nucleus and enter the cytoplasm for translation.

A triplet of bases in a mRNA string is referred to as 'codon'—that can be viewed as the information unit for transfer of information from DNA to RNA.

The message of mRNA codon string encodes a sequence of amino acids. The message of mRNA codon string encodes a sequence of amino acids. Before we conclude this section, we present the codon-amino acid table displaying the amino acids encoded by different codons (Fig. 1.17).

First Letter	Second Letter				Third Letter
	U	**C**	**A**	**G**	
U	phenylalanine	serine	tyrosine	cysteine	**U**
	phenylalanine	serine	tyrosine	cysteine	**C**
	leucine	serine	stop	stop	**A**
	leucine	serine	stop	tryptophan	**G**
C	leucine	proline	histidine	arginine	**U**
	leucine	proline	histidine	arginine	**C**
	leucine	proline	glutamine	arginine	**A**
	leucine	proline	glutamine	arginine	**G**
A	isoleucine	threonine	asparagine	serine	**U**
	isoleucine	threonine	asparagine	serine	**C**
	isoleucine	threonine	lysine	arginine	**A**
	(start) methionine	threonine	lysine	arginine	**G**
G	valine	alanine	aspartate	glycine	**U**
	valine	alanine	aspartate	glycine	**C**
	valine	alanine	glutamate	glycine	**A**
	valine	alanine	glutamate	glycine	**G**

Codon to amino acid relationship showing degeneracy

Fig. 1.17 Mapping of 64 codons to 20 amino acids

There are 64 combinations of 4 nucleotides taken three at a time (referred to as codon) and only 20 amino acids. As a result in most cases, there are more than one codon per amino acid. This phenomenon is referred to as 'codon degeneracy'. As an example, the simplest amino acid glycine (gly) is encoded by codons GGU, GGC, GGA and GGG.

1.4.4 Translation Process

We next discuss a highly complex molecule called ribosome that translates the information in the mRNA string to a polypeptide chain of amino acids. For example, if it reads a base triplet, say GGU in mRNA string, the amino acid glycine gets attached to the growing peptide chain. Ribosome is a highly specialized molecule of great importance because it is the biological entity that acts as a bridge between Proteomics and Genomics (Fig. 1.18).

A ribosome builds a protein. Hence, the flow of genetic information is

$$DNA \rightarrow mRNA \rightarrow Protein$$

The ribosome is a central entity to the process of life. Ribosomes occur both as free particles within cells and also as particles attached to the membranes inside of the cells. A ribosome is made of about 4 nucleic acid molecules and 70 different types of proteins. Ribosomes are several in numbers within a cell and account for a

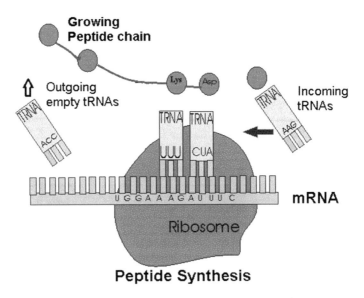

Fig. 1.18 Process of translation

Fig. 1.19 Entities involved
in the process of translation

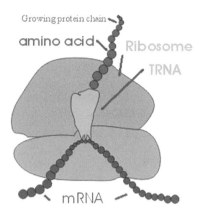

Details of entities involved
in translation process

large proportion of its total nucleic acid. The reason for such huge quantity of ribosomes is because the translation process is extremely slow. Every second, in an animal cell, around 1 million amino acids are added to growing proteins. However, each ribosome can add only between 3 and 5 amino acids to the growing peptide chain of protein each second.

A schematic of the ribosome seems to be analogous to a scanner + translator + printer. The scanner section is smaller in size and the translator + printer section is bigger as shown in Fig. 1.19 with larger section above the smaller one.

The RNA and proteins that make ribosomes in bacteria are different from that of eukaryotes. Also, these vary between different species of eukaryotes. However, their basic structure and function are always the same.

Where are the amino acids coming from for protein synthesis in ribosome? Molecules called transfer RNAs (tRNAs) bring amino acids to the ribosome. Ribosome has three slots (Fig. 1.20) for tRNAs: the A site, P site and E site. tRNAs move through these sites (from A to P to E) as they deliver amino acids during translation.

The small and large subunit of ribosome come together to provide three locations for tRNAs to bind (the A site, P site, and E site). In the diagram, the sites appear in A-P-E order from right to left. After the initial binding of the first tRNA at the P site, an incoming charged tRNA would bind at the A site. Peptide bond formation transfers the amino acid of the first tRNA, and methionine to the amino acid of the second tRNA (in this case, Tryptophan). This chain of two amino acids would be attached to the tRNA in the A site. The ribosome would then move along the mRNA template by one codon. The tRNA in the A site (with the polypeptide chain) would shift to the P site (Fig. 1.20), and the empty tRNA previously in the P site would shift to the E site (where it would exit the ribosome). A new tRNA in this case, one bearing phenylaniline (Phe), would bind to the newly exposed codon in the A site, and the process could then repeat.

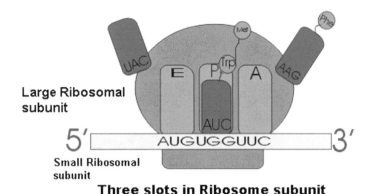

Three slots in Ribosome subunit

Fig. 1.20 tRNA slots in ribosome

Fig. 1.21 tRNA as enabler connecting mRNA codon to amino acid

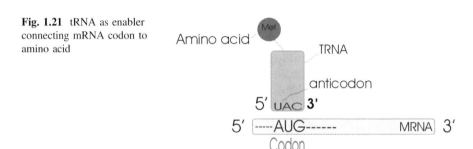

A transfer RNA (tRNA) is a special kind of RNA molecule. Its job is to match a mRNA codon with the amino acid it codes for. Each tRNA contains a set of three nucleotides called an anticodon (Fig. 1.21). The anticodon of a given tRNA can bind to one or a few specific mRNA codons. The tRNA molecule also carries an amino acid, the one encoded by the codons that the tRNA binds. Information flow has a direction, and the nucleotide chain in DNA and RNA has also a direction because the two ends of a strand of DNA or RNA are different from each other. Many processes, such as DNA replication and transcription, can only take place in one particular direction of a DNA or RNA strand. A tRNA loaded with amino acid 'Met' for the codon AUG is illustrated in Fig. 1.21.

Figure 1.21 shows a tRNA acting as an adapter connecting a mRNA codon to an amino acid. At one end of the tRNA, it has an anticodon of 3'-UAC-5' that binds to a codon in a mRNA that has a sequence of 5'-AUG-3' through complementary base pairing of first two bases followed by wobble pairing for third base. The other end of the tRNA carries the amino acid methionine (Met), which is the amino acid specified by the mRNA codon AUG.

There are many different types of tRNAs floating around in a cell, each with its own anticodon and matching amino acid. Some tRNAs bind to multiple codons through 'wobble' pairing, as elaborated next.

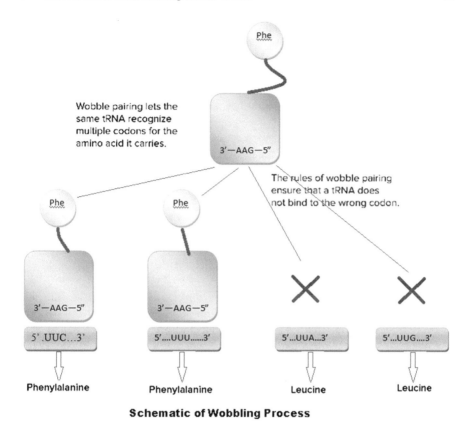

Wobble pairing lets the same tRNA recognize multiple codons for the amino acid it carries.

The rules of wobble pairing ensure that a tRNA does not bind to the wrong codon.

Schematic of Wobbling Process

Fig. 1.22 Wobbling process

Some tRNAs can form base pairs with more than one codon. A typical base pair between nucleotides other than A–U and G–C can form at the third position of the codon, a phenomenon known as wobble. Wobble pairing does not follow normal rules, but it does have its own rules. For instance, a G in the anticodon could pair with a C or U (but not an A or G) in the third position of the codon. Wobble pairing lets the same tRNA recognize multiple codons for the amino acid it carries elaborated in Fig. 1.22 with reference to tRNA for amino acid 'Phe'.

The tRNA for phenylalanine (Phe) has an anticodon of 3′-AAG-5′. It can pair with a mRNA codon of either 5′-UUC-3′ or 5′-UUU-3′ (both of which are codons that specify phenylalanine—Phe). The tRNA can bind to both codons because it can form both a normal base pair with the third codon position (5′-UU**C**-3′ codon with 3′-A**A**G-5′ anticodon) and an atypical base pair with the third codon position (5′-UU**U**-3′ codon with 3′-A**A**G-5′ anticodon).

The rules of wobble pairing ensure that a tRNA does not bind to the wrong codon. The tRNA for phenylalanine has an anticodon of 3′-AAG-5′, which can pair with two codons for phenylalanine but not with 5′-UU**A**-3′ or 5′-UU**G**-3′ codons.

These codons specify leucine, not phenylalanine, so this is an example of how the rules of wobble pairing allow a single tRNA to cover multiple codons for the same amino acid, but does not introduce any uncertainty about which amino acid will be delivered to a particular codon. Wobble pairing allows fewer tRNAs to cover all the codons of the genetic code, while still making sure that the code is read accurately. This is a computational method nature has adopted to increase accuracy and efficiency in information translation.

A tRNA molecule has an 'L'-shaped structure held together by hydrogen bonds between bases in different parts of the mRNA sequence. One end of the L shape has the anticodon, while the other has the attachment site for the amino acid. Different tRNAs have slightly different structures, and this is important for making sure they get loaded up with the right amino acid.

1.4.5 Loading a tRNA with an Amino Acid

How does the right amino acid get linked to the right tRNA to make sure that codons are read correctly. Enzymes called aminoacyl-tRNA synthetases perform this very important job. There is a different synthetase enzyme for each amino acid, one that recognizes only that particular amino acid and its tRNAs (and no others). Once both the amino acid and its tRNA have attached to the enzyme, the enzyme links them together, in a reaction fuelled by the 'energy currency' molecule Adenosine TriPhosphate (ATP). The active site of each aminoacyl-tRNA synthetase fits an associated tRNA and a particular amino acid like a 'lock and key'. ATP is then used to attach the amino acid to the tRNA.

Occasionally, an aminoacyl-tRNA synthetase may commit a mistake—it might bind to the wrong amino acid (one that 'looks similar' to its correct target). For example, the threonine synthetase sometimes grabs serine by accident and attaches it to the threonine tRNA. Luckily, the threonine synthetase has a proofreading site, which pops the amino acid back off the tRNA if it is incorrect.

DNA sequences that encode protein sequences are referred to as 'coding DNA'. There exists non-coding DNA that does not encode protein. Non-coding RNA transcribed out of non-coding DNA is introduced in next section.

1.4.6 Non-coding RNA

There are RNA molecules that are not translated into protein; these are non-coding RNAs **(ncRNA)**.

A few varieties are small nuclear RNA (snRNA), non-messenger RNA (nmRNA) and functional RNA (fRNA). The DNA sequence from which a functional non-coding RNA is transcribed is named RNA gene. The RNA molecules tRNA (transfer RNA) and rRNA (ribosomal RNA) are associated with the translation process. In addition, certain ncRNAs perform important controlling function.

The number of ncRNAs encoded within the human genome is greater than thousands. Some of them are also described as junk RNAs which are likely to be the product of spurious transcription. A large part of them are interpreted as 'microfossils' of primitive life forms. These may be viewed as the codes in storage in hard disk but no longer useful or executable in the present operating system. The 'unused code' analogy also forces us to think—should not some of these unused RNAs have roles in causing deleterious effect on normal functioning of life?

In fact, ncRNAs do show abnormal expression patterns in cancerous tissues. Precursors appear to be much more frequent in patients with chronic lymphocytic leukaemia. The protein 'p53' is a tumour suppressor; it is a guardian of the genome. The same protein is sometimes inactivated and more importantly promotes human cancer. The p53 (also known as TP53) protein functions as a transcription factor and plays a crucial role in cellular stress response. Besides its crucial role in cancer, p53 has also been found to be associated with other diseases such as diabetes, cell death after ischaemia, and various neurodegenerative diseases like Alzheimer, Huntington and Parkinson's.

At this stage, it is proper to introduce the biological 'swiss knife' RNA as a precursor to present DNA centric life.

1.4.7 RNA World

When we started with nucleic acid, we mentioned that they are polymeric macromolecules assembled from nucleotides. The bases A and G are purine bases, while pyrimidine base are C, T and U. Purines and pyrimidines have been synthesized in vitro (in glass with other micromolecules). DNA has a complex structure; its double helix structure gives it a structural rigidity required to store the information for inheritance. They have an inherent redundancy—if one strand breaks, it could be recreated from the complementary strand. However, such a stable structure triggers the question—are there precursors of DNA in a primitive life form that got replaced with the current form of life, as we see today? This hypothesis is important for our model design. Though there is no conclusive proof, it is easier to think of the two basic nucleotide groups, purine and pyrimidine, as the starting point, out of which 4 bases evolved—A, G from purine and U/T, C from pyramidine. As a result, single-strand RNA with two bases got replaced with more complex RNA string with four bases A, U, G, C similar to the one of current life form. The big problem is—in order to sustain primitive life form of RNA world, from where can we get the protein enzymes? For current life form, these enzymes are transcribed from DNA. Let us briefly present the viewpoint of the proponents of 'RNA World' hypothesis. The hypothesis was proposed independently by Carl Woese, Francis Crick and Leslie Orgel [6] in the 1960s.

Sustaining of life requires that molecules have a crucial property: the fundamental ability to catalyse reactions that lead to the production of more molecules similar to themselves. The first 'biological' molecules may have been created by metal-based catalysis on crystalline surfaces of minerals. It has been proposed that an elaborate system of molecular synthesis and breakdown could have persisted on these surfaces long before the first living cell came into existence. These catalysts having special self-promoting properties could ideally use raw materials to reproduce and replenish themselves and thus divert these same materials from the production of other substances. In living cells, the most flexible catalysts are polypeptides which are composed of different amino acids having versatile and chemically diverse side chains. Hence, they are able to take on diverse three-dimensional structures that rise with reactive chemical groups. But, although polypeptides are versatile as catalysts, there is no recognized way in which they can reproduce themselves.

Polynucleotides have one unique property that contrasts with those of polypeptides. Polynucleotides can direct the formation of exact copies of their own sequence. This unique capability is because of complementary base pairing that enables one polynucleotide to act as a template for the creation of another.

Efficient synthesis of polynucleotides by such template mechanisms requires catalysts or enzymes to promote the polymerization reaction. Without these catalysts or enzymes, this synthesis is extremely slow, inefficient and error prone. Nucleotide polymerization is catalysed by catalysts like DNA and RNA polymerases.

In 1982, these missing enzymes were identified. It was discovered that the RNA molecules themselves can act as catalysts. Further, it was found that RNA is the catalyst for the peptidyl transferase reaction that takes place inside the ribosome. The discovery of ribozymes lends strong support to the RNA world hypothesis. In the early 1980s, research groups led by Sidney Altman and Thomas R. Cech independently discovered that catalyse RNAs—ribozymes. These are capable of catalysing specific biochemical reactions similar to that of protein enzymes.

So according to the 'RNA World Hypothesis', life later evolved to use DNA and proteins due to RNA's relative instability and comparatively poorer catalytic properties. As a result, ribozymes, as we see it today, can be viewed as remnants of an ancient world that existed prior to evolution of proteins. Although self-replicating systems of RNA molecules have not been found in nature, scientists are hopeful of constructing such molecules in the laboratories. Even though laboratory demonstration would not prove the role of self-replicating RNA molecules in the origin of life, it would definitely suggest such a possibility. RNA molecule acts both as an information carrier and as well as a catalyst, which forms the basis of the RNA world hypothesis.

The hypothesis provides us with a starting point to initiate the model building process. Chapter 3 of this book focuses on this aspect. Next two sections of the current chapter lay the foundation for the 'New kind of Computational Biology' proposed in this book.

1.4.8   The Tools for a Model Maker

The primary ingredients to make a model—protein, RNA and DNA. The overall behaviour of these entities is dependent on their constituent parts. The atoms making an amino acid and its position in 3-D space impart certain properties to it. Amino acids linkage imparts different variations of such properties. So things get complicated as amino acids interact; even interconnected pair of identical amino acids displays the behaviour that is different from its constituents. The underlying reason for such wide variations is due to the interaction of charge difference in space filled with water. Figure 1.23 displays an instance of charge distribution in an amino acid molecule. In this context, let us examine the nature of three major forces dominating the molecular interaction.

- **van der Waal's force**: This type of force arises from induced dipole and the interaction is weaker than the dipole–dipole interaction. In general, heavier the molecule, the stronger is the van der Waal's force of interaction. Each interaction has a characteristic optimal distance of twice the atomic radius. Van der Wall's sphere is related to atomic radius. Although individually it is a weak force, they collectively play an important role in determining biomolecular structure and interactions. This force is extremely important in molecular dynamic modelling.

Fig. 1.23 Charge distribution in amino acid proline

- **Hydrogen bond**: Hydrogen bonds affect the molecular structure of organic compounds such as alcohols, acids, amines, and amino acid chain of proteins, nucleotide chain of DNA, RNA, etc. We have mentioned hydrogen bonding in earlier sections since it is the most dominant and important force in biological science. The hydrogen bond is an attractive force between a hydrogen atom from a molecule or a molecular fragment Z–H in which Z is more electronegative than H. Hydrogen bonding is easily found in liquid water, it is the reason for its higher boiling point. The hydrogen atom in a water molecule is attracted towards the oxygen atom of the neighbouring water molecule as the oxygen atom is partially negatively charged while the hydrogen atom is partially positively charged. It is also described as an electrostatic dipole–dipole interaction.
- **Covalent bonding**: Covalency of an atom refers to the number of electrons it shares with other atoms. It is an intramolecular force between a pair of atoms, rather than an intermolecular force like H-bond. It shapes the molecule itself rather than affecting the shape of a polypeptide chain.

In addition to the above three, there are other electrostatic interactions leading to sulphur bond, salt bridge and hydrophobic core due to hydrophobic forces. Figure 1.23 illustrates charge distribution in a computer synthesized proline amino acid having 17 atoms; the reddish colour indicates positive charge distribution, while faint bluish represents negatively charged region. It can be seen that the positive charge gets concentrated on a region rather than balancing negative charge (bluish) which covers larger area.

The dominant tools of a model maker in biology are the equations representing the molecular forces in play between the molecules of a macromolecule. We could then run computer programs to simulate the behaviour of the macromolecule based on the molecular attraction and repulsion. This is the basis of Molecular Dynamics (MD) modelling.

Biology is data/information-centric science. Cosmology and high energy physics also generate voluminous data. However, there is a significant difference between biology and other branches like cosmology and high energy physics. While it is much easier to filter the essential information in other branches, it is impossible to do so in biology due to very high interdependencies of information. Consequently, large-scale approximation is a necessity in the mathematical model derived out of the forces acting on molecules inside a medium.

At this stage, it is appropriate to mention a few dominant methodologies used in laboratories to get the data we use for modelling a biomolecule. To start with, we explain the standard Latin phrases used for identifying the environment or mode where the experiment is being conducted.

In vitro (Latin for within the glass) is a technique of doing a procedure in a controlled environment but outside of a living organism. This is the way majority of experiments in cellular biology are conducted. This is the oldest method but its weakness is that it fails to replicate the precise cellular environmental conditions of an organism, say a microbe. It is estimated that 99% of the species in the human microbiota have not been characterized through in vitro techniques since the cellular environment of the microbiota could not be replicated in vitro. Further details are noted in Annexure 1.2 that reports a brief introduction to a biological cell.

One important in vitro analysis is cell line experiment or immortalized cell lines. These are widely used as a simple model to study biological cells. The primary advantage of using an immortal cell line for research is that the cell line is immortal; they can be grown indefinitely in culture. This simplifies analysis of the biological process of cells.

Immortalized cell lines can also be cloned and hence can be propagated indefinitely. Now an analysis could be repeated many times over on genetically identical cells, an amazingly desirable property for repeatable scientific experiments. The cell line experiments are hence a cost-effective way of growing cells similar to those found in a multicellular organism in vitro. This method is commonly used for testing toxicity of drugs and its impact in the production of proteins in living organisms.

In vivo (Latin for 'within the living') refers to performing an experiment within a living organism as against on a partial or dead organism. Two forms of in vivo research are active—animal studies and clinical trials. In vivo testing is preferred over in vitro wherever possible as it is more suited for observing the real effects of an experiment and its overall impact within a living subject. This method of study is relatively expensive and also animal studies are being partially or totally banned in a few countries.

In situ frequently means that you are looking at something in its natural context but perhaps not under natural conditions (in vivo). One of the most common examples is in situ *hybridization*, where you take a slice of a living specimen, fix all of its contents in place on a microscope slide, and then determine where a molecule (say, a specific RNA) are within the fixed tissue. This retains the context, but not the living processes that determine and drive the localization of the RNA. Hybridization, in general, refers to the process of generating a hybrid out of different varieties.

The expression In silico means 'performed on a computer', silicon being the primary material for constructing semiconductor devices that constitute the active parts of any computing system. The expression 'in silico' was first coined in public in 1989 in the workshop 'Cellular Automata: Theory and Applications' in Los Alamos, New Mexico. Pedro Miramontes, a mathematician from National Autonomous University of Mexico (UNAM), presented the report 'DNA and RNA Physicochemical Constraints, Cellular Automata and Molecular Evolution'. In his talk, Miramontes used the term 'in silico' to characterize biological experiments carried out entirely in a computer [7].

In silico studies are used widely to predict how drugs interact with the body and with numerous other pathogens. For example, software simulations are now being used to predict how certain drugs already on the market could potentially treat multiple-drug-resistant strains of tuberculosis. In silico experiments are also developed for the following applications.

1. **Bacterial sequencing techniques**

This technique has been developed as a support to in vitro methods for identifying various bacterial DNA and RNA. The standard method is known as Polymerase Chain Reaction (PCR). In this process, a few copies of a piece of DNA are taken and increased by several orders of magnitude, hence generating millions of copies of the same DNA sequence. Using this method, we can detect bacteria associated with a variety of conditions with increasingly high sensitivity. In silico PCR analysis is an efficient and useful method for bacterial sequencing.

2. **Molecular modelling**

This modelling method is used to demonstrate how drugs and other drug-like substances may interact with the nuclear receptors of cells. As an example, Trevor Marshall [8] has used computer-based molecular modelling techniques to demonstrate that 25D, one of the vitamin D metabolites, and capnine, a substance produced by bacteria, turns off the vitamin D receptor.

3. **Whole cell simulations**

Researchers have now built a computer model of the interior of a bacterial cell. In a test to demonstrate its response to sugar in its environment, it was able to accurately simulate the behaviour of living cells. We will discuss a little more on in silico methods in Sect. 1.4.9.

Extracting structural data of a biomolecule is a critical step. Major techniques and methodologies used in determining the structure of biomolecules are:

X-ray Crystallography is the oldest and the most favoured technique currently in use for structure determination of proteins and biological macromolecules. This is primarily because the method is least expensive. X-rays having much shorter wavelength than visible light give a definite pattern of diffraction when passed through crystal lattice; trained experts can form the structure of the molecules from X-ray plates containing the images. This method has been improved through using solid sensors and mathematical techniques and regularly used in structure-based drug design, elucidation of enzyme mechanisms, specificity of protein–ligand interactions and site-directed mutagenesis.

After obtaining the crystals of suitable size the subsequent step is to analyse their X-ray diffraction behaviour. Trial exposures of new crystals can be performed on X-ray generators. X-ray generated from synchrotron has the advantage of an extremely intense X-ray beam with high-quality optical characteristics. To arrive at the final image from the complex diffraction pattern, computer image processing techniques are used.

Nuclear Magnetic Resonance (NMR) spectroscopy is used to obtain information about the structure and dynamics of proteins and nucleic acid. The process is not unlike MRI techniques used in medical imaging.

The NMR spectroscopy process involves the quantum mechanical properties of atomic nucleus. These quantum mechanical properties depend on the local molecular environment. Their measurements provide a comprehensive map of how the atoms are linked chemically and their proximity in special domain and how rapidly they move with respect to each other. The 3D structure itself is derived from complex computer calculations. Information derived from the behaviour of nuclear spin (quantum number) and the signals generated from them are employed to synthesize an image. The process is performed in an aqueous medium and thus a more appropriate image of living organelle can be visualized.

The sample is placed inside a powerful magnet, radio frequency signals are sent through the sample, and the degree of absorption of these radio frequencies by the sample is measured. Individual atoms within the protein absorb different radio frequencies depending on their environment. The absorption signals of different nuclei may be perturbed by adjacent nuclei. Sensors capture the above-mentioned signal and images formed through computer processing.

Electron Cryomicroscopy [9] is a form of Transmission Electron Microscopy (TEM). TEM itself is a dominant instrument in biology and to find the structure of micro-materials. An electron gun beams out on a processed material (light falling on an object) however unlike ordinary light, the wavelength of the falling electrons is extremely small, these are called de-Broglie waves. In cryo-electron microscopy, the sample is studied at liquid nitrogen temperatures. Appropriate precaution is taken so that ice crystal does not form on the sample.

Cryo-electron microscopy allows the observation of specimens in their native environment. An enhanced version of this equipment is Cryo-Electron Tomography (CET) that enables 3D reconstruction of a sample. Tomography is a process where multiple images are taken of the sample from different angles, and using computer image processing, a three-dimensional image of the object can be formed.

The processes mentioned above are only three of the multitude of techniques used to get information about biomolecules. An important part in experiments is to get the required samples from living organisms in vivo (say from a patient under some treatment and administered with drugs) and to change them in vitro (in test tube in the laboratory).

To sum up the methods, NMR is limited to smaller proteins, while X-crystallography has no limit.

Protein dynamics can be studied using NMR, while X-ray crystallography gives static structural information. No single technique is strong enough to solve a problem in biology.

1.4.9 More on In Silico Methods

All the experimental setups discussed earlier, even a non-specialist can realize that they are extremely time consuming and expensive. Even though costly and time consuming, these are essential for design of a target molecule—new drug or a product for different applications. Further, design of new drugs demands testing of the target drug on animal model followed by its testing on human volunteers. The associated cost and delay force us to think for a viable alternative.

Ideally, we could try the earlier experimental set-up in silico, i.e. in a computational machine. These are relatively cheaper and faster, depending on the methodology and algorithms employed. The in silico experimental results are next validated through in vivo and in vitro tests. In the process, the in silico experiment could filter out unnecessary trials of target molecules for in vivo test and thereby reducing cost and time. The inherent challenge for the researchers developing in silico experimental set-up is to achieve the accepted level of prediction quality. This issue has been given prime importance in the design proposed in this book.

Modelling *in silco* is dominated by Molecular Dynamics (MD) simulation in computers. MD has its roots in theoretical physics but now commonly used in several domains including material science, biochemistry and biophysics. MD is frequently used to refine the three-dimensional structures of the proteins based on experimental constraints from crystallography or NMR spectroscopy. Thus, molecular dynamics starts from first principle, i.e. molecular forces and interaction. The first simulation is performed on models where data found through other methods are known, this is the tuning phase. Ultimately and through incremental steps larger biosystems, even whole biological cells are simulated.

Limitations of the MD method are related to the parameter sets used, and also the underlying molecular mechanics force field. Hydrogen bonds, as discussed earlier, play a dominant role in cellular biology. One of the biggest problems lies in incorporating hydrogen bonds which have to be described as Columbic interactions of atomic point charges and hence at best a crude approximation. Furthermore, the surrounding aqueous solution has a much higher dielectric constant which is not given due consideration.

Design of a molecular dynamics simulation is dependent on the available computational power even though computing power has become exceptionally cheap. Simulation size, time step and total time duration must be selected so that the simulation can finish within a reasonable time period. However, to make statistically valid conclusions from the simulations, the time span simulated should match the kinetics of the natural process.

MD research, the de facto standard simulation tool, received strong industrial support. Significant progress has been achieved with the advent of refined methodologies and availability of distributed computing environment. However, the basic issue of exponential computational complexity will continue to be a major hurdle, particularly for handling large size molecules.

If we analyse the global behaviour of proteins, RNA and DNA, we would find that it would be extremely helpful if there is a method which is able to mimic their behaviour and is not dependent on depicting how the various forces interact to derive the observed behaviour. Viability of such a methodology is worth exploring provided availability of results of detailed experimental data is ensured to make the in silico experimental set-up. Fortunately, the availability of experimental data is very high and growing significantly. Further, diverse research groups around the globe have focussed their attention on organizing the databases to access the relevant information from this vast repository of wet lab experimental data. In this background, we concentrate on the design of a methodology for behavioural modelling of the building blocks of life sciences—DNA, RNA, protein.

We need an in silico methodology which will be designed to capture the behaviour of the ingredients involved—codons of mRNA string, nucleotides of DNA and RNA string, and amino acids of protein chain. Next, we let a given set of these entities to evolve according to certain rules derived for mimicking the physical properties of the ingredients. The entities will evolve as per the given rules and finally will attain a steady state. Next, we will try to map the observed behaviour of these entities to the features of the snapshot of life (protein + nucleotides) acquired experimentally through in vivo, in vitro or cell line study. Let us call it as cLife (our coined short form of cellular automata based life, we will come back to it later). A cLife is not a simulation of life; it is an analogy of life running in a different way. The cLife itself has to run on a computer, so we call it in silico method. However, it is different from all traditional in silico methodologies proposed so far. The mathematical framework used for the method was proposed long ago. Next section briefly depicts the historical background of this proposition employing a new kind of computational biology.

Prior to switching our discussions on our in silico framework we have designed to model biomolecules, let us point to a recent trend in computational biology—Artificial Intelligence (AI) driven methodology for drug design.

The branch of computer science named AI aims to generate programs those can execute the intended functions as intelligent as human beings. In the 1980s while John McCarthy [10] proposed this branch as—'The science and engineering of making intelligent machines, especially intelligent computer programs', it generated considerable interest in academia and industry. The faith deposed on the discipline got belied due to its failure to meet the minimum level of expectation of user community. The primary reason was the limited computing power available in that period. Recent surge of interest in AI is primarily driven by phenomenal increase in computing power and decrease of storage cost. In parallel, the associated support for AI has come forward from the disciplines of computer science, mathematics, psychology and cognitive science in respect of reasoning, learning and problem solving. In addition, fine-tuning of AI platform demands availability of large-scale data sets for the specific application domain that was also lacking at that period.

In the above background and with the availability of large-scale clinical data available from hospitals, AI community in association of drug designers has generated high hope of developing AI-driven drug design methodologies to address different classes of disease.

Compared to bottom-up approach of MD simulation, the AI-driven platform bank on top-down modelling of observed behaviour extracted from laboratory and hospital clinical data and knowledge derived out of experience of drug designers. AI-driven approach tries to extract semantic meaning of a sentence purely from the sentence itself without any semantic knowledge of words used to frame the sentence. At best, such semantic knowledge is retrieved from the experience of drug designers. Further, identification of a consistent pattern out of voluminous data is an extremely time consuming and computing intensive, and error prone due to diverse sources of data, and their validation for different age, sex and race of geographically dispersed targeted patient community around the globe.

1.5   Introduction to Cellular Automata (CA)

The concept of Cellular Automata (CA) was originally proposed in the 1940s by John von Neumann [11] and Stanislaw Ulam [12] while they were at Los Alamos National Laboratory. Neumann's initial design referred to as the kinematic model was based on the notion of a robot building its own copy. As he continued to develop his design, Neumann realized the fundamental difficulty of building a self-replicating robot, along with its high cost of providing the robot with a 'sea of parts' to build its own replica. In 1948, Neumann reported his ideas of self-replicating system in the paper entitled 'The General and Logical Theory of Automata'. Ulam suggested using a *discrete* system for creating a reductionist model of self-replication. Nils Aall Barricelli also performed many of the earliest explorations of these cellular automata models of artificial life supporting self-replication.

It was only after the publication of John Conway's Game of Life [13] (a two-dimensional cellular automation) in the 1970s, interest in the subject expanded beyond academia. Stephan Wolfram engaged in the systematic study of one-dimensional cellular automata while his research assistant Mathew Cook demonstrated that one of these cellular automata rules is 'Turing complete'. We suggest readers to use 'Conway Game of Life program' [13] to get the initial excitement of CA.

Prior to undertaking study of CA, we will take a short discourse on discreteness, computation and computability.

A 'Turing complete' is defined as a system in which a program can be written which will find an answer to any given problem. The closely related term 'Turing equivalence' defines two computers A and B as equivalent if B can simulate A and vice versa. Turing completeness is significant for the fact that every real-world design of a computing device can be simulated by a universal Turing machine. Alan Turing, Alanzo Church, and independently in another way, Kurt Gödel went on to prove that there exists a set of simple instructions, which, when put together, are able to produce any possible computations.

At the end of Sect. 1.2.2, we mentioned that the fundamental entities or building blocks of the known universe and their interactions can be treated as representation of a set of numbers (quantum numbers). We do not see the subnuclear or subatomic particles, everything we say that exists are due to their interactions. These interactions are guided by a set of rules. The building blocks of atoms could be represented in numbers; what we interpret as force, mass/ energy are just information processing as per a set of rules. In fact, that is what we do in Molecular Dynamics modelling (MD). Common molecular dynamics simulations scale by $O(n^2)$, big 'O' notation is used in computer science to describe the performance or complexity of an algorithm. 'O' specifically represents the worst-case scenario, and is used to describe the execution time required or the space used by an algorithm for a problem size n which is an extremely large number for biomolecules. For example, for modelling a macromolecule, n refers to the number of atoms covered by the molecule. This is the scenario after taking approximations which are essential to complete the computation within reasonable time. It is conjectured that to simulate the behaviour of a single proton (made of the quarks) with modern supercomputers without imposing any approximation would take greater than a million years. Life intrinsically is more complicated.

We are not sure if life as an information processor is Turing complete. If we consider the availability of any amount of time and memory, could we create its computational equivalent? To add to the problem, life is changing in macro-molecular level and we have no direct mapping of this to their constituent atomic or molecular behaviour. Life is an evolving system inside a universe which has its relatively simpler dynamics.

Having finite memory and computational power and assuming that the behavioural model of life is computable, we have to find a method in silico which would help us to build the model. The great theoretical physicist John Archibald Wheeler [14] enriched this idea in a concept he called '*It from Bit*'. In his own words:

"I, like other searchers, attempt formulation after formulation of the central issues and here present a wider overview, taking for working hypothesis the most effective one that has survived this winnowing: It from Bit. Otherwise put, every it—every particle, every field of force, even the space time continuum itself — derives its function, its meaning, its very existence entirely—even if in some contexts indirectly—from the apparatus-elicited answers to yes or no questions, binary choices, bits"

"It from Bit symbolizes the idea that every item of the physical world has at bottom—at a very deep bottom, in most instances—an immaterial source and explanation; that what we call reality arises in the last analysis from the posing of yes–no questions and the registering of equipment-evoked responses; all things physical are information-theoretic in origin and this is a participatory universe".

In his book *Einstein: An Intimate Study of a Great Man*, the author Dimitri Marianoff, states:

"No account of existence can ever hope to rate as fundamental which does not translate all of continuum physics into the language of bits. We will not feed time into any deep-reaching account of existence. We must derive time—and time only in the continuum idealization—out of it. Likewise with space" [15]

If matter, energy and space are informatics in origin, so are the molecules of life. As first demonstrated by Gregor Mendel, the genetic information is passed through generation by discrete means, though he did not know about DNA back then. We have seen the discreteness of life as established through DNA.

The earlier discussions emphasize the fact that if the universe is an information processing machine then so is biological life. There are infinite ways to interpret and store information, in words, in pictures and in numbers. In biological strings, this is done using chemical combinations.

The thoughts mentioned here are not established, so far there is no experimental confirmation of either binary or quantized nature of the universe. This book is about a method dealing with information based on behaviour of the blocks of life. Our method itself is independent, if any or all 'discreteness' hypothesizes are false. But if proven true it will definitely vindicate most of the conjectures.

On the lighter side, we quote Neil deGrasse Tyson, director of the Hayden Planetarium, who put the odds at 50–50 that our entire existence is a program on someone else's hard drive. 'I think the likelihood may be very high,' he said. He noted the gap between human and chimpanzee intelligence, despite the fact that we share more than 98% of our DNA. Somewhere out there could be a being whose intelligence is that much greater than our own. "We would be drooling, blithering idiots in their presence," he said. "If that's the case, it is easy for me to imagine that everything in our lives is just a creation of some other entity for their entertainment". For scientists, such an entity is 'Mother Nature' being studied since prehistoric age.

Now we will show that CA is a universal computer (in a trivial manner) and introduce this unique branch of computational method for building our model.

One could question why has not CA dominated over traditional mathematical techniques. There are two important reasons—first, CA study demands a computer, it is not analytical. For hundreds of years, traditional mathematical techniques are applied very successfully. Computers are used to solve the problem numerically where analytic methods fail. Second, similar to a mathematical framework, CA just like mathematics should be viewed as an independent subject, a universe on its own. Only in the very recent years when we are finding standard mathematical techniques failing, we are revisiting the CA universe. CA techniques have been used effectively in fluid dynamics instead of using numeric solutions of Navier–Stokes equation. We have reported a brief introduction in Annexure 1.3—on CA applications in other branches of science.

1.5.1 Cellular Automata

One-dimensional CA is a linear array of cells (Fig. 1.24a).

Figure 1.24b displays a ten cell representation numbered from 9 to 0, where the 0th and 9th are pseudo cells representing CA boundary. For null boundary, these two pseudo cells are permanently set at zero value as shown in Fig. 1.24. Each square block represents a CA cell that can store binary information—'0' or '1'. The information stored is referred to as 'state' of a cell. Instead of binary, a CA cell can store more than two states.

Let us discuss the simplest CA with two states per cell, three-neighbourhood and null boundary. Each cell in Fig. 1.24 has two neighbours—left and right; in addition, the cell itself is assumed as its neighbour, thus making three-neighbourhood of a CA cell. Each cell of a CA evolves in discrete time steps —also known as generations. Prior to evolution of CA cells, each cell is initialized

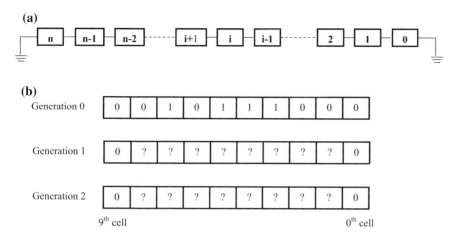

An 8- Cell boundary CA: each cell changes its state according to the CA rule, the two cells marked '9' and '0' define the cell boundary

Fig. 1.24   Cellular Automata (CA) cell and evolution

with state '0' or '1'. This binary input string is referred to as 'seed' to initiate CA evolution. The two terminals cells of one-dimensional CA have only two neighbours instead of three—leftmost cell has no left neighbour, and rightmost cell has no right neighbour. If we assign state 0 (null) for the missing neighbours, the resulting CA is referred to as 'null boundary' CA. The CA is called as periodic boundary if rightmost cell is the neighbour of leftmost one and vice versa. The CA model proposed in this book only employs null boundary CA.

Each ith cell in a cell array can have one of the two state values 1 or 0, sometimes shown as blackened box or empty (white box). Consequently, there can be $2^3 = 8$ combinations of current states of cell triplets $<(i - 1) \, i \, (i + 1)>$ marked as 7 (111), 6 (110), 5 (101), 4 (100), 3 (011), 2 (010), 1 (001), 0 (000) noted in Fig. 1.25. The last row shows the evolution of next state of ith cell for each of the 8 possible combinations of the current states of its neighbours. The decimal counterpart of the 8 bit string of cell next state is referred to as 'CA rule' for the ith cell. Figure 1.25 displays the CA rule 90: $(01011010)_2 = (90)_{10}$. In the leftmost position (the 7th place), the current state of three cells $<(i - 1) \, i \, (i + 1)>$ is $<111>$ for which the next state of ith cell is 0. Similarly, for the 6th place the current states of three cells have values $<110>$ for which next state value of ith cell is 1, and so on for all the 8 places.

Next let us analyse an interesting rule noted in Fig. 1.26: $(110)_{10} = (01101110)_2$

If we work in the same manner for Rule 110, as we did using Rule 90 (Fig. 1.25), we can justify the value of central cell (01101110) as shown in Fig. 1.26. As per this rule, for input states $<000>$, next state is 0s, while the output is 1 for input state triplets as $<001>$, and so on.

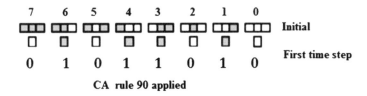

CA rule 90 applied

Fig. 1.25 States of a three-neighbourhood CA and evolution with CA Rule 90

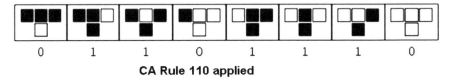

CA Rule 110 applied

Fig. 1.26 States of a three-neighbourhood CA and evolution with CA Rule 110

The interesting point is the evolution of the 32 cell CA, with rule 110 assigned to each cell. If this CA is operated for 14 steps after initializing with value 1 for 14th cell and 0's for all other cells, we shall arrive at the pattern shown in Fig. 1.27.

It is definitely tedious to determine the state of cells after a couple of generations, hence we require computers. In order to design an n-cell CA, we need to fix the value of n, set the rules for each cell, set the boundary (null/periodic), fix the seed (initialization value 0 or 1 for each cell) along with the number of steps of evolution. We may terminate after generating reasonable number of evolution steps. We have provided computer codes to run and generate behaviour of CA in Annexure 1.4.

Let us present a small example of linking behaviour of CA with real-world logic. Logic gates are essential parts of digital systems; using logic gates in correct combination, we can perform logical and arithmetic operations. The NAND or

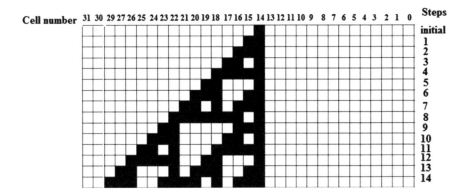

Fig. 1.27 Evolution of 32-bit CA cell with Rule 110

NOR logic gates are called 'universal logic' gates, as any other gates can be constructed from them. Let us see the output of a two input NAND gate with inputs A and B and output Q.

A	B	Q
0	0	1
1	0	1
0	1	1
1	1	0

Let us revisit Rule 110 to check how this rule mimics NAND logic (Fig. 1.28).

Figure 1.28 shows that using Rule 110, universal gate like NAND can be constructed. In other words, we can extrapolate that Rule 110 is Turing complete (obviously, the above method is not a proof). As rule 110 and other rules results in Turing completeness, then we can say 'any problem that needs a universal computer to solve, its simulation can be done with CA'. If the protein and nucleotide base life is Turing complete, then cLife (CA based Life) should be able to mimic biological life provided we use appropriate rules, correct seed, employ proper viewing option in respect of what CA evolution patterns we should look for behavioural modelling of a biomolecule.

In the above background, let us summarize the essential requirements of building CA model. In the subsequent discussions, CAM (CA Machine) and CA are used interchangeably.

1. A CAM (CA Machine modelling a macromolecule—RNA, DNA, protein) has a rule assigned to each cell of the CA. The cell models a building block of macromolecules. As discussed in earlier sections, these are—sugar–phosphate molecule and nucleotide base of DNA/RNA, codon base triplet of mRNA string, amino acid of protein peptide chain. The rule is designed in such a manner that the physical and chemical properties of a micromolecule get reflected in the rule structure.

2. In each time step of CAM evolution, next state of ith cell depends on the current state of $(i − 1)$, i, $(i + 1)$ cells. Consequently, after k time steps, next state of ith

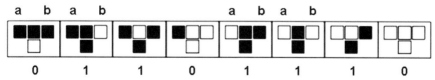

We have chosen those states where initial state of central cell = 1. Depending on value of a, b inputs, next state of central cell follows NAND gate rules

Fig. 1.28 CA rule 110 mimics NAND Logic

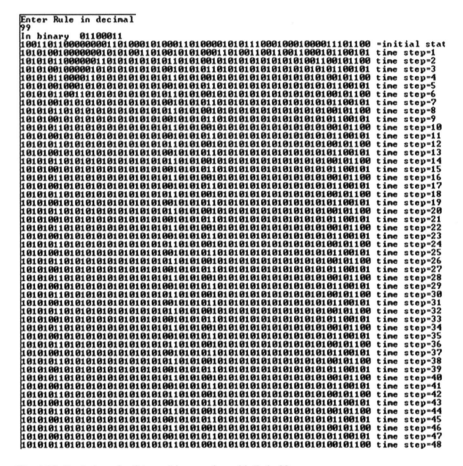

Fig. 1.29 Evolution of a CA machine running with Rule 99

cell will depend on the current states of all the cells in the region $(k - i)$ to $(k + i)$ positions. A CAM evolves allowing dependence of a cell state based on the state of any other cell over the evolving time steps—this is referred to as 'global evolution' of a CA, as shown in Figs. 1.29 and 1.30. At the end of its traversal through transient states, each CA cell settles at its attractor cycle where the cell cycles through a specific set of states.

3. The variation of basic properties of the micromolecule (e.g. amino acid may be polar or non-polar) should get reflected in the initial state value assigned to a cell, called seed, modelling the building block.

Based on this simplest CA system, it is interesting to see how a behavioural model can be generated out of CA evolution. A CAM evolves in two discrete phases—transitory and cyclic. It is worth noting the fact that the duration of transitory states of different cells may differ from one another. For the example 64

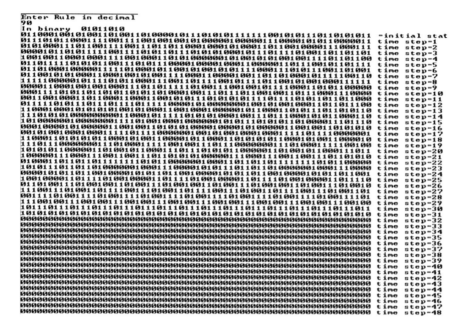

Fig. 1.30 Evolution of a CA machine running with Rule 90

cell CA in Fig. 1.29, while for cells 48, 49, the transition phase ends at time step 2, the same for 50, 51, 52, etc. transitions phase end at end of evolution step 3, 4, 5 respectively.

The first phase is referred to as Transitory State (TS) while a cell transitions through non-repeating states. The end of TS is marked as CS (Cycle Start). The second phase is called cyclic, while each cell runs through a set of cyclic states. Figure 1.29 displays the screenshot of the output of a CA machine using Rule 99 for each cell. The program included in Annexure 1.4 will prompt the user to enter the CA rule applied on 64 cells; a random initial value is given to each cell and the CAM runs for 48 steps. We can see that after maximum of 7 steps (transitory states), each cell goes through a cyclic state of length 1 or 2. If we run the same program with different rule and seed, we will get different results. Figure 1.30 displays the results of running the program with rule 90 for each cell. For this case, the length of transitory states is 31 (that is, CS = 31) and length of cycle is 1 (CL = 1)—that is, each cell runs through one cyclic state.

If we study Fig. 1.30, we can see a pattern in the transitory phase that ends at evolution step 31. The shape visible during the transitory state has a time-space diagram in the form of a triangular shape referred to as Sierpinski triangle. On the other hand, each cell runs through a single cycle with state 0 from evolution step 32 onwards.

When a CAM is in a cyclic state, we say it has settled in its 'attractor basin'. The CAM represents a nonlinear dynamic system, and therefore the result of a

CAM evolution can be ascertained only by running the machine. It might be possible to ascertain CA evolution pattern only for a specific sequence of rules assigned to CA cells. This aspect of CA evolution will be elaborated in Chap. 2.

CA has been used in many experiments ranging from modelling forest fire, architectural growth, generating different complex patterns which appear in biological systems and even in financial modelling. However, if we build the model with complex CA with higher dimension, dynamic rule change, multivalued state cells and larger neighbourhood, the complexity of CA evolution becomes equal or may be more than that of the target problem. The rudiments of biology we depicted are simple in nature, an aggregation of basic building blocks in chains storing encrypted information. The blocks interact with its neighbours in successive steps to output the end results as encrypted information. The only way to decrypt the input and output information is through wet lab experimental set-up. Fortunately, due to significant effort put forward around the globe on study of life sciences, enormous volumes of data are available in different databases. We shall take help of data/information available in different databases for validation of CA model output.

We will take the basic building blocks of life—proteins, RNA, DNA and design equivalent CAM for each of these blocks. However another major problem crops up, how do we visualize the result to compare with the real world? This point has been addressed in next section subsequent to presentation of the basic framework of designing CA rules for our application.

1.5.2 Design of CA Rules

Each CA rule is represented as a string of 8 bits; consequently, there are $2^8 = 256$ CA rules for three-neighbourhood two-state (0 and 1) CA cell. A rule having four 0's and four 1's is defined as a Balanced Rule (BR). There are $^8C_4 = 70$ such balanced rules, out of which 64 non-trivial rules are selected to model 64 codons of a mRNA codon string (refer to Sect. 1.4.3).

We will describe the CA modelling of protein, RNA and DNA in their respective chapters. In the current chapter, we present an overview of formulation of the CA rules used for building the model.

1-Major/0-Major Representation of a CA Rule

The eight-bit string of rule 141 (1000 1101) of CA cell (say, marked as ith cell) represents its next state corresponding to the current states of $(i - 1)$, i, and $(i + 1)$ cells. The triplet of current states is represented as—0 (000) to 7 (111), 6(110), 5 (101), 4(100), 3(011), 2(010), 1(001), 0(000). The triplets for 7, 6, 5, 3 have two 1's and termed as 1-major, while 4, 2, 1, 0 are referred to as 0-major. On rearranging the decimal values of 8 triplets <7 6 5 3 4 2 1 0> as 1-major and 0-major groups, the rearranged next state bit string leads to the following table for a sample CA rule 141:

Table 1.2 Rule equivalence CA rule 141 and CA rule 169

CA Rule	Normal	Rearranged bit string <1-major 0-major>	Equivalent CA rule
141	<1000 1101> (Decimal: 141)	<1001 0101>	Reversal of <1001 0101> to <1010 1001> Rearrange to original switch <7654 3210> Rule number of decimal value of string <1011 0001> = CA Rule 177

By convention, we evaluate the decimal value of a binary string on assigning highest/lowest weight to left/rightmost bit. A transform derived on reversing this convention (that is, highest/lowest weight to right/leftmost bit) leads to rule 177, as noted above. Hence, rule pairs 141 and 177 are equivalent under that transform. Such equivalent rule groups are derived by using different transforms. For subsequent discussions throughout this book, we shall only use the rearranged bit string of a rule (shown in column 3 of Table 1.2), while its decimal counterpart is always evaluated out of normal bit string (as shown on column 2), as per normal convention with highest/lowest weight assigned to left/rightmost bits. Design of CA rules is reported in Chap. 2, while modelling micromolecule with such rules is discussed in subsequent chapters.

1.5.3 Introduction to CA-Based Life (cLife)

Conway's 'Game of Life' [13] proposed in 1970 generated considerable interest. Game of life employs two-dimensional CA and complex rules for its evolution. However, CA-based life, cLife is entirely different. For the design of CA-based Life, we employ simplest one-dimensional CA—three-neighbourhood two-state per cell. Rather than employing uniform CA rule for each cell, we employ both uniform and hybrid (non-uniform) CA that do not employ the same rule for each of its cells. As per our study, we observed that even such a simple CA display exponentially wide variation of patterns in its evolution; such patterns, if selected properly, consistently represent relevant features of biomolecules. Further, as we proceed with our model, we observed that three neighbourhood CA does not suffice. At that point, we used one-dimensional five-neighbourhood CA. Throughout this book, we employ both three- and five-neighbourhood CA. Rules of cLife are designed based on the analysis of micromolecules.

The constituents of the basic building blocks of DNA, RNA, and protein are—sugar–phosphate backbone, nucleotide bases, codon representing a nucleotide base triplet and amino acids. CA rules are designed based on the data/information of these constituents. While nucleotide base, amino acid are modelled from the information of covalent bonds connecting a pair of atoms, a codon is modelled from the information of the specific base triplet it contains. Once CA rules are designed, a biological string of DNA, RNA, peptide chain of protein gets represented with a CA. Subsequent steps of cLife modelling do not model the environment or associated forces of molecular interaction. It is taking the information from established results and behaviour from in situ, in vivo and in vitro experiments. The cLife does not model hydrogen bonds, van der Waal's force, the pH of the liquid in which the protein, RNA and DNA are working. Because such physical features are associated in the laboratory set-up out of which results are derived. We extract relevant information/behaviour retrieved from the wet lab experiment to set up the in silico cLife model.

Thus cLife model can be viewed as bottom-up model to design CA rules for constituents (micromolecules) of the macromolecules—DNA, RNA, protein, followed by behavioural top-down modelling of the macromolecules. Once CA model conforms to the established results, the model is employed for extrapolation and predicting behaviour of any macromolecule, its interaction with other macro- and micromolecule (ligand, drug, other proteins, etc.). The prediction results will enable the experimentalists to narrow down the wet lab experimental study resulting in significant reduction of cost and time.

We have coined the word **cLife** (CA-based Life) because it is different from conventional simulation methodologies for biological systems. The proposed cLife only utilizes the data/information/knowledge embedded in the building blocks of life and their interactions derived from wet lab experiments. The cLife approach is not simulation, as it does not model the inherent details in time steps such as MD modelling.

To view the results of any *in silico* experiment, we generally try to produce in silico experimental results representing as close as possible to the relevant data/information/knowledge retrieved from wet lab experiment. A CA evolves in discrete time steps. As noted earlier, each CA cell runs through transitory states followed by settling in its attractor cycle. These two pieces of information of transitory and cyclic states are referred to as CS (Cycle Start) and CL (Cycle Length). In its cyclic evolution, the cell runs through a set of states. Specific parameters retrieved from CA evolution for each cell can be represented in Y-axis while representing serial cell position on X-axis to generate a graph referred to as 'signal graph'. On visualization, we extract specific patterns of signal graph that represent relevant features of a physical system. This step is referred to as 'signal graph analytics' elaborated in different chapters.

1.6   Summary

The objects in nature are made out of simple building blocks; each block has properties which can be expressed in numbers. The numbers dictate how they self-organize or interact with other blocks; a set of rules guide this processing of data/information contained in the object. This happens with subatomic particles, the building blocks of everything in the universe. Based on this universal information processing paradigm, we build behavioural model of the macromolecules of life— DNA, RNA, protein in different chapters of this book.

At a certain level of complexity, the blocks are formed of organic entities based on 'carbon' and a few other elements. They can store, modify information and self-replicate. Behaviour of these known macro-blocks of life is modelled using Cellular Automata (CA). Rather than bottom-up approach of traditional in silico simulation techniques, like Molecular Dynamic (MD) simulation, we concentrate on a combination of both—(i) bottom-up mathematical model of constituents (micromolecules) of DNA, RNA, protein with CA rules, followed by (ii) top-down behavioural model with CA evolution patterns.

Codons or amino acids in biological strings are comparable to words in a sentence we write to convey our feelings. Individually, a word does not carry the idea contained in the sentence. The words arranged in a specific sequence convey the meaning/feeling expressed in the sentence. The words can be arranged in many ways to convey different meaning to a sentence. So rather than concentrating on study of individual words in a sentence, we could start exploring the meaning of a sentence from the semantic knowledge of how the words are normally arranged in a sentence as per a given set of grammatical rules. Following this analogy, we build the behavioural model for biomolecules based on the information of its constituents —nucleotides, codons and amino acids. Fortunately unlike natural language, the grammatical rules to model the behaviour of constituent parts are limited. It is true that the context under which a biomolecule is analysed can vary significantly. In building the in silico model we have proposed, we have assumed availability of wet lab experimental data/information/knowledge on considering all such contextual variables. Our assumption is based on the available repository of biological data-bases that have been growing significantly in recent years. In this background, we concentrate on building the in silico model using the mathematical framework of Cellular Automata.

Cellular automata provides a unique method which can be used to solve any problem that a universal computer can solve. The simplest CA we have used has three basic properties—one dimension, three/five-neighbourhood, and two states per cell. On assigning appropriate rules to CA cells representing the constituents of a macro-block, we can build a dynamic behavioural model of the macro-blocks of life (RNA, DNA and protein). This model could be used to probe *in silico* and the results derived from the model can be validated from in vitro, in vivo experimental results available in public domain or available from wet lab experiments. On completion of this in silco experimental set-up, it can be used for prediction of results prior to proceeding for costly and time-consuming wet lab analysis. The sole purpose is to cut down the significant cost and time of wet lab study by filtering out case studies which are not relevant.

Questions

1. Why do you think this chapter started with the basic elements for introducing a concept in computational biology?

   (a) It is essential to know the laws of the physical domain to model biological entities.
   (b) Scientific investigation needs both top-down and bottom-up approach. So it is easier to start with something tangible at the top to introduce a discipline.
   (c) We are building up our biological model from the very smallest constituents of nature.

2. All the molecules in life are organic in nature, but all organic molecules are not. Give an example.
3. What is the heaviest element required and found in human body?
4. In DNA phosphates favour helicity. For a stable helix there must be an opposing force, what part opposes helicity?
5. An estimated 1 min 20 s is required to transcribe and translate 600 nucleotides into a 200 amino acid protein. Which organelles are responsible to make the translation process faster?
6. Computer simulation is used to model galaxy formation and subatomic particles, but what is the main difficulty faced in modelling macromolecules in life?

   (a) Biomolecules are more in numbers than stars in galaxies.
   (b) The environment of leaving system (water) is computationally difficult to model.
   (c) We know nothing about life forming molecule but a lot about subatomic particles.

7. Why the $n-1$ cell and 0th cell are null terminated in a one-dimensional CA?
8. All arithmetic and logical operation can be performed by using CA, justify the statement as true or false.
9. Without consulting any other chapter/resources indicate which CA rule represents the exclusive OR (XOR) logic.

Annexure

Annexure 1.1: The Standard Model

In Sect. 1.3.1 of this chapter, we have mentioned 'The Standard Model' without any elaboration. Hence a brief introduction to this model follows. First, let us go through the scale of the universe to get the idea where we stand in the grand plan from the smallest to the largest dimensions of the things we know of.

Naturally, there are dimensional uncertainties in both directions, from the smallest to the largest. Dimensions are in metres. The highlighted entries are associated directly with biomolecules of life.

- 10^{-24}—low energy neutrinos
- 10^{-17}—up/down quarks
- 10^{-15}—proton
- **10^{-12}—carbon nuclei**
- **10^{-10}—hydrogen atom**
- **10^{-9}—alpha helix**
- **10^{-8}—DNA**
- **10^{-5}—cell nucleus**
- $10^{-2.4}$—grain of rice
- 1.71—human
- 10^{7}—Earth
- 10^{9}—Sun
- 10^{21}—Milky Way (our galaxy)
- 10^{23}—Local cluster of galaxy
- 10^{26}—As far we can see of the universe

Most of us are deceived by the exponents. Suppose we have a sufficiently big paper of 0.1 mm thickness. We first tear the paper in halves and put the two pieces one over the other. We then take the two papers together and tear them and again put the two pair over the previous pair. Now we have 4 layers of paper and the total thickness is 0.4 mm. If we can proceed in this manner for total 94 times what will be thickness? Well, the final number is 2×10^{24} metres which is out of our local cluster of galaxies.

Particle physics describes what the universe (matter, energy and forces) is made of. Forces are defined by the interaction of particle–particle or particle–energy. Scientists have created a Standard Model which defines the properties of the particles and forces. The quest is to find the fundamental or unification of all kind of forces, a kind of single equation or at least a set of co-related relations which will define everything we can measure. The oldest of forces felt by us is gravity. We are not aware of it as on Earth we are under its influence all the time unless we are falling off from sufficiently tall structure, which is called 'free fall'. During the free fall, we realize that we are weightless. The gravitational attraction between masses is universal. This force is not described by the Standard Model. It is described in general theory of relativity.

The next two forces are electric and magnetic. Electricity was discovered when we constructed galvanic cells. Magnetic forces were discovered with natural occurring lodestones. We found that relative movement of magnets and electrical conductors generates current in the conductors. James Clark Maxwell unified two forces, the electric and magnetic and we now call it electromagnetic. After the discovery of subatomic particles, we found that other forces are required which

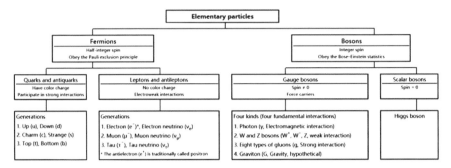

Fig. 1.31 Family tree of elementary particles

obey different laws. The quest of such unification of law of forces and discovery of subatomic particles gave us what is known as 'The Standard Model'.

The theory describes three of the four known fundamental forces in the universe, the electromagnetic, weak and strong interactions, as well as classifying all known elementary particles. It was developed in stages throughout the latter half of the twentieth century, through the work of many scientists around the world. The current formulation was finalized in the mid-1970s upon experimental verification of the existence of quarks. The discovery the top quark the tau neutrino and Higgs boson have added further credence to the Standard Model. The family tree of all the particles is shown in Fig. 1.31.

For most of us, we are familiar with the electron, photon (a lepton and a boson). The proton and the neutron are not fundamental as they are made of three quarks as mentioned in Sect. 1.2.2: Below the atoms of this chapter. That the neutron may not be fundamental was conjectured in the second decade of twentieth century. Outside the atomic nucleus, the neutron is unstable and decayed within 10 min. Both matter and the mediator of forces behaves as particle (they also behave as waves but we are not going through all that here). All particles that carry or mediate forces are called bosons and their spin must be non zero. For electromagnetic we have photons, for the weak (required in radioactive decay), we have three particles W+ , W− and Z.

James Clark Maxwell showed that electric and magnetic field can be expressed in unity as electromagnetic field such as radio waves, visible light and X-rays. It was found in 1961 that the weak and electromagnetic can also be unified in so-called electroweak. The strong force which holds the quarks together to make the hadrons like the proton took years of research to combine with electroweak force. Today, we have a model of high confidence level from which we could form the chart Fig. 1.2 of this chapter. The only vital force that remains isolated is the gravity.

The Standard Model includes 12 elementary particles with ½ integer spin known as fermions. Fermions obey Pauli's Exclusion principle. The principle states that two or more particles with half-integer spin cannot occupy the same quantum state within a quantum system simultaneously. For example, if two electrons reside in the same orbital, and if their n (principal quantum number), ℓ (angular momentum quantum number), and $m\ell$ (magnetic quantum number) values are the same, then their *ms* (spin quantum number) must be different. This quantum number dictates the structure of atoms, the physical and chemical behaviour of elements.

The fermions of the Standard Model are classified according to how they interact. The quarks interact through strong interaction. This works for very short distances but is extremely strong. There are six quarks and six leptons. They are grouped into three generations. We have no idea why this is so. The three quarks are combined in such a way that they are colour neutral (just like charge neutrality). Quarks are the building blocks of hadrons. Hadrons made of a quark–antiquark is called a meson, those made from three quarks are baryons, like proton and neutron. One can just see combination rules and quantum number results in hundreds of subatomic particles; most very short lived (they decay into more stable particles and give out energy) and some like proton which effect everyday life, probably do not decay. Quarks also carry electrical charge and weak isospin and interact with leptons electromagnetically and weak interaction.

The six fermions do not carry colour charge and are called leptons. The three neutrinos do not carry electric charge so their motion is directly influenced only by the weak nuclear force. The electron, muon and tau all interact electromagnetically.

Each member of a generation has greater mass than the corresponding particles of lower generations. The first generation charged particles do not decay; hence all ordinary (baryonic matter = the constituents of matter making life) matter is made of particles of the first generation.

All particles that have rest mass must move at the velocity less than light (photons) in vacuum. The force carrier particles are called gauge bosons each mediating different forces. The gauge boson for electromagnetic field is the photon. At a microscopic level of life, electromagnetism allows (photons and electrons) particle to interact with each other. As far as we know the very small world of the quantum has little influence on the behaviour of life.

As matter particles, the gauge bosons of the Standard Model all have spin. The value of the spin is 1 and Pauli's exclusion principle cannot be applied to them. The bosons do not have a theoretical limit on spatial density, one can pack as many bosons in a unit volume of space. Photons mediate the electromagnetic force between electrically charged particles. The photon is massless and its interaction with fermions is described in quantum electrodynamics.

The W+, W− and Z gauge bosons mediate the weak interaction between quarks and leptons. They are massive. The W carries an electric charge of +1 and −1 and couples to the electromagnetic interaction. The electrically neutral Z boson interacts with both left-handed particles and antiparticles. These three gauge bosons along with the photons are grouped together, as collectively mediating the electroweak interaction.

The eight gluons mediate the strong interaction between colour charged particles, the quarks. Gluons are massless. The eightfold multiplicity of gluons is labelled by a combination of colour and anti-colour charge. Because the gluons have an effective colour charge, they can also interact among themselves. The gluons and their interactions are described by the theory of quantum chromodynamics.

Colour charges are named so as to differentiate the name of a type of charge and the kind of housekeeping name. It has nothing to do with actual colour, just as the spin is not analogous to a spinning top, else how could one explain +1/2 and −1/2 spin.

Lastly, the Higgs particle is a massive scalar elementary particle. It has no spin. The Higgs boson plays a unique role in the Standard Model, by explaining why the other elementary particles are massive, while the photon and gluon have no mass. In particular, the Higgs boson explains why the photon has no mass, while the W and Z has. The Higgs boson generates the masses of the leptons (electron, muon, and tau) and quarks. As the Higgs boson is massive, it also interact with itself.

In quantum field theory particles are manifestation of an excited state of a field, photon is to electromagnetic as is Higgs to Higgs field. More a particle interacts with the Higgs field, more massive it becomes.

The fundamental matter–force description of the basic constituents making our universe was developed from mathematical principles based on rules, symmetries and conservation. That by doing mathematics, we can predict the existence of matter particles and forces is by itself a deep mystery. We have come very close to finding out the basic building blocks, this gives us confidence that from all the tables, charts and rules we can make sense and life will be able to find its own basics.

Although the Standard Model is theoretically self-consistent and has demonstrated huge successes, it falls short of being a complete theory. Among others, it does not fully explain gravitation, matter–antimatter asymmetry and neutrino oscillation. We have a long way to go and that brings hope that our life will remain interesting for a long time, we have so many unanswered questions.

Annexure 1.2: Biological Cells

We referred to biological cell in Sect. 1.4 of this chapter without any details. This annexure reports a brief introduction on biological cell.

The cell is the basic structural, functional unit of all known living organism. A cell can replicate independently. Cells are the building blocks of life. Organism consisting of a single cell is called unicellular, for example, bacteria. Higher organism such as plants contains multiple cells. Human contain more than 10 trillion cells. Cell's dimension varies from 1 to 100 micrometres.

Cells consist of a membrane enclosing a disordered colloidal solution called cytoplasm. The material properties of the cytoplasm remain an ongoing investigation. Recent measurements reveal that the cytoplasm can be likened to an elastic solid, rather than a viscoelastic fluid.

The simplest cell does not have a nucleus inside the cell membrane; these are called prokaryotic cells and were the first form of life on earth which emerged at least 3.5 billion year ago. The DNA of the prokaryotic cell consists of a single chromosome that is in direct contact with the cytoplasm. The dimension of prokaryotic cell ranges from 0.5 to 2.0 μm in diameter. Bacteria fall in the domain of prokaryotic cellular organism.

Enclosing the cell is the plasma membrane covered by a cell wall, but in some microorganism, the cell wall is absent. The cell wall gives rigidity to the cell and serves as a protective filter. Inside the cell is the cytoplasmic region that contains DNA, ribosome, etc. There are variations in the internal elements of a cell depending on different unicellular life forms. This shows the diversity and evolutionary dynamics that even the simplest cell went and is going through.

Plants, animals, fungi, slime moulds, protozoa and algae belong to eukaryotic group. Eukaryotic cells are 15 times larger than prokaryotic (and hence thousand times more in volume) cells. The various organelles that perform specific metabolic function are compartmentalized inside the cell membrane. The most important among these is cell nucleus, an organelle that houses the cell's DNA. The plasma membrane resembles that of the prokaryotes, i.e. physical communication with the environment for nourishment and waste disposal.

The eukaryotic DNA is organized in one or more linear molecules, called chromosomes. All chromosomal DNA is stored in the cell nucleus, separated from the cytoplasm by a membrane. Some eukaryotic organelles such as mitochondria also contain some DNA. We have mentioned the existence of two different kinds of genetic material, DNA and RNA in this chapter.

Cells use DNA for their long-term information storage. The biological information contained in an organism is in its DNA sequence. RNA is used for information transport (mRNA). Transfer RNA (tRNA) is used to add amino acids during protein translation. The organelle 'ribosome' is the organic body which connects genomics to proteomics.

On Cell Genetic

A human cell has genetic material contained in the cell nucleus and in the mitochondria. In humans, the nuclear genome is divided into 46 linear DNA molecules called chromosomes, including 22 homologous chromosome pairs and a pair of sex chromosomes. The first 22 pairs are called autosomes. The chromosomes of the 23rd pair are called allosomes consisting of two X chromosomes in women, and a Y chromosome in men.

The X chromosome is always present as the 23rd chromosome in the ovum, while either an X or Y chromosomes can be present in an individual sperm. Early in female embryonic development, in cells other than egg cells, one of the X chromosomes is randomly and partially deactivated; in some cells, the X chromosome inherited from the mother is deactivated, while in others, the X chromosome from the father is deactivated. This ensures that both sexes always have exactly one functional copy of the X chromosome in each cell.

The mitochondrial genome is a circular DNA molecule distinct from the nuclear DNA. It is very small compared to nuclear chromosomes and it codes for 13 proteins involved in mitochondrial energy production and specific tRNAs. In human, mtDNA is inherited solely from the mother.

Foreign genetic material (most commonly DNA) can also be artificially introduced into the cell by a process called transfection. Transfection is transient, if the DNA is not inserted into the cell's genome. Transfection is stable if it is inserted. Certain viruses also insert their genetic material into the genome. There are various methods of introducing foreign DNA into a eukaryotic cell including electroporation, cell squeezing and nanoparticles. Several chemical materials and biological particles (viruses) are used as carriers. In Chap. 4 of this book, we have discussed how similar processes are used by biotechnology and in silico processes are attempting to simulate them.

Anatomy of a cell structure is noted below following a brief description of its organelles. Analogous to the different functions of organs in a human body, organelles are parts of the cell which are adapted and/or specialized for carrying out one or more vital cell functions. Both eukaryotic and prokaryotic cells have organelles; prokaryotic organelles are generally simpler and are not membrane-bound.

There are several types of organelles in a cell. A few such as the nucleus and Golgi apparatus are typically solitary in number. Others such as mitochondria, lysosomes exist in large numbers.

Mitochondria are self-replicating organelles that occur in various numbers, shapes and sizes in the cytoplasm of all eukaryotic cells. The biological processes of a cell are often referred to as its 'respiration' that occurs in the mitochondria, generate the cell's energy, using oxygen to release energy stored in cellular nutrients. Mitochondria multiply through binary fission like prokaryotic cells. They are known as the powerhouses of the cell. They are organelles that act like a digestive system which takes in nutrients, breaks them down and creates energy-rich molecules for the cell. The biochemical processes of the cell are known as cellular **respiration**.

The **Endoplasmic Reticulum** (ER) is a transport network for molecules targeted for certain modifications and specific destinations, as compared to molecules that float freely in the cytoplasm. The ER has two forms: the rough ER, which has ribosomes on its surface that secrete proteins into the ER, and the smooth ER, which lacks ribosomes. The smooth ER plays a role in calcium sequestration and release. The primary function of the **Golgi apparatus** is to process and package proteins and lipids that are synthesized by the cell.

Lysosomes contain enzymes that digest excess or worn-out organelles, food particles and engulfed viruses or bacteria. **Peroxisomes** secrete enzymes to get rid of toxic peroxides in the cell. The **centrosomes** produce the microtubules of a cell. It directs the transport through the ER and the Golgi apparatus. Centrosomes are

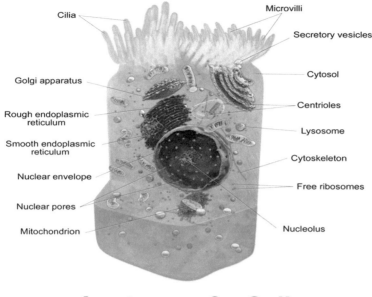

Anatomy of a Cell

Created by Blausen.com staff (2014) and contributed to Wikimedia
Projects (open access publications) of Wikimedia Foundation

Fig. 1.32   Anatomy of a cell

composed of two centrioles, which separate during cell division and help in the formation of the mitotic spindle. A single centrosome is present in the human cell. Vacuoles provide structural support, as well as serving functions such as storage, waste disposal, protection and growth.

A vital member of all the organelle is the **ribosome**, we have already described this unit in details in this chapter (Fig. 1.32).

Cell division involves a single cell (called a *mother cell*) dividing into two daughter cells. Such procreation of cells results in growth of multicellular organism. Prokaryotic cells divide by binary fission while eukaryotic cells usually undergo a process of nuclear division, called mitosis, followed by division of the cell, called cytokinesis. A diploid cell (cells that have two homologous copies of each chromosomes) may also undergo meiosis (specialized type of cell division that reduces the chromosome by half) to produce haploid cells, usually four. Haploid cells serve as gametes in multicellular organisms, fusing to form new diploid cells.

In mitosis, DNA replication always happens when a cell divides into two. In meiosis, the DNA is replicated only once, while the cell divides twice. DNA replication only occurs before meiosis I. DNA replication does not occur when the cells divide the meiosis II. Cell division is an extremely complex and essential process.

In complex multicellular organisms, cells specialize into different cell types that are adapted to execute specific function. In mammals, major cell types include skin cells, muscle cells, neurons, blood cells and others. Cell types differ both in appearance and function, yet they are genetically identical. Cells are able to be of the same genotype but of different cell type due to the differential expression of the genes they contain.

Cells emerged at least 3.5 billion years ago. The early cell membranes were probably more simple and permeable than modern ones, with only a single fatty acid chain per lipid. Lipids are known to spontaneously form bi-layered vesicles in water, and could have preceded RNA, but the first cell membrane could also have been produced by catalytic RNA, or even have required structural proteins before they could form.

In silico study of cell demands approximately whole cell computational modelling. This is essential since the study of detailed behaviour of the various organelles cannot be done in vitro. The brief description of this annexure points to the following properties of a biological cell—it has a digestion system (energy input from external environment), an excretory system to get rid of waste and toxic product, an organelle to store its characteristics and a mechanism to create the required proteins for survival and damage repair. Most important, it can make a replica of itself. We have not yet created cLife for whole cell modelling, but we aim to do so in the future after fine-tuning the models of RNA, DNA and proteins.

Annexure 1.3: Application of Cellular Automata in Different Branches of Science

A Cellular Automata (CA) is a discrete dynamical system. In Sect. 1.5.2 of this chapter, we introduced simplest one-dimensional two states per cell CA with three neighbourhood and null boundary. It evolves in discrete time steps with next state value of a cell dependant on the states of its three neighbours. We have employed such one-dimensional three or five-neighbourhood CA. By convention, the CA dimension is referred to as its grid. In different chapters of this book, we have employed one-dimensional (1D) simple CA modelling biological strings—DNA, RNA and protein.

Majority of CA modelling of physical phenomenon, including Conway's 'game of life', uses two-dimensional grid operated with complex rules. Generic definition of a cellular automaton includes the following parameters:

- A grid is defined along with its boundary conditions. In many applications, periodic boundary conditions are used; consequently, a two-dimensional (2D) grid represents the surface of a torus.
- Grid cells can have finite values; a cell storing k states is referred to as k state per cell CA; every CA cell evolves as a state machine.

- Cell neighbourhood defines which nearby cells affect the next state of a grid cell.
- A local rule, by which a cell's next state evolves, may be either deterministic or probabilistic. In the case of stochastic cellular automata, a probabilistic rule is employed, where the rule is defined according to some probability distribution.
- For all real-life applications, hybrid CA is used for modelling, where the same rule is not employed for the grid cells. A uniform CA, on the other hand, employs the same rule for each cell—a few illustrations are noted in Sect. 1.4.7 of this chapter; these are mostly of theoretic interest

In Chap. 2, we have introduced one-dimensional (1D) 5 neighbourhood hybrid CA. Limited real-life applications have been reported in published literature with this class of simple 1D three/five-neighbourhood CA proposed in this book. On the other hand, volumes of CA research have been reported in different application domains using CA with multiple states per cell, higher dimensional grid, and rules of higher complexity.

In the **physical sciences**, CA simulation has been used to model the environment of spatial dynamics. CA is capable to simulate nonlinear and discontinuous processes. Solution of problems in the field of complex fluid dynamics or similar applications, demands very large computational support; CA modelling has been used in such cases using much less computational power. The first lattice-gas cellular automaton was proposed in 1973 by Hardy, de Pazzis and Pomeau (HPP model). This model uses 2D grid along with the following characteristics:

- All gas particles have the same mass, which is set equal to 1 for simplicity, and are indistinguishable.
- The evolution in time proceeds by an alternation of collision and propagation.
- The collision should conserve mass and momentum while changing the occupation of the cells.
- The collisions are strictly local, i.e. only particles of a single cell are involved.

Refinement of the above model and the associated rules has led to the development CA model for air flow patterns (streamlined and vortices) created by different objects in an airflow which are analogous to wind tunnel experiments. In this case, CA has become a complement to solutions derived out simultaneous partial differential equations.

One of the major focus of CA application in **biology** is the study of tumour growth. There are a number of attributes of a tumour's environment that are difficult to model in discrete terms, such as nutrient levels, toxin levels and pH levels of the extracellular environment of a cell. However, it is possible to define CA rules from the following characteristics associated with tumour growth.

- In view of its need for more energy, tumour cells prefer to move towards regions with high nutrient levels and away from regions with higher toxin levels.
- Higher acidity of the extracellular environment aids in tumour invasion.

- Low pH triggers the productions of enzymes that help in the breakdown of the extracellular matrix.
- Low pH is also detrimental to the intercellular adhesion and communication of normal cells, aiding the tumour cells in their invasion between normal cells.
- Cancer cells seem to rely on anaerobic rather than aerobic respiration involving oxygen.

Defined on a large two-dimensional grid, the state of each cell of that grid is a vector of four components:

- One of four discrete grid cell states: Tumour cell, normal cell, micro-vessel (small blood vessels), or vacant.
- The continuous H+ ion concentration, of 27 that grid cell (represented by H+, for the duration of this chapter).
- The continuous glucose concentration, of that grid cell.
- For micro-vessels only: a set of values, four (for the four walls of the vessel) for each of the two chemical concentrations, to enforce gradient boundary conditions.

In 2001 Patel, Gatenby et al. published a paper describing an agent-based model with the purpose of studying this problem. It mimicked the observation of clinical tumour growth to a large extent.

Annexure 1.1 introduces an overview of 'The Standard Model'—the fundamental theory of the structure of the Universe. We also touched upon the fact—how Quantum Theory (QT), which is being successfully exploited in everyday life, is incompatible with the large structure of the universe. Quantum theory has in itself so many inherent ambiguities. Even in this fundamental domain attempts are being made to bring in CA methodology. Such an attempt by Nobel Laureate physicist Gerard't Hooft has been reported in (https://arxiv.org/abs/1405.1548). In this chapter, we mentioned the viewpoint of scientists that the universe itself is an information processing machine and fundamentally it is discrete. Based on this realm, Prof. Hooft has conjectured a CA model of a quantized discrete interpretation of all the phenomena in the universe which is too complex. The fundamental particles, as reported in Annexure 1.1, do follow certain rules. Take Pauli's exclusion principle: the inability of two fermions with the same quantum number to occupy the same space—such a characteristic can be modelled with a CA rule. One of the effects of discreteness and its ability to evolve in a nonlinear manner frees CA from infinities, singularities and normalization problems which plague high energy theoretical physics.

In this chapter, we mentioned Conway's game of life. Another game was conceived which gave birth to CA application to model socio-economic problem. The game is played in a toroid grid. A toroid can be formed with a 2D grid and then connecting the top of the page to the bottom (this makes a cylinder) and connecting the left of the page to its right at the same time. So in a flat screen the object vanishing upwards from top of the screen will appear from the bottom of the screen and vice versa; likewise, object going out of the screen from the left/right of the screen and reappears from right/left. The game called Wa-Tor was designed by Alexendar Keewatin Dewdney. There are two inhabitants of Wa-Tor—sharks and fishes.

The fish movement is guided by the rules which include survival for unit time in cells while it moves randomly in each unoccupied cell in successive time steps. If it survives certain unit of time, it reproduces leaving its offspring in its previously occupied cell and its own survival time reset to zero. On the other hand, the shark movement is guided by the rules as follows—it moves randomly; if it moves to a cell occupied by a fish it gains energy and survives for more movement; if there is no fish it loses energy, and at zero energy level it dies. On the other hand for a certain value of energy, it reproduces.

This game of Wa-Tor CA evolution has been made more elaborate with movement of CA cell inhabitants guided by different rules. For a given set of rules and population densities of CA cells in the grid, the game would represent the behavioural pattern that can mimic real-life socio-economic situation. It can be observed that this game represents a model for two nonlinear first-order differential equations. This forms the base of many CA applications in physical and socio-economic nonlinear dynamic systems.

Can we use CA for weather prediction, or to predict the effect of deforestation or forest fire? Theoretically yes, researchers have tried to use CA in such application domains and many more. But like any mathematical tool, we need to design the CA rules and the associated framework (dimension, number states/cell, neighbourhood, boundary) to solve the problem. We would like to conclude this annexure while repeating the point that if we use complicated rules and complex framework for building the model, the CA will evolve and may even generate interesting visual pattern, but interpreting the result will need more resource than needed to solve the original problem.

Annexure 1.4: Example of CA Program

This is the code used in this chapter to create many of the Cellular Automata (CA) evolution patterns. Compilation of the code can be done with any c/c++ compiler like DevC++. The reader is advised to change the #define values for both the number of iterations and size of the CA cells to experiment.

```c
#include "stdio.h"
#include "stdlib.h"
#include "math.h"
#include "sys/types.h"
#include "time.h"

#define NITERATIONS 60
#define LENGTH 125

int enter_rule;
int rules[8];

void DisplayState(int *,int);

void get_rule ()
{
        int z,p;
        printf("Enter Rule in decimal \n");
        scanf("%d",&z);
        printf("In binary ");
        for (int k=8;k>0;k--) {
                p=z%2;
                z=z/2;
                rules[k-1]=p;
        }

        for (int k=0;k<8;k++) {
                printf("%d",rules[k]);
        }
        printf("\n");
}


int main() //int argc,char **argv
{
        int i,j,k,d;
        long secs;
        int state[LENGTH],newstate[LENGTH];
        get_rule();
        time(&secs);
        srand(secs);

        for (i=0;i<LENGTH;i++) {
                state[i] = rand() % 2;
        }
        DisplayState (state,LENGTH);
        printf("\n");
        for (i=0;i<NITERATIONS;i++) {
                for (j=0;j<LENGTH;j++) {
                        k = 4*state[(j-1+LENGTH)%LENGTH] + 2*state[j] + state[(j+1)%LENGTH];
                        newstate[j] = rules[k];
                }
```

```
                    for (j=0;j<LENGTH;j++) {
                            state[j] = newstate[j];
                            DisplayState(state,LENGTH);
                    }
            }
    }

    void DisplayState(int *s,int len)
    {
            int i;
            for (i=0;i<len;i++) {
                    if (s[i] == 1)
                            putchar('1');
                    else
                            putchar('0');
            }
            printf("\n");
    }
```

References

1. Biography of John Dalton. www.biography.com/people/john-dalton-9265201
2. Bensaude-Vincent, B.: Dmitri Mendeleev. Encyclopædia Britannica, Encyclopædia Britannica, Inc., 2 Jan 2018. www.britannica.com/biography/Dmitri-Mendeleev
3. Andrew, E.: What Are Fundamental Particles? IFLScience, 20 Mar 2018. www.iflscience.com/physics/what-are-fundamental-particles/
4. Croft, S.: Ingredients for Life. A place for life: Understanding evolution. University of California Museum of Paleontology. evolution.berkeley.edu/evolibrary/article/0_0_0/astrobio_life_01
5. Structure of Nucleic Acids.: SparkNotes. www.sparknotes.com/biology/molecular/structureofnucleicacids/section3/
6. Exploring Life's Origins: Ribozymes & the RNA World. Exploring Origins. exploringorigins.org/ribozymes.html
7. Definition of In silico. https://en.wikipedia.org/wiki/In_silico
8. Introduction to the Marshall Protocol. Marshall Protocol Knowledge Base. mpkb.org/home/patients/protocol_overview
9. Cressey, D., Callaway, D.: Cryo-electron microscopy wins chemistry nobel. Nature News **550** (7675), 167 (2017). https://www.nature.com/news/cryo-electron-microscopy-wins-chemistry-nobel-1.22738
10. John McCarthy (Computer Scientist). Wikipedia, Wikimedia Foundation, 28 Mar 2018. en.wikipedia.org/wiki/John_McCarthy_(computer_scientist)
11. The Embryo Project Encyclopedia. John Von Neumann's Cellular Automata | The Embryo Project Encyclopedia. embryo.asu.edu/pages/john-von-neumanns-cellular-automata
12. Rucker, R.: Classical Era: Von Neumann to Gosper. The Origins of Cellular Automata, Primordial Soup Kitchen. University of Wisconsin - Madison, psoup. math.wisc.edu/491/CAorigins.htm
13. A Discussion of Game of Life. Game of Life, Stanford.edu. web.stanford.edu ~ cdebs/GameOfLife/
14. John Wheeler.: Important Scientists, The Physics of the Universe. www.physicsoftheuniverse.com/scientists_wheeler.html
15. Einstein an Intimate Study of a Great Man Dimitri Marianoff, Palma Wayne by Dimitri Marianoff (Author), Palma Wayne (Author)

Chapter 2
Design of Cellular Automata Rules for Biomolecules

"I believe that new mathematical schemata, new systems of axioms, certainly new systems of mathematical structures will be suggested by the study of the living world".

—Stanislaw Ulam

2.1 Introduction

In Chap. 1 we presented the viewpoint that the universal logic function of NAND can be realized with the simple one-dimensional Cellular Automata (CA) evolution and CA can be used to design a universal computing machine. We also emphasized the discreteness of the fundamental behaviour of physical objects and yet there is a requirement of smoothness. This means that while at the bottom of the construction there are individual entities like elements, molecules bound together by forces, the macro level behaviour is smooth and continuous. The behaviour of discrete constructs is also rarely reflected on their combined macro-behaviour. For example, the properties of the compound sodium chloride is different from either sodium or chloride.

Cellular automata is discrete in both space and time. Utilizing the structural evolution through time, exponentially large number of CAs can be designed employing different rule sets. However, the evolving pattern of a CA may be meaningful or utterly meaningless to model the behaviour of a physical system including biomolecules of current life form. This happens because we know only one form of life, a life based on nucleotides and proteins. The possibility of the existence of other forms of life has been debated for long and is a topic of active interest through exploration of outer space.

We introduced the concept of cellular automata-based Life (cLife) in Sect. 1.5.3 of Chap. 1. The current chapter presents the methodology to design CA rules and different variants of CA that will bring cLife as close as possible to life as we know it. We will try to produce Cellular Automata Machines (CAM) variation used for cLife so that a uniform relationship exists among the determinants of CA rule structure. We lay down the process to represent variants of CAM as it may be used

© Springer Nature Singapore Pte Ltd. 2018
P. P. Chaudhuri et al., *A New Kind of Computational Biology*,
https://doi.org/10.1007/978-981-13-1639-5_2

to mimic the behaviour of essential biomolecules of life. The evolving pattern of CAMs may lead to unidentifiable set of numbers, beautiful but not of any use to model biological entities. The most challenging task is to design the structure and neighbourhood of CA rules for CAMs to model behaviour of a physical entity.

In the concluding remarks of Chap. 1, we emphasized the concept of the bottom-up approach of designing CA rules, followed by its validation from top-down behavioural model of the physical systems as obtained from microbiological experiments. This chapter presents the logical steps of the iterative process of bottom-up approach for rule design, followed by top-down validation in subsequent chapters. The process of validation demands consistent representation of the behavioural features of biomolecules (retrieved out of experimental results of molecular biologists) with the parameters derived out of CAM evolution. In the event of inconsistent representation, we go back to redesign of the CA rule structure and associated modification on CAM to start the next cycle of iteration. While the current chapter presents the methodology of CA rule design for different biomolecules, the top-down validation has been followed in subsequent three chapters dealing with three major biomolecules RNA, DNA and protein.

Nature has failed many a time in millions of years and through the process of experimentation it has formed and arrived at an organic conglomerate of current life form. We learn from nature through experiments and micro/macro observations and attempt to design corresponding rules for cLife. The basic concept introduced is elaborated further in the CAM fundamentals presented next.

2.2 Revisiting CA Machine Fundamentals

Chapter 1 reviews the fundamentals of biological units—the macromolecules— DNA, RNA, protein and reports the basics of Cellular Automata Machine (CAM). The current chapter reports the methodologies formulated in the design of CA to mimic those entities. We design the CA rules for CAM based on the alternative representation of the building blocks (referred to as micromolecules) of three macromolecules. CA rule design explores both the biological information contained —(i) in a micromolecule, and also (ii) its molecular structure. Thus, a CAM is built from bottom-up, with the basic knowledge of biological entities. The CAM runs and we have results of its evolution. Once a CAM evolution is terminated after the desired number of evolution steps, we now have to map CAM evolution parameters to the features extracted from the knowledge base of biological experiments, generally known as the wet lab results. The database of different wild and mutated RNA, DNA and proteins are incredibly large, which are used to validate the CA model. If the results derived out of CAM differ much from wet lab results, then we modify the CA rules designed for the CAM. We build cLife bottom-up and evaluate top-down to estimate how closely cLife mimics the current form of life.

Since the visualization of CA by Neumann and Ulam in 1950s, a large volume of CA research has been documented in the published literature, which cover a wide variety of CA rule types reported in the book (Stephen Wolfram, *A New Kind of Science*). As indicated in Sect. 1.5.1 of Chap. 1, our CA model employs the simplest one-dimensional CA rule. In this chapter, we will introduce the basic steps of designing CA rules based on the analysis of the building blocks (micromolecules)—sugar–phosphate molecule, nucleotide bases, amino acid. Different CA modeling techniques are employed to design CA rules along with the underlying principle followed to mimic those essential building blocks. The details of the evolution of CAMs to model the biomolecules are reported in subsequent chapters—RNA in Chap 3, DNA in Chap. 4 and protein in Chap. 5.

2.3 Three-Neighbourhood CA Machine

In Chap. 1 we have introduced a single CA rule in the design of a CA. We reported the evolution of CA cells as per the rule with the change of its state values. Each cell is initialized with binary value referred to as 'seed'. The machine runs according to the applied rule. We have used three-neigbourhood CA using a single rule. We formally stated that such a Cellular Automata Machine (CAM) is a three-neigbourhood one-dimensional machine evolving with a single CA rule and is referred to as uniform CAM. A hybrid CAM, on the other hand, employs different CA rules for its cells.

The three-neighbourhood CA (3NCA) cell structure is reproduced in Fig. 2.1. It is a two-state per CA cell. The cell structure has been detailed in Fig. 2.1b. The next

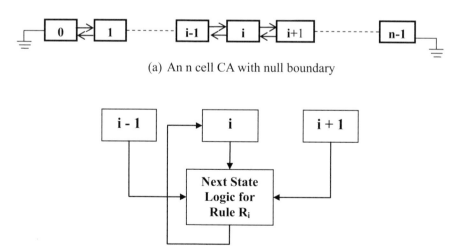

(a) An n cell CA with null boundary

(b) Rule R_i employed on the ith cell

Fig. 2.1 A basic CA cell

state function of ith cell depends on the current states of its left neighbour, the cell itself, and its right neighbour. Hence, the next state function could be viewed as a three-variable Boolean function [1].

For a three-variable Boolean function, there are eight combinations of current states marked as 0(000), 1(001), 2(010), 3(011), 4(100), 5(101), 6(110), 7(111). Each of these decimal numbers 0–7 are termed as 'Min Term'. Since for our application, these min terms are associated with CA rules, we refer to the min terms of a Boolean function as (Rule Min Term (RMT) in different chapters of this book.

When a CAM evolves, each of its cells goes through non-repeating state transitions prior to reaching its cyclic evolution. A CA cell runs through a set of (binary) states, which repeat after (say) q time steps ($q = 1, 2, 3, \ldots$). The value q is referred to as the Cycle Length (CL) of the cell. The cycle is also referred to as its 'attractor cycle'. Prior to reaching its cycle, a cell runs through (say) p transitory states ($p = 0, 1, 2, \ldots$). The parameter p is referred to as Cycle Start (CS); prior to reaching its cyclic state, a cell runs through p number of transitory states after initialization with its seed. The parameter pair (CS, CL) conveys the characteristics of CA evolution. At any time step of the cyclic evolution of ith cell of a 3NCA, its state could be also evaluated as a triplet of current states of its neighbours $(i - 1)$, i, and $(i + 1)$, that is the associated RMT of the cell at that time step. Hence, a CA cell generates a set of RMTs through its cyclic evolution.

In each time step, a CA cell evolves to its next state as per its rule of evolution. This evolution step is the time step elapsed for each cell to generate its next state from the current states of its neighbours. Since each cell evolves at the same time, such a CA is referred to as 'synchronous CA'. For an asynchronous CA, time of evolution differs for different CA cells.

Several studies have implemented asynchronous models and found that their behaviour differs from that of synchronous ones. Most importantly, an interesting behaviour of synchronous CA may disappear in the asynchronous version. Harvey and Bossomaier [2] pointed out that stochastic updating result in no repeatable cyclic behaviour, although they introduced the concept of loose cyclic attractors. Kanada [3] has shown that some one-dimensional CA models that generate non-chaotic patterns when updated synchronously generate edge of chaos patterns when randomized. Orponen [4] has demonstrated that any synchronously updated network of threshold logic units could be simulated by a network that has no constraints on the order of updates. There are different types of asynchronous updating. Throughout this book, we use synchronous CA, where all cells are updated in parallel at each time step.

We first construct the model of a wild (natural) molecule. In order to study the effect of specified changes in the behaviour of the molecule, we reconstruct its mutated version by introducing changes in the rule structure of the corresponding cell locations. Next, we mark the differences in the evolution pattern of the wild and mutated version to compare with the real-world (wet lab) data. This general principle is implemented in different chapters to construct CA analogy of different biomolecules.

The CA time steps are the time steps taken to construct the CA analogy of the concerned biomolecule. It is worthwhile to clearly note the point that the time step of a CA evolution, unless mentioned otherwise, does not have any corelation with the time steps of a physical system being simulated with CA. The information retrieved from a CAM evolution is next introduced. Each of these parameters captures specific characteristics of a CAM.

The parameters extracted out of CAM evolution are the following:

1. The 0/1 binary evolution pattern—a two-dimensional matrix of binary pixels in its transitory state or its cyclic evolution state.
2. The parameter pair CS (Cycle Start) and CL (Cycle Length) that capture the switching point from transitory to cyclic evolution—this parameter pair plays a dominant role in the design of CA model.
3. The set of RMTs in the cycle of a cell.
4. The sum of RMT states in its cycle is referred to as CycTotal.
5. Difference between a wild and mutated version of a biomolecule is captured with the earlier parameters. The difference is evaluated from the output of CAMs representing wild and mutated biomolecules with an XOR (or any other) operation on the binary output of each cell at each evolution step for two versions of the CAMs.

Note: logical NAND operation on a pair of bits has been explained in Sect. 1.5.1 of Chap. 1; logical XOR operation outputs 1 if input bit pair differs (that is either (0, 1) or (1, 0), else output is 0 (that is the input bit pair is either (0, 0) or (1, 1)).

Design of a CA model for any physical system proceeds through the following basic steps:

- Step 1: Based on the detailed study of the physical system, design the CA rules to represent the building blocks of the physical system.
- Step 2: Run the CA Machine (CAM) with the CA rules designed in Step 1.
- Step 3: Extract the parameters discussed earlier.
- Step 4: Consistent mapping of relevant CAM parameters with the features observed for biomolecules either from wet lab experiment or clinical study.
- Step 5: If consistent mapping fails, the steps 1–4 are iterated until the desired goal is reached.

In Table 2.1, we have shown the three-neighbourhood CA (3NCA) rule in tabular form. The eight RMTs of 3NCA ith cell refer to the current state values of left neighbour, cell itself, and its right neighbour is denoted as $<i{-}1, i, i + 1>$. The binary strings of three variables, as noted in row 2 of Fig. 2.1 are 111, 110, 101, 100, 011, 010, 001, 000. Their decimal values are noted in row 1. Decimal counterpart of the 8-bit pattern of the next state values of 8 RMTs 0 (000) to 7 (111) is referred to as CA rule. The next state value of a cell, denoted as b_k, for a specific input combination (of current states) $k = 0$ to 7 is noted on the third row along with an illustration of two CA rules 45 and 105 on last rows of Table 2.1.

Table 2.1 Three-neighbourhood CA rule

Current state values	7	6	5	4	3	2	1	0
$K = <a_{i+1}\ a_i\ a_{i-1}>$	111	110	101	100	011	010	001	000
Next state b_k	b7	b6	b5	b4	b3	b2	b1	b0
Rule 45	0	0	1	0	1	1	0	1
Rule 105	0	1	1	0	1	0	0	1

The decimal value of the eight-bit pattern is referred to as a rule of a 3NCA. By convention, the decimal value of a bit string is evaluated with left most bit as the MSB (Most Significant Bit) and right most bit as LSB (Least Significant Bit). For a 3NCA rule, we deal with an eight-bit binary string. Since it is a two-state per cell CA, there are $2^8 = 256$ rules for 3NCA cell.

Out of eight, four RMTs, 7, 5, 2, 0 are termed as PRMT (Palindromic RMT) because on reversing their three-bit pattern, there is no change in the pattern and so the decimal value remains identical. On the other hand, on reversing the three-bit pattern for 6 (110) the decimal value becomes 3 (011) and vice versa; these two RMTs (6, 3) are referred to as CoP (Conjugate Pair). Similarly, RMT pair (4, 1) is another CoP. RMTs (7, 6, 5, 3) have two '1's and are referred to as 1-major, while 0-major RMTs (4, 2, 1, 0) have two '0's.

Figure 2.2 shows the two CA rules 45 and 105 with their eight-bit pattern rearranged with RMTs in 1-major and 0-major format. Throughout this book, a CA rule is represented in 1-major and 0-major format. Even though the format of RMT sequence of a rule has been modified along with the corresponding modification of its eight-bit pattern, the decimal value of the rule is evaluated as per the convention

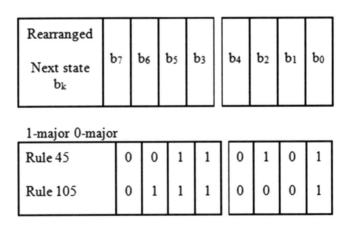

Rule 45 and 105 expressed in 1-major and 0-major format

3NCA rules in 1-major and 0-major format

Fig. 2.2 Rule 45 and 104 in 1-major and 0-major format

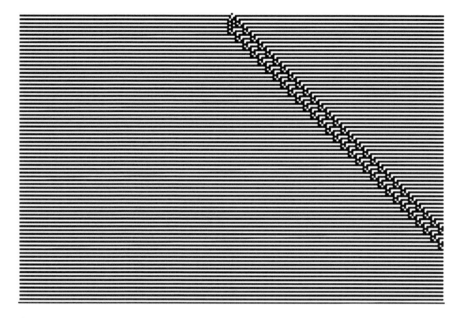

Fig. 2.3 Evolution of three-neighbourhood uniform CA with Rule 107

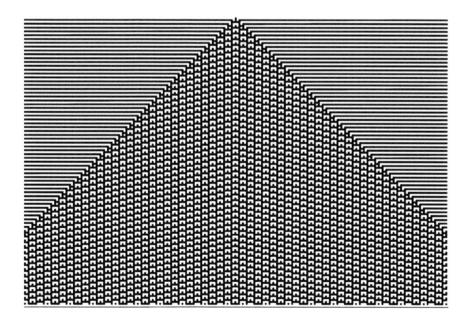

Fig. 2.4 Evolution of three-neighbourhood uniform CA with Rule 109

Fig. 2.5 Evolution of
three-neighbourhood uniform
CA with Rule 110

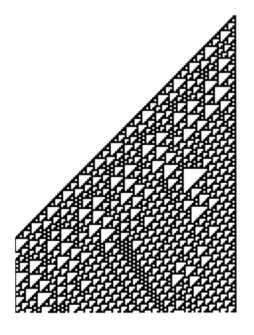

with normal arrangement of RMTs in sequential order (7, 6, 5, 4, 3, 2, 1, 0) from
left to right with the corresponding 8 bit pattern of cell next state. As will be evident
from subsequent discussions, this representation of CA rules in 1-major/0-major
format provides the foundation of designing CA model for biomolecules.

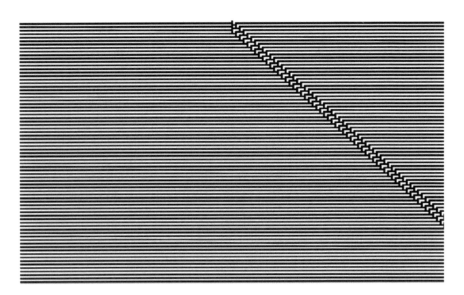

Fig. 2.6 Evolution of three-neighbourhood uniform CA with Rule 111

Table 2.2 shows the 8-bit patterns of two 3NCA rules in 1-major/0-major format. A CA rule is denoted by the decimal value of its conventional 8-bit pattern.

A CA employing the same rule for each cell is referred to as Uniform CA. In Chap. 1 Sect. 1.5.1, we have shown an example of the evolution of uniform CA. Figures 2.3, 2.4, 2.5 and 2.6 illustrate the evolution of a uniform CAM designed with different CA rules.

2.4 Uniform Three-Neigbourhood CA

Figure 1.29 of Chap. 1 showed the evolution of uniform CA Rule 99, where cells are initialized with different seed values 0 or 1. In this section, we present examples of CAs evolving with a middle cell of the cell array initialized with seed value 1 and all the remaining cells initialized with 0 values. In the graph displaying CA evolution of binary patterns, the CA cell state 1 and 0 is represented in black and white, respectively. All the CAs depicted here is working with seed [....000000001000000....]. Also, one can see that in Fig. 2.3 the value of CS is 5 time steps; that is after 5 steps, the evolving pattern continuously shifts to the right due to the uniform CA Rule 107 (01101011) with the CL value of 2 for each cell.

Figure 2.3 shows the evolution of uniform CA with Rule 107. Next, we show the evolution of Rule 109, 110 and 111 in Figs. 2.4, 2.5 and 2.6, respectively. Note the similarity of evolution patterns for uniform CA with rules 107 (0110 1011) and 111 (0110 1111).

From the analysis of evolving patterns of Fig. 1.29 of Chap. 1 and those of Figs. 2.3–2.6 of the current chapter, we can conclude that the CA evolution pattern generally depends on the CA Rule and CS value is dependent on the seed (initialization). Design of rule and seed for each cell of a CA are based on detailed analysis of the physical system being modelled with CA framework.

2.5 Three-Neigbourhood Hybrid CA Machine (CAM)

For modeling each of the 64 nucleotide base triplets of DNA and RNA, we employ a non-trivial three-neighbourhood CA rule. Such a CA is referred to as hybrid CA.

A hybrid CAM employs multiple CA rules; the rule design depends on the model requirement of the biomolecule. Figure 2.7 shows a six-cell three-neigbourhood hybrid CA (3NHCA) with its rules noted in a rectangular box. This rule vector <112 14 58 58 147 92> refers to the first six rules of Table 2.2 column 7. Table 2.2 displays a 42-cell 3NHCA of a hypothetical DNA strand having 42 nucleotide base triplets, each modelled with the 3NCA rule noted on column 7.

Figure 2.8 displays the evolution pattern of a 42-cell null boundary 3NHCA with 7 blocks of 6 cells. Each block employs the 6 cells shown in Fig. 2.7. Sample

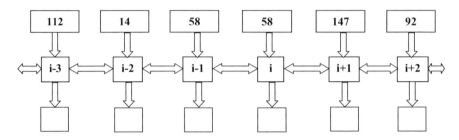

Fig. 2.7 Six-cell three-neigbourhood hybrid CA

Table 2.2 Evolution of a hypothetical DNA strand having 42-nucleotide base triplets

Nucleotide triplet	Initialization seed	Cycle start		Cycle length		CA rule representing triplets
ATG	0	CS	6	CL	14	Rule [112]
ATC	0	CS	6	CL	14	Rule [14]
ACC	1	CS	6	CL	5	Rule [58]
ACC	1	CS	6	CL	5	Rule [58]
AGC	1	CS	7	CL	5	Rule [147]
ACG	1	CS	9	CL	5	Rule [92]
ATG	0	CS	9	CL	5	Rule [112]
ATC	0	CS	9	CL	5	Rule [14]
ACG	1	CS	8	CL	5	Rule [92]
ACC	1	CS	8	CL	5	Rule [58]
AGC	1	CS	7	CL	5	Rule [147]
ACG	1	CS	9	CL	5	Rule [92]
ATG	0	CS	9	CL	5	Rule [112]
ATC	0	CS	9	CL	5	Rule [14]
ACG	1	CS	8	CL	5	Rule [92]
ACC	1	CS	8	CL	5	Rule [58]
AGC	1	CS	7	CL	5	Rule [147]
ACG	1	CS	9	CL	5	Rule [92]
ATG	0	CS	9	CL	5	Rule [112]
ATC	0	CS	9	CL	5	Rule [14]
ACG	1	CS	8	CL	5	Rule [92]
ACC	1	CS	8	CL	5	Rule [58]
AGC	1	CS	7	CL	5	Rule [147]
ACG	1	CS	9	CL	5	Rule [92]
ATG	0	CS	9	CL	5	Rule [112]
ATC	0	CS	9	CL	5	Rule [14]
ACG	1	CS	8	CL	5	Rule [92]
ACC	1	CS	8	CL	5	Rule [58]

(continued)

Table 2.2 (continued)

Nucleotide triplet	Initialization seed	Cycle start		Cycle length		CA rule representing triplets
ATG	0	CS	6	CL	14	Rule [112]
AGC	1	CS	7	CL	5	Rule [147]
ACG	1	CS	9	CL	5	Rule [92]
ATG	0	CS	9	CL	5	Rule [112]
ATC	0	CS	9	CL	5	Rule [14]
ACG	1	CS	8	CL	5	Rule [92]
ACC	1	CS	8	CL	5	Rule [58]
AGC	1	CS	7	CL	5	Rule [147]
ACG	1	CS	9	CL	5	Rule [92]
ATG	0	CS	9	CL	5	Rule [112]
ATC	0	CS	9	CL	5	Rule [14]
ACG	1	CS	8	CL	8	Rule [92]
ACC	1	CS	8	CL	8	Rule [58]
AGC	1	CS	7	CL	8	Rule [147]
ACG	1	CS	8	CL	8	Rule [92]

Fig. 2.8 Evolution of a 42-cell null boundary 3NHCA with 7 blocks of 6 cells

001111001111001111001111001111001111001111

3NHA Evolution for DNA

evolution of $7 \times 6 = 42$ cell 3NHCA is shown in Fig. 2.8 for 53 evolution steps. The black boxes in the 0/1 evolution pattern represent '1' and white represent '0'. Horizontally, there are six blocks of identical patterns while pattern for first and last three cells differ. Each of the six blocks evolves with identical rule vector noted in Fig. 2.7 and identical seed noted below. It is worth noting the point that null boundary of first and last cell resulted in the difference in evolution patterns for first and last three cells. So, the obvious conclusion is that null boundary affects the evolution of terminal cells. This observation should be taken into consideration while designing CA for modeling a physical system.

The initialization seed for 42-cell null boundary 3NHCA is

001111001111001111001111001111001111001111

The 3NCHCA evolution shown in Fig. 2.8 runs for 53 time steps. How do we decide to stop the CAM evolution? As soon as a CAM cell enters in its attractor cycle, the cell RMT values keep repeating after elapse of CL (Cycle Length) number of time steps. Consequently, we allow evolution of a CA until each cell enters in its cyclic evolution. It is worthwhile to note the point that the cyclic evolution of a 3NCA is evaluated with the repetition of RMT patterns in a cycle rather than binary output of a cell. Thus, CAM evolution stops after attaining the Maximum CL value (MaxCL) observed for any cell. There are instances of CA evolution where MaxCL value is very high. That is, a cell does not enter in its cyclic evolution within a specified evolution step. For such cases, we analyse the CA evolution pattern for a specific time step, where some cells may enter in cyclic evolution, while a few others continue in their transitory evolution.

Table 2.2 column 1 shows the 42 nucleotide base triplet of a hypothetical DNA strand. Figure 2.8 displays the evolution of this 42-cell 3NHCA. The seed for each cell is shown in column 2 with CS and CL values of each cell reported on column 4 and 6, respectively. The seed for 3NHCA modeling a base triplet of a DNA string is derived out of the following consideration. Figure 1.17 of Chap. 1 presents the codon–amino acid table valid for any living organism; there are certain codons that represent non-polar amino acids, while other groups of codons represent polar amino acids. The seed for cells modeling codons representing non-polar (polar) amino acid is set as 0(1). Chapter 3 presents the CA model for nucleic acid RNA. It reports the details of CA model for a codon string translated in a biological system for synthesis of amino acid chain of a protein. Graphical representation of the two parameters CS and CL is shown in Fig. 2.9.

The Cycle Start (CS) and Cycle Length (CL) parameters represent relevant features of each triplet and its neighbourhood in a DNA strand. The 42-cell 3NHCA representing a DNA strand of 42 nucleotide triplets are evolved and shown in Fig. 2.8 using initialization seed value: 001111001111001111001111001111001111001111 11001111.

The CS and CL values are tabulated in Table 2.2. The CS and CL values noted in the table are graphically represented in Fig. 2.9. We will refer to this type of graph as a signal graph in Chaps. 3 and 4.

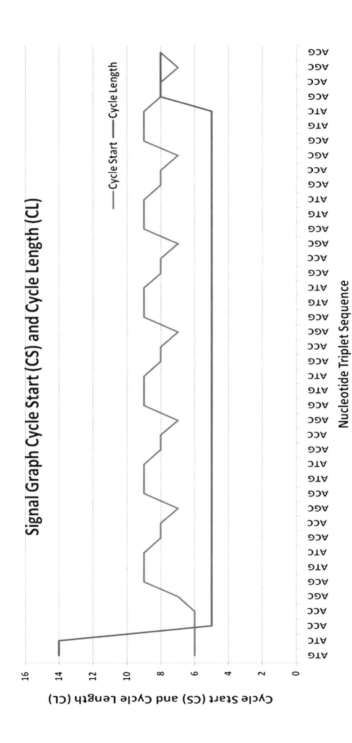

Fig. 2.9 Signal graph of Cycle Start (CS) and Cycle Length (CL) of DNA strand having 42 nucleotide base triplets

Fig. 2.10 CA evolution for
the codon sequence of a RNA
strand

3NHA Evolution for RNA

We extend the illustration of hypothetical DNA strand for the design of CA for a
RNA strand, where the model is similar to that of DNA except the base Thiamine
(T) replaced by Uracil (U). Consequently, wherever a 'T' appears in DNA base
triplet, the CA rule changes for the corresponding RNA triplet. By convention, the
nucleotide base triplet of an mRNA (Messenger RNA) transcripted out of DNA is
referred to as 'codon'. CA rule design details are reported in Chap. 3 for RNA,
while Chap. 4 covers the CA model for DNA.

Evolution of the codon string of RNA in Fig. 2.10 employs the same 'seed' used
for DNA strand—00111100111100111100111100111100111100111111. The evolution pattern shown in Fig. 2.10 for RNA is different from those of DNA strand
(Fig. 2.8) due to thiamine of DNA replaced with the base uracil for RNA.

2.6 Five-Neigbourhood Cellular Automata (5NCA)

The earlier three sections deal with the design of 3NCA rules for the biological entity of codon that represents nucleotide base triplets of an mRNA (messenger RNA) string transcribed out of a gene in a DNA strand. A codon is basically an information-level representation of nucleotide base triplets out of which an amino acid is synthesized by the ribosome. The translation process in ribosome is presented in Sect. 1.4.4 of Chap. 1. There are a total of 256 (two hundred and fifty-six) 3NCA rules. The information of 64 codons, is reported in Chap. 3 and adequately represented in the 8-bit pattern of 64 (sixty four) 3NCA rules. However, to design CA rule to represent the physical structure of a biological entity, we need to consider the information of covalent bonds of different atoms (hydrogen and non-hydrogen) in its molecular structure. It can be observed that 8-bit pattern of 3NCA rules fails to adequately represent the information derived out of the physical structure of nucleotide bases of purine (A, G) and pyrimidine (C, U, T). The molecular structures of each of these micromolecules cover 12–16 atoms—carbon, oxygen, nitrogen and hydrogen.

The inadequacy of 3NCA rule structure to represent structural information forced us to consider 5NCA rules to represent micromolecules. Each 5NCA rule is represented by $2^5 = 32$-bit pattern. The 32-bit pattern has the flexibility of representing the information content of wide varieties of the molecular structure of micromolecules. Since our major focus is to study biomolecules DNA, RNA, we concentrate in the design of 5NCA rules to represent their building blocks—sugar–phosphate molecule, and nucleotide bases (A, C, G, T, U). The basic framework of 3NCA rule structure is extended for the design of 5NCA rules. For ease of reference, the 3NCA rule structure is recapitulated below with reference to a grouping of its 8 RMTs.

3NCA rule has 8 RMTs 0(000) to 7(111) out of which 7, 6, 5, 3 are 1-major RMTs, and 0,1,2,4 are 0-major RMTs. The RMTs 7, 5, 2, 0 are PRMT (Palindromic RMT), while (6, 3) and (4, 1) are referred to as CoP (Conjugate Pair). The 4 RMTs of 1-major/0-major blocks could be further divided, as noted in Table 2.3a, into two sub-blocks. The left most sub-block covers a single RMT with three 1's/0's which is palindromic, while next block covers 3 RMTs, each having two 1'/0's. The three RMTs in second sub-block cover a conjugate pair (6, 3)/(1, 4), while the middle RMT 5/2 is palindromic. The underlying principle of such an arrangement of 3NCA rules has been elaborated in Chap. 3 while designing rules for 64 codons. This format of RMTs of a CA rule structure has been extended in next section for similar analysis of 32 RMTs (from 31(1111) to 0(00000)) of a 5NCA rule.

Table 2.3 Uniform structure of RMT grouping of 3NCA and 5NCA rule

(a) 1-major/0-major format of 8 RMTs of a 3NCA rule structure		
<7>	<6 5 3>	1-major RMTs
<0>	<1 2 4>	0-major RMTs
(b) 1-major/0-major format of 32 RMTs of a 5NCA rule structure		
PRMT	CoP (30, 15), PRMT 27, CoP(29, 23)	CoP (28, 7), CoP (26, 11), PRMT (21, 14), CoP (25, 19), CoP (22, 13)
<31>	<30 29 27 23 15>	<28 26 25 22 21 14 13 19 11 7> (1-major RMTs)
<0>	<1 2 4 8 16>	<3 5 6 9 10 17 18 12 20 24> (0-major RMTs)
PRMT	CoP (1, 16) PRMT 4 CoP (2, 8)	CoP (3, 24), CoP (5,20), PRMT (10, 17), CoP (6, 12), CoP (9, 18)

2.6.1 Analysis of 32 RMTs of a Five-Neighbourhood CA (5NCA) Rule

Figure 2.11 shows the structure of a 5NCA cell. A 3NCA rule for ith cell is derived out of current states of $<(i − 1), i, (i + 1)>$ cells. A 5NCA rule for ith cell is derived out of current states of $<(i − 2) (i − 1), i, (i + 1) (i + 2)>$ cells.

A 5NCA rule has $2^5 = 32$-bit pattern and 32 RMTs marked as 31(11111) to 0 (00000). Similar to the grouping of RMTs of 3NCA rule, the 1-major/0-major RMTs of a 5NCA rule are grouped as shown in Table 2.3b. The second and third rows show the three sub-blocks of 1-major and 0-major RMTs. Left most block covers single Palindromic RMT (PRMT) 31/0 with five 1's/0's. The second sub-block covers five RMTs, each having four 1's/0's. The third sub-block covers 10 RMTs, each having three 1's/0's. The PRMTs and Conjugate Pairs (CoPs) are marked on first and fourth row of Table 2.3b for 1-major and 0-major RMTs. The PRMTs for second and third sub-block is shown at the middle with CoPs on its two sides. Reversal of bit pattern of RMT of a CoP generates its conjugate pair. For

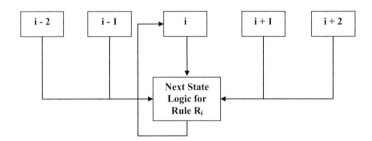

Fig. 2.11 Structure of a five-neighbourhood CA (5NCA)

example, reversal of 5-bit pattern (01001) of RMT 9 generates RMT 18 (10010) generating the CoP (9, 18) for the third sub-block of 0-major RMTs. Similarly, 22 (10110) and 13 (01101) form a CoP of third sub-block of 1-major RMTs.

2.6.2 Design of 5NCA Rule for Sugar–Phosphate Backbone of DNA and RNA

The DNA backbone structure with sugar–phosphate and nucleotide base molecules is shown in Fig. 2.12. It shows the molecular structure of sugar–phosphate backbone with two adjacent phosphate molecules covalently bonded (phosphodiester bond) to the intermediate sugar molecule. Figure 2.13 shows the molecular structure of deoxyribose sugar of DNA backbone and ribose sugar of RNA backbone.

The uniform backbone structure of a sugar–phosphate molecule of a nucleic acid is represented with a uniform 5NCA. The 5NCA rule for each CA cell is designed as an alternative representation of the information noted in the molecular structure of sugar–phosphate molecule in respect of connectivity of its atoms. For example, the Phosphate (P) of the molecule is interconnected to four Oxygen (O) atoms.

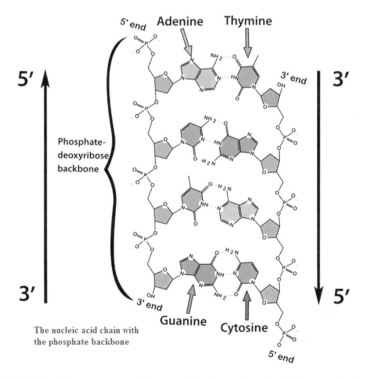

Fig. 2.12 DNA backbone structure with sugar–phosphate and nucleotide base molecules

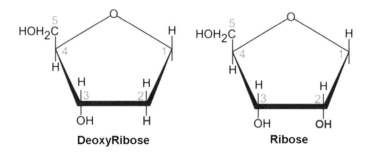

Fig. 2.13 Molecular structure of deoxyribose sugar of DNA backbone and ribose sugar of RNA backbone

The O atom of pentose sugar is connected to two carbon atoms and so on. A procedure is next reported for the design of 5NCA rule for DNA/RNA sugar–phosphate molecule. It is to be noted that the rule designed for executing the procedure is not unique. Alternative designs of CA rule exist for representing the information of the molecular structure of sugar–phosphate molecule. This is evident from the steps noted in the procedure. As the bottom-up approach of CA rule design for sugar–phosphate molecule, we derived a large number of alternative designs. Next, for top-down validation—(i) we derived the relevant parameters out of the evolution of the 5NCA representing nucleic acid backbone, and (ii) mapped the derived parameters on to the features observed in the structure and function of the molecule reported in wet lab results. Two 5NCA rules reported in Table 2.4 for DNA and RNA sugar–phosphate molecule have been derived through such an iterative procedure. These rules are used in Chaps. 3 and 4 for the study of RNA and DNA molecules. The procedure for CA rule design is illustrated below for design of 5NCA rule for sugar–phosphate molecule.

Algorithm: CA rule design

Inputs:

 (i) 32 RMTs of 5NCA rule structure divided into 1-major and 0-major format (as reported in Table 2.2)
(ii) Molecular structure of sugar–phosphate molecule (Figs. 2.13 and 2.14).

Output: 32-bit 5NCA rule in 1-major/0-major format for DNA/RNA (Table 2.4)

Steps

1. Analyse molecular structure of input molecule and arrange atoms with descending order of covalent bonds it makes with other atoms.
2. A 1 is assigned to a Palindromic RMTs for representing atoms with a high number of covalent bonds. A pair of 1's assigned to a Conjugate Pair

(CoP) RMTs represents a pair of atoms with a covalent bond between them.

Note: In the design of CA rule structure for a micromolecule, an atom represented by a palindromic RMT is assumed to represent bond, wherever it exists, with an atom represented by another RMT.

3. Hydrogen (H) atoms of molecules are represented by assignment of 1's for 0-major RMTs, while atoms for non-hydrogen atoms (P, C, O, N) are represented by assignment of 1's for 1-major RMTs.

4. Pair of non-H atom and H atom with covalent bond between them are so assigned to a 1-major RMT (say T) and 0-major RMT (say T′) so that T + T′ = 31.

5. Execution of earlier steps generates the 32-bit 5NCA rules in 1-major/0-major format.

The 32-bit pattern of 5NCA rule for DNA/RNA sugar–phosphate molecule is tabulated in Table 2.5 in the descending order of RMTs. The bit patterns are identical except the bit for RMT 19 having 0 for DNA and 1 for RNA representing

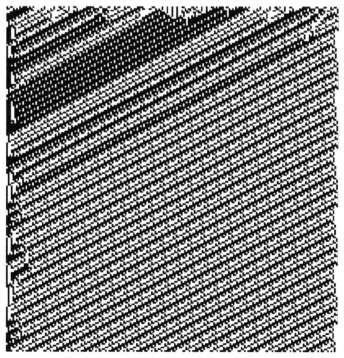

Evolution of 0/1 Pattern of BBDCAM

Fig. 2.14 Evolution of CA for Backbone for DNA CA Machine (BBDCAM)

Table 2.4a 5NCA rule for DNA sugar–phosphate

Five 1's	Four 1's RMTs (5)	Three 1's RMTs (10)			
31	30 29 27 23 15	28 26 25 22	21 14	13 19 11 7	Sixteen 1-major RMTs
1	0 1 0 1 1	1 1 0 0	1 1	1 0 1 1	11 non-H: 1P, 5C, 5O atoms assigned on 1-major RMTs placing 1's
0	1 0 1 0 0	1 1 0 0	1 1	0 0 0 1	7 H atoms assigned on 0-major RMTs on placing 1's
0	1 2 4 8 16	3 5 6 9	10 17	18 12 20 24	Sixteen 0-major RMTs
Five 0's	Four 0's RMTs(5)	Three 0's RMTs (10)			

Table 2.4b 5NCA rule for RNA sugar–phosphate

Five 1's	Four 1's RMTs (5)	Three 1's RMTs (10)			
31	30 29 27 23 15	28 26 25 22	21 14	13 19 11 7	Sixteen 1-major RMTs
1	0 1 0 1 1	1 1 0 0	1 1	1 1 1 1	12 non-H: 1P, 5C, 6O atoms on 1-major RMTs on placing 1's
0	1 0 1 0 0	1 1 0 0	1 1	0 0 0 1	7 H atoms assigned on 0-major RMTs on placing 1's
0	1 2 4 8 16	3 5 6 9	10 17	18 12 20 24	sixteen 0-major RMTs
Five 0's	Four 0's RMTs (5)	Three 0's RMTs (10)			

Table 2.5 32-bit pattern of DNA and RNA sugar–phosphate molecule

Biomolecule	32-bit binary pattern of DNA/RNA sugar–phosphate molecule	Physical information in CA model
DNA	10110101101000010111011001011101	18 1's for 18 atoms
RNA	10110101101010101011011001011101	19 1's for 19 atoms

extra O atom of RNA ribose sugar. CAMs designed with such uniform 5NCA rules are referred to as BBDCAM (BackBone DNA CA Machine) and BBRCAM (BackBone RNA CA machine). Evolution of 0/1 pattern for BBDCAM and BBRCAM are reported in Figs. 2.14 and 2.15.

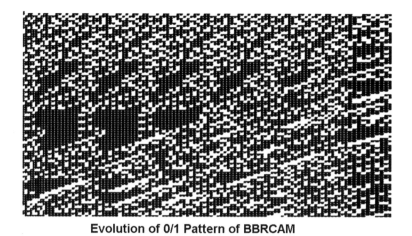

Evolution of 0/1 Pattern of BBRCAM

Fig. 2.15 Evolution of CA for Backbone for RNA CA Machine (BBRCAM)

2.6.3   Design of 5NCA Rules for Nucleotide Bases and Hybrid 5NCA for DNA/RNA

Figure 2.16 shows the molecular structure of the five nucleotide bases associated with DNA and RNA. The 32-bit patterns of 5NCA rules for five bases of nucleic acids are designed by executing the 'algorithm for CA rule design' reported in Sect. 2.6.2 and illustrated for the design of 5NCA rules for sugar–phosphate molecule. The algorithm outputs the rules in the 1-major/0-major format of RMTs out of which the format (of descending order of RMTS) noted in Table 2.5 have been derived. Alternative 32-bit rule designs exist for nucleotide bases. Subsequent to experimentation with different alternative designs, we picked up these rules through top-down validation. The nucleotide bases are represented by the 32-bit rules reported in Table 2.6 for bases A, G, C, T for (DNA) and U (replacing T) for RNA in the format for RMTs (31, 30, …, 1, 0).

For a DNA strand with bases A, C, G, T, a hybrid 5NCA is designed on assigning the corresponding rules for each base of the strand. Such a CAM is referred to as Nucleotide Base DNA CA Machine (NBDCAM). The corresponding CAM for RNA is referred to as Nucleotide Base RNA CA Machine (NBRCAM) designed with bases A, G, C, U. An example of NBDCAM of hypothetical DNA and RNA strings are shown in Figs. 2.17 and 2.18

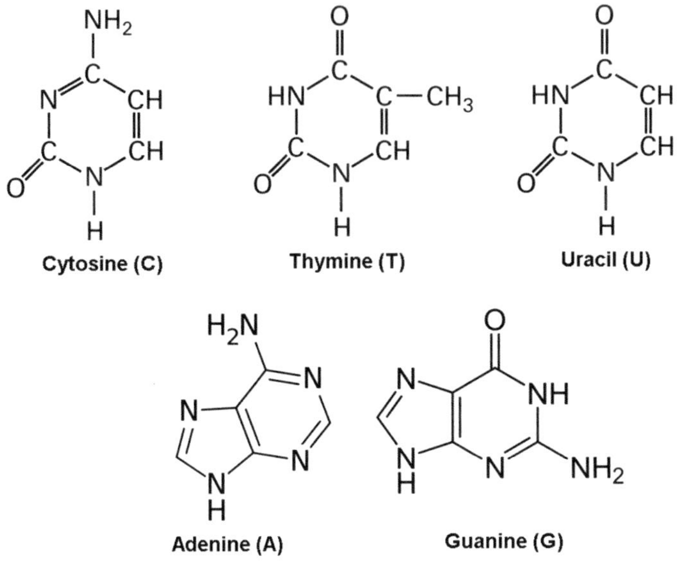

Fig. 2.16 Structures of nucleotide bases

Table 2.6 5NCA rule for nucleotide bases in the format of descending format of RMTs

Nucleotides	32-bit value (31, 30, …, 1, 0)	Number of atoms
Adenine	01011001011010101110000010110001	15 atoms
Guanine	01011011011010101110000010110010	16 atoms
Cytosine	01011001011000101100000010110010	13 atoms
Thiamine	11011001001000101110000010110011	15 atoms
Uracil	01011001001000101110000010010010	12 atoms

2.6.4　Design of Composite DCAM and RCAM

Earlier two sections report the design of four independent CAMs—(i) BBDCAM and NBDCAM for a DNA strand, and (ii) BBRCAM and NBRCAM for a RNA strand. Independent evolution of each of these CAMs is reported for hypothetical DNA/RNA strand. However, the bases are covalently bonded to sugar–phosphate molecule of backbone. Consequently, to develop the CA model for nucleic acid molecule, it is necessary to derive a composite DCAM/RCAM through co-evolution of patterns of BBCAM and NBCAM. Such CAMs are referred to as Composite DNA CA Machine (ComDCAM) and Composite RNA CA Machine (ComRCAM). These are designed by implementing the following steps. A CAM has n cells if it models a string of n bases.

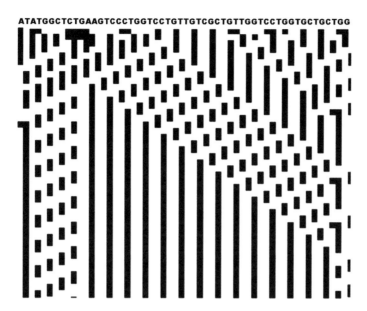

Fig. 2.17 Evolution pattern for NBDCAM for a hypothetical DNA base sequence (ATATGGCTCTGAAGTCCCTGGTCCTGTTGTCGCTGTTGGTCCTGGTGCTGCTGG)

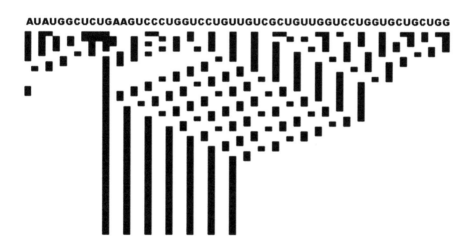

Fig. 2.18 Evolution of NBRCAM, the nucleotide base remains the same (except that T has replaced U)

Algorithm: Design a composite CA machine

Steps

1. The cell of a CAM is initialized with '1' if it models a purine base (A or G) and it is initialized with seed '0' if it models a pyrimidine base (C, T or U).
2. Each of the two CAMs (BBCAM and NBCAM) evolves one time step.
3. The ith cell ($i = 0, 1, 2, \dots (n - 1)$) 0/1 output of BBCAM and NBCAM are merged with logical NAND function to derive the next state bit for ith cell of NBCAM. The ith cell output of BBCAM continues without any modification. The output of logical NAND function defines the output of the Composite CAM (ComCAM).
4. Steps 2 and 3 are iterated for the number of evolution steps till each cell runs through its cycle.

Example of Composite CAM (ComCAM) using a hypothetical sequence noted below is shown in Figs. 2.19 and 2.20 for DNA (with bases A, G, C, T) and RNA (with bases A, G, C, U): **GTCCAGCCTTCCCTGGGCAAGGAAACTGC AGCAGCCAAGTTTGAGCGGCAGCACATGGAC**

In the ComRCAM evolution shown in Fig. 2.20 for comRCAM, the same nucleotide base chain is used as in ComDCAM except 'T' has been replaced by 'U'.

Detailed analysis of Composite DCAM and Composite RCAM evolution illustrated in Figs. 2.19 and 2.20 are reported in Chaps. 4 and 3 respectively.

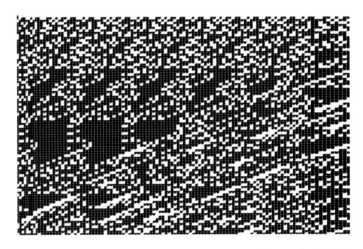

Fig. 2.19 Composite DNA CAM evolution derived out of BBDCAM and NBDCAM

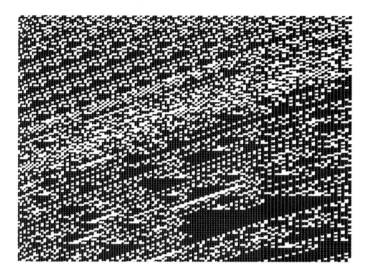

Fig. 2.20 Composite RNA CAM evolution derived out of BBRCAM and NBRCAM

In addition to other parameters, we shall focus on the analysis of signal graph derived out of Cycle Start (CS) values for successive cells of the CAMs.

For the mutational study of DNA and RNA strand, we shall evaluate the difference of the—(i) Cycle Start (CS) signal graphs, and (ii) 0/1 evolution pattern of wild and mutant version of NBCAM and ComCAM. Parameters derived out of the evaluation of the difference of NBCAM and ComCAM are next mapped to the features extracted from the wet lab results for top-down validation. This procedure for mutational study based on NBCAM/ComCAM is a necessity in view of the following observation: due to the strong sugar–phosphate backbone of DNA/RNA reflected in ComCAM, the effect of mutation, specifically SNP ((Single Nucleotide Polymorphism)) on a DNA and RNA strand does not always get represented in the difference in evolution of the wild and mutated version of ComCAM. The reverse has been also observed to be true for some cases, where NBCAM fails to detect the difference in wild and mutant probably due to the similarity of rule structure of the bases and the SNP location. Failure to detect the difference between Wild and Mutant by a CAM cannot be predicted without running the CAMs for Wild and Mutant.

2.7 Design of Protein Modeling CA Machine (PCAM)

Earlier sections present the design of 5NCA rules for the building blocks of nucleic acids—DNA and RNA. Based on a detailed study of physical domain features, two different CAMs are designed—one for backbone and other for a string of nucleotide bases, each bonded to a sugar–phosphate molecule of the backbone. Next, the third

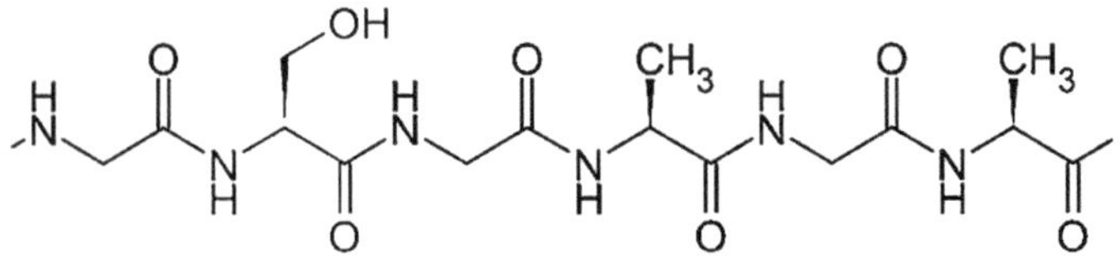

Fig. 2.21 Backbone peptide chain of amino acid residues

CAM (referred to as Composite CAM) is designed out of co-evolution of these two CAMs to represent the behaviour of a nucleic acid molecule.

Another biomolecule of primary importance is protein—the workhorse of biological functions in a living organism. Amino acids are the building blocks of a protein chain. Similar to the building block of DNA/RNA, each of the 20 common amino acid molecules displays the molecular structure of uniform backbone with differing side chain bonded to the C-alpha atom of its backbone. However, major differences exist between these two biomolecules (nucleic acid and protein) from the point of view of their information content. There is limited variation of four bases in nucleic acids, while variation in a protein chain is much wider with 20 widely diverse amino acids. Further, sugar–phosphate backbone of nucleic acid is much stronger compared to the peptide backbone chain of a protein. All these lead to a much wider variation of structure and function of a protein compared to nucleic acids. Consequently, the design of CA rules for amino acids and PCAM representing a protein chain significantly deviate from that of the nucleic acid molecule.

A peptide is a short chain of amino acids. A sequence of amino acids in a peptide is connected by peptide bonds between their backbones. Typically, peptides are distinguished from proteins by their shorter length, although the cut-off number of amino acids for defining a peptide and protein could be arbitrary. Figure 2.21 shows the peptide chain of amino acid residues of a protein. Figure 2.22 shows the structures of the 20 amino acids. Each amino acid covers a common backbone with side chain attached to the carbon atom of backbone referred to as C-alpha and modeled with 3NCA rules.

2.7.1 Design of CA Model for Amino Acids

Procedure to design CA rule to represent the molecular structure is reported in Sect. 2.6.3 with an illustration of the design of a 5NCA 32 bit rule for sugar–phosphate molecule. Two basic steps of the procedure are—(i) assign an atom displaying a high number of bonds in the PRMT (Palindromic RMT), and (ii) a pair of atoms with the bond between them are assigned to the CoP (Conjugate Pair) RMTs. As reported in Sect. 2.5 of this chapter, out of 8 RMTs of 3NCA rule, there are four PRMTs (7, 5, 2, 0) and other four RMTs forms two CoPs (6, 3) and (4, 1).

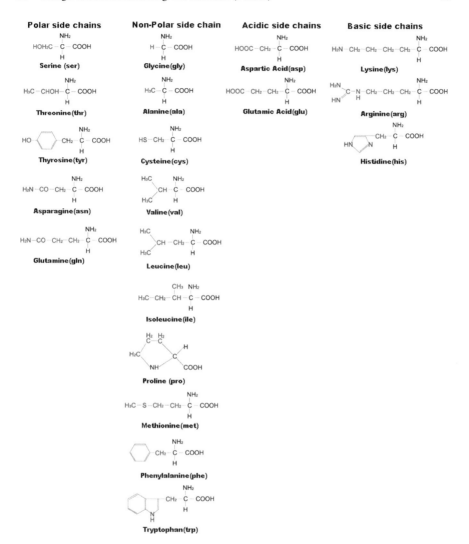

Fig. 2.22 Structure of 20 amino acids

The 'Algorithm: CA rule design' illustrated for the design of 5NCA rule for sugar–phosphate molecule is implemented to map the information of the molecular structure of—(i) backbone and (ii) side chain in the 8-bit structure of 3NCA rule. 64 alternative rules are designed for uniform backbone and 20 different rules are designed for side chain of 20 common amino acid residues. As pointed out in Sect. 2.6.3, alternative designs of 3NCA rules for side chains of amino acid residues exist. Subsequent to experiment with different alternatives, we have picked up the set of rules reported in this section. For design of PCAM for a protein chain, we utilize all the 64 alternatives of backbone rule to analyse possible variation of the features of a protein for interaction with other biomolecules.

For the design of CA rules for amino acid residue (i) backbone and (ii) side chain, we consider only heavy atoms (C, N, O) and exclude hydrogen atom (H). On exclusion of H atom, the molecular structure of the backbone has 4 atoms—O, N and two C atoms. We design 8-bit pattern of 3NCA rules for backbone. There are ($^{8}C_4$) seventy (70) 3NCA rules with four 1's and four 0's. Out of which we have chosen 64 non-trivial CA rules noted below to represent the molecular structure of backbone:

[141,197,177,163,75,89,210,154,45,101,180,166,58,114,92,78,139,209,153,195, 57,99,156,198,46,116,60,102,77,178,90,165,135,149,225,169,27,83,216,202,39,53, 228,172,30,86,120,106,147,201,150,105,15,85,240,170,29,71,184,226,51,204,54,108]

Similarly, on considering only non-H atoms of side chain, twenty (20) 3NCA rules are designed to represent information content of side chain molecular structure for 20 amino acids. In order to restrict the number of 1's in the 8-bit pattern of a 3NCA rule, we avoid populating more than six '1's in the design of a rule. The only exception is the design of Arg (R) side chain having a long aliphatic chain with 7 atoms. Further, an aromatic ring present in a side chain (trp (W), tyr (Y), phe (F)) is represented by assigning 1 to RMT 7 which is not used to represent any other atom. The structure of his (H) displays 6 non-H atoms including those in its ring; consequently, for the design of rule for its side chain, RMT 7 is not assigned even though it has an aromatic ring. The amino acid side chain encoding is reported in details in Chap. 5 (Sect. 5.4.1).

Based on the earlier considerations, 3NCA rules for side chains of 20 amino acid residues are designed as noted in Table 2.7 with second column displaying single letter code for amino acid and third column showing 8 bit pattern for 8 RMTs of a residue—7, 6, 5, 4, 3, 2, 1, 0.

The methodologies of designing sixty four (64) 3NCA rules for amino acid backbone and 8-bit pattern representing information of the molecular structure of side chain are elaborated in Chap. 5 on proteomics.

2.7.2 PCAM (Protein Modeling CA Machine) Design

In order to model a protein chain, we implement the following two steps:

1. An amino acid residue of a protein chain is represented by 8 CA cells which are initialized with the 8-bit seed derived out of its side chain 3NCA rule.
2. The backbone is designed with a uniform 3NCA with one of the 64 CA rules.

For an input protein chain, we have 64 different PCAMs. Based on the similarity of evolution patterns of different PCAMs, the 64 PCAMs for a protein are grouped into (say) k number of classes. Each of these k classes represents different features of a protein for interaction with different micro and macromolecules under normal temperature and pH value. Any variation of these two environmental parameters (temperature and pH) adds further complexity for analysis of protein structure and function.

Table 2.7 The 8-bit pattern of 3NCA rules for side chain of 20 amino acid residues noted with single letter code on first column with first row displaying 8 RMTs

Amino acids	Abbreviations	RMTs 7 6 5 4 3 2 1 0
Glycine	G (gly)	0 0 0 0 0 0 0 0
Alanine	A (ala)	0 0 0 0 0 1 0 0
Proline	P (pro)	0 0 1 0 0 1 1 0
Valine	V (val)	0 0 0 1 0 1 1 0
Methionone	M (met)	0 0 1 1 0 1 1 0
Typtophan	W (trp)	1 0 1 1 0 1 1 0
Phenylalaline	F (phe)	1 0 0 0 0 1 0 0
Isoleucine	I (ile)	0 0 0 1 1 1 1 0
Leucine	L (leu)	0 0 0 1 0 1 1 1
Serine	S (ser)	0 0 1 0 0 1 0 0
Cysteine	C (cys)	0 1 0 0 0 1 0 0
Threonine	T (thr)	0 0 1 1 0 1 0 0
Asparagine	N (asn)	0 0 1 0 1 1 1 0
Glutamine	Q (gln)	0 0 1 0 1 1 1 1
Tyrosine	Y (tyr)	1 0 1 0 0 1 0 0
Histidine	H (his)	0 1 1 1 1 1 1 0
Lysine	K (lis)	0 0 1 1 0 1 1 1
Arginine	R (arg)	0 1 1 1 1 1 1 1
Aspartic acid	D (asp)	0 1 1 1 0 1 0 0
Glutamic acid	E (glu)	0 1 1 1 0 1 1 0

A protein chain with n amino acid residues is represented by a $8n$ cell uniform 3NCA. Each 8-bit block of cells is initialized with the 8-bit seed derived out 8-bit pattern of the side chain 3NCA rule of each amino acid. Figure 2.23 illustrates a 32-cell 3NCA representing a peptide chain with 4 amino acid residues represented by single letter code M, R, P, S. The 8-bit pattern of residue side chain 3NCA rule reported in Table 2.7 is assigned to 8 cells representing a residue.

An application detailed in Chap. 5 addresses the mutational study of a protein chain. The target goal is to specify one out of 64 PCAMs for which the wet lab results get mapped to a specific parameter of the difference in evolution pattern

$$4 \times 8 = 32 \text{ Cells}$$

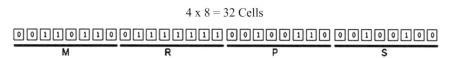

32 cell CA representing a peptide chain of four amino acids M, R, P, S (single letter code) with corresponding 8-bit side chain rule as shown in Table 2.7

Fig. 2.23 32-cell CA representing peptide chain of 4 amino acids

between two PCAMs (wild and mutant version). Difference in evolution patterns could be measured by bitwise XOR (or other) operation on the evolved pattern for each cell at each evolution step for the two PCAMs employing the same backbone rule.

An illustration of mutational study follows for an example protein with three 3NCA backbone rules 54, 105 and 135. Single letter amino acid sequence for the protein chain is noted below.

**MALKSLVLLSLLVLVLLLVRVQPSLGKETAAAKFERQHMDSSTSAA
SSSNYCNQMMKSRNLTKDRCKPVNTFVHESLADVQAVCSQKNVACKN
GQTNCYQSYSTMSITDCRETGSSKYPNCAYKTTQANKHIIVACEGNPY
VPVHFDASV**

The above protein chain has the UniProt ID RNAS1_bovine.

For this study, the PCAMs run through 1000 evolution steps. Figures 2.24, 2.25 and 2.26 display the results of XOR operation of PCAM evolution for wild and its mutated version employing three different CA rules for backbone—54, 105 and 135, respectively. Figure 2.26 is derived employing 3NCA rule 135 for backbone which confirms the limited change in a specific region around the mutational site. Mutation site for all the three cases is K67A (residue location 67 mutated from K to A).

The XOR-ing of the wild and mutated protein is done only to visualize the subtle difference when the 'seed' changes for a CA rule on insertion of mutation at residue location 67 (marked on the line displaying 0th evolution step). It will be shown in Chap. 5 how the changes are detected and how the location of protein–protein interaction site is predicted. A brief highlight follows.

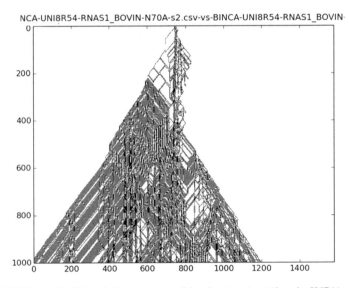

Fig. 2.24 Difference in CA evolution between wild and mutant (mutation site K67A) running CA Rule 54

CA-UNI8R105-RNAS1_BOVIN-K67A-s2.csv-vs-BINCA-UNI8R105-RNAS1_BOVIN

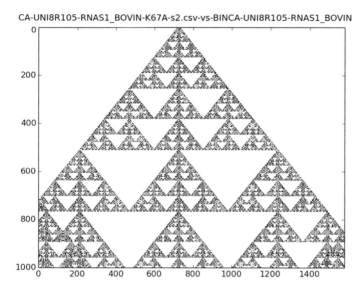

Fig. 2.25 Difference in wild and mutant (mutation site K67A) running CA Rule 105

CA-UNI8R135-RNAS1_BOVIN-N70A-s2.csv-vs-BINCA-UNI8R135-RNAS1_BOVIN-

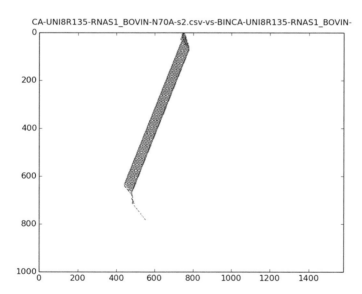

Fig. 2.26 Difference in wild and mutant (mutation site K67A) running CA Rule 135

The PCAM generates CA Evolution Image known as CA-image. We have seen in Fig. 2.26 that there are limited differences when we run selected rules on different protein sequences (wild and mutant).

We will show in Chap. 5 how we can get a $n \times m$ binary matrix where n is the length of CA cells and m is time step. For the current study $m = 1000$. A statistical method of examining texture that considers the spatial relationship of pixels is the Gray-Level Co-occurrence Matrix (GLCM) [5], also known as the grey-level spatial dependence matrix. The GLCM functions characterize the texture of an image by calculating how often pairs of the pixel with specific values and in a specified spatial relationship occur in an image, creating a GLCM, and then extracting statistical measures from this matrix. Our CA texture analysis method is based on GLCM method. The following steps are implemented to analyse the texture of the CA evolution.

Algorithm: Analyse the texture of CA evolution

Input: Binary matrix of CA evolution
Output: CA evolution of texture features
Steps

1. Generate Gray-Level Co-occurrence Matrix (GLCM)

   - Row-wise GLCM (rGLCM)—Count the occurrence of four transitions (00, 01, 10, 11) from left to right and right to left for each of the rows. We get a row-wise GLCM matrix of $t \times 4$ size from $t \times n$ CA evolution matrix, where t is the CA evolution time step
   - Column-wise GLCM (cGLCM)—Count the occurrence of four transitions (00, 01, 10, 11) from top to bottom and bottom to top for each of the columns. We get a column-wise GLCM matrix of $4 \times n$ size from $1000 \times n$ CA evolution matrix, where n is the size of CA

2. Normalize rGLCM and cGLCM
3. Calculate feature functions from rGLCM and cGLCM: Four feature functions Angular Second Moment (ASM), Contrast (CON), correlation (COR) and Entropy (ENT) are calculated from the matrices. Thus, we get 4 numeric values (ASM, CON, COR, ENT) for each of the matrices, forming a set of 8 features.

Row-wise GLCM (rGLCM) is sequence length independent, which means any length of protein sequence will form a $t \times 4$ rGLCM matrix with t the time step of CA evolution. Thus, rGLCM will be particularly useful in comparing different length sequences for an experiment like classification of proteins. Column-wise GLCM (cGLCM) is length dependent and particularly useful in comparing effect of mutation on a specific protein sequence, where length is the same for PCAM matrices.

2.8   **Summary**

In this chapter, we went through the basic design architecture of cellular automata machines referred to as CAMs. We described the procedure and steps through which we designed the CAMs.

For representation of information of nucleotide base triplets (codons), we used 64 rules. The nucleotide triplets of RNA and DNA are replaced by corresponding rule and thus we have a CA with such rules to represent a codon string. The rules are three neigbourhood and as multiple rules are employed in the design, it is a hybrid CA referred to as 3NHCA.

Rather than dealing with codon information, when we tried to design CA rule to represent information content of the sugar–phosphate molecule and nucleotide bases, we concluded that the 8-bit structure of 3NCA rules does not suffice. Consequently, we designed five-neighbourhood CA rules to represent the building blocks of DNA/RNA. The sequence of nucleotides is represented by a 5NHCA— five-neighbourhood hybrid CA. The backbone is represented by a uniform 5NCA. Finally, to represent the composite structure of a DNA/RNA strand, it is necessary to co-evolve the five-neighbourhood hybrid CA and 5N uniform backbone CA. These co-evolved CAMs are called ComRCAM for RNA and ComDCAM for DNA. These CAMs are employed in Chaps. 3 and 4 for cLife representation of RNA and DNA.

The behaviour and property of proteins are substantially different from nucleic acids. Protein behaviour change significantly based on environment/interaction with other biomolecules (micro or macro). Consequently, the design of PCAM (Protein modelling CA Machine) deviates from that of CAMs designed for DNA/RNA. A PCAM designed for protein chain having n residues has $8n$ cells. Its peptide backbone is designed with a uniform 3NCA employing one of the sixty four (64) 3NCA rules. A block of 8 cells representing a residue is initialized with the 8-bit pattern of the side chain rule of the residue. Consequently, there are 64 different variations of PCAM for a protein chain, each representing characteristics of the protein for different interaction/environment.

For the mutational study of a protein chain, we design PCAMs for wild and mutant versions for each of the 64 backbone rules. Parameters are derived out of the difference in the evolution patterns for wild and mutant for each of the 64 backbone rules. One PCAM for which the parameters get consistently mapped to the features extracted from the wet lab experimental results is selected to represent the interaction/environment of the mutation.

Questions

1. In the design of 3NCA rule for codon, which of the following statements are correct?

   (a) The CA rules represent nucleotide base triplets

    (b) A single 0/1 seed input to a CA cell corresponds to the middle base of the codon triplet

    (c) A single 0/1 seed input to a CA cell corresponds to a nucleotide triplet

2. Explain the term 'Rule Min Term (RMT)'. What are '0-major' and '1-major' RMTs? How these RMTs are used in the design of 3NCA and 5NCA rules? What is 'Palindromic RMT'?

3. A 3NCA rule is used to represent nucleotide triplet, while a 5NCA rule is used to represent a single nucleotide base—explain the reason for such difference in the CA rule design?

4. The 5NCA for RNA and DNA backbone differs by a single bit in a position—this difference is due to which atom of sugar—phosphate molecule?

5. What is the difference between uniform and hybrid CA? Why is a uniform CA used to represent the backbone of nucleic acid, while a hybrid CA represents the chain of nucleotide bases?

6. What is a co-evolved CA? How a co-evolved CA is designed to represent a DNA/RNA strand?

7. Design BBDCAM and BBRCAM for a nucleic acid strand having 32 bases, where the sequence (A, T, C, G) repeats eight times. Note Cycle Start (CS) and Cycle Length (CL) values of each CA cell after each cell settles down in its cycle.

8. Design of NBDCAM and the corresponding NBRCAM for the example DNA strand of Question 7.

9. Explain the steps implemented in the design of PCAM (Protein modeling CA Machine). How many cells have to be used to represent a 6 amino acid peptide chain?

10. 'CAM design for protein chain differs from CAM design for nucleic acid'—True or False; if 'True' explain the underlying reasons from analysis of structure and function of these.

References

1. https://books.google.com › Computers › Programming › Algorithms
2. Harvey, I., Bossomaier, T.: Time out of joint: attractors in asynchronous random boolean networks. In: Proceedings of the Fourth European Conference on Artificial Life. MIT Press, Cambridge (1997)
3. Kanada, Y.: Asynchronous 1D cellular automata and the effects of fluctuation and randomness. In: Proceedings of the fourth Conference on artificial life (A-Life IV). MIT Press. 1994
4. Orponen, P.: Computing with truly asynchronous threshold logic networks. Theoret. Comput. Sci. **174**(1–2), 123–136 (1997)
5. Haralick, R.M., Shanmugam, K.:. Textural features for image classification. IEEE Trans. Syst. Man Cybern. **6**(1973), 610–621

Chapter 3
Cellular Automata Model for Ribonucleic Acid (RNA)

"I'm fascinated by the idea that genetics is digital. A gene is a long sequence of coded letters, like computer information. Modern biology is becoming very much a branch of information technology".

—Richard Dawkins

3.1 Introduction

What led to the origin of life on earth?—it continues to be one of the deepest and unsolved mysteries in modern science. The basic molecules available on earth millions of year ago might have given rise to two compounds, namely purines and pyrimidine—chemicals that are the key ingredients of DNA and RNA. The key early event in this process, as per 'RNA First' hypothesis introduced in Chap. 1, was the formation of RNA, a long chain-like molecule which is responsible to convey genetic information and speed-up chemical reactions within the cells.

In the current life form, Ribonucleic Acid (RNA) is generally viewed as the bridge between Genomics and Proteomics. The mRNA carries the message from the DNA, which in turn controls and coordinates the cellular activities within a cell. If a cell requires a particular protein to be synthesized, the corresponding gene is 'turned on' and then, the mRNA is synthesized through the process of transcription. In the subsequent translation process, the mRNA interacts with the ribosome and other mechanisms within the cell to direct the synthesis of the protein which it encodes. Compared to double-stranded DNA, single-stranded mRNA, in general, is relatively unstable and short-lived in the cell, to ensure that the proteins are only made when needed, especially in prokaryotic cells.

Ribosome is composed of rRNA and protein. As referred by its name, rRNA (Ribosomal RNA) is a major constituent of ribosomes, constituting about 60% of the ribosomal mass and specifying the location where the mRNA binds. The rRNA ensures the proper alignment of the mRNA, tRNA (Transfer RNA) and the ribosomes. Further, the rRNA is responsible to perform an enzymatic activity (peptidyl

© Springer Nature Singapore Pte Ltd. 2018
P. P. Chaudhuri et al., *A New Kind of Computational Biology*,
https://doi.org/10.1007/978-981-13-1639-5_3

transferase) to catalyze the formation of the peptide bonds between two aligned amino acids during elongation of the peptide chain for protein synthesis.

tRNA (Transfer RNA) is the third main type of RNA and one of the smallest, usually only 70–90 nucleotide bases long. It is responsible to carry the correct amino acid to the site of protein synthesis in the ribosome. Nature has designed a fail-safe design of base pairing between the tRNA and mRNA that allows the correct insertion of amino acid in the polypeptide chain being synthesized. Consequently, a mutation in the tRNA or rRNA can result in severe problems for the cell. Both the molecules are hence necessary for ensuring proper protein synthesis as the proteins control the functions of living cells.

siRNAs (Small Interfering RNAs) have a well-defined structure—a short usually 20–24-bp (base pair) derived out of a longer RNA molecule referred to as Double-Stranded RNA (dsRNA). A dsRNA has phosphorylated 5 prime (5′) end and hydroxylated 3 prime (3′) end with two overhanging nucleotides. An enzyme referred to as dicer catalyzes production of siRNAs and Small Hairpin Loop (HPL) RNAs from long dsRNAs. For wet lab experiments, siRNAs may be introduced into cells by a process named 'transfection'. In principle though, any gene could be knocked down by a synthetic siRNA which has a complementary sequence. Consequently, in post-genomic era, siRNAs have become a principal tool for validating gene function and drug targeting.

Similar to siRNAs, miRNAs (micro RNA) resemble another small RNA in the RNA interference (RNAi) pathway. While miRNAs are derived from regions of RNA transcripts that fold back onto themselves to form short hairpin; the siRNAs are usually derived out of longer regions of double-stranded RNAs. The human genome may encode over 1000 different miRNAs. They are abundant in many mammalian cell types and target more than 60% of the genes of mammals including humans.

Lastly, we note an important information from the study of viruses; some of the viruses, for example, retrovirus utilize RNA to retain their hereditary functions. HIV (Human Immunodeficiency Virus) is a retrovirus.

From the study of the relevant research publications, we mentioned in Chap. 1 that RNA world hypothesis is getting more positive results from wet lab experiments and also from information complexity (entropic) viewpoint. So, we start our CA model introduction from RNA itself. RNA in its different variants plays an important role in the current life form, but it may have played an initialization role in starting life. This Chapter focuses on the study of RNA molecules transcribed from both coding and non-coding DNA strand. The messenger RNA (mRNA) derived out of gene (coding DNA) and functional RNAs derived from non-coding DNA are briefly introduced in Chap. 1. In Sect. 1.4.7 of Chap. 1, it has been proposed in 'RNA World' hypothesis that RNA precedes DNA, and it was the main building block in the early life form. Consequently, it is necessary to undertake a detailed study of this macromolecule, as noted below, along with the design of CA model for its structure and function.

3.2 Cellular Automata (CA) Model for Ribonucleic Acid (RNA)

This chapter focuses on the study of RNA molecules transcribed from both coding and non-coding DNA strand. While Chap. 1 introduces RNA fundamentals, design of CA rules for the building blocks of RNA is reported in Chap. 2. Rules are designed on extraction of the underlying 'information' of the molecular structure of the building blocks—sugar–phosphate, nucleotide base and nucleotide base triplet referred to as codon of mRNA. The different sections of this chapter are noted below.

- A Mathematical Model for Codon Degeneracy.
- CA Model for Ribosomal Translation.
- Mutational Study on mRNA Codon String—Case Studies.
- Modelling Co-Translational Folding of Nascent Peptide Chain in Ribosome.
- Prediction of Secondary Structure of RNA Molecules—tRNA, miRNA Precursor and miRNA (micro RNA).
- Modelling siRNA and its Target Gene in cLife (CA-based Life).
- RNA-Binding Protein (RBP).

3.3 A Mathematical Model for mRNA Codon Degeneracy

A degenerate code points to the fact that several code words have the same meaning. The genetic code is degenerate because there are multiple instances in which different codons specify the same amino acid. This has been shown in Chap. 1, Fig. 1.17 where the simplest amino acid glycine (gly) is encoded with four codons GGU, GGC, GGA and GGG with first two bases as GG followed by any one of the four bases in the third base location. It has been suggested that the degeneracy makes the DNA more tolerant to point mutations—that is, replacement of the third base with another in the base triplet of a codon. Degeneracy is associated with the translation of an mRNA codon to an amino acid in a protein chain. In order to present the underlying principle of CA model for degeneracy, we next present a few definitions along with illustrations.

A position of a codon is said to be an *n*-fold degenerate site if only $n(n = 4, 3, 2)$ of four possible nucleotides (A, C, G and U) at this position specify the same amino acid. A nucleotide substitution at an n-fold degenerate site is referred to as a synonymous mutation. Substitutions resulting in the change of the amino acid are denoted as non-synonymous mutations.

Table 3.1 Codon degeneracy table for 20 amino acids and start/stop codons

Amino acid	Codons	Amino acid	Codons
Ala/A	GCU, GCC, GCA, GCG	Leu/L	UUA, UUG, CUU, CUC, CUA, CUG
Arg/R	CGU, CGC, CGA, CGG, AGA, AGG	Lys/K	AAA, AAG
Asn/N	AAU, AAC	Met/M	AUG
Asp/D	GAU, GAC	Phe/	UUU, UUC
Cys/C	UGU, UGC	Pro/P	CCU, CCC, CCA, CCG
Glu/E	GAA, GAG	Thr/T	ACU, ACC, ACA, ACG
Gly/G	GGU, GGC, GGA, GGG	Trp/W	UGG
His/H	CAU, CAC	UAU, UAC	UAU, UAC
Ile/I	AUU, AUC, AUA	Val/V	GUU, GUC, GUA, GUG
START	AUG	STOP	UAA, UGA, UAG

A codon triplet has three base sites—left, middle and right. A position of a codon is defined as a non-degenerate site, if any mutation at this position results in amino acid substitution. There is only one threefold degenerate site where changing to three of the four nucleotides have no effect on the amino acid, while changing to the fourth possible nucleotide results in an amino acid substitution. This is the third position of isoleucine (ile) codons: AUU, AUC and AUA all encode isoleucine, but AUG encodes Methionine (Met). Hence for computation, this condition is often treated as a twofold degenerate site. We present the inverted genetic code in Table 3.1 displaying each of the 20 amino acids and their associated codons.

3.3.1 Mathematical Model for Design of CA Rule Groups Representing Codon Groups

In this section, we present the CA rule design for the corresponding code. In Table 3.2 (column 2 and 3), we present the CA rules (both in decimal and binary 1/0 major format) for different 16×4 codon groups associated with 20 common amino acids. The underlying principle of our design is noted followed by the details on how we selected CA rules (of Table 3.2) using symmetry and a set of transformations applied on the 8-bit binary pattern of three-neighbourhood CA rules.

In addition to introduction of the basic macromolecules (DNA, RNA and protein) in Sect. 1.5.1 of Chap. 1, we reported the fundamentals of three-neighbourhood two-state per cell null boundary CA. Each CA rule is represented by its 8-bit pattern.

We have assumed that RNA-based primitive life was initiated with two bases—purine and pyrimidine. The primitive codons of base triplets (with two bases) can be modelled with three-neighbourhood two-state per cell CA rules. So, there were

Table 3.2   64 CA rules—16 × 4 rule group representing 16 × 4 codon amino acid groups

1	2-3rd base	3-3rd base	4	5	6	7
ID	[R1] = [C] [R3] = [G]	[R2] = [U] [R4] = [A]	$\sum$(ID)	N1,N0,δ	y-ERG	First two bases of the codon/ amino acid
1	141 (1001,0101) 177 (1010,1001)	197 (1100,0101) 163 (1010,0011)	678	2,2,1 2,2,1	4	GG/Gly
2	75 (0101,0011) 210 (1100,1010)	89 (0101,1001) 154 (1001,1010)	528	2,2,1 2,2,1	4	GU/Val
3	45 (0011,0101) 180 (1010,1100)	101 (0110,0101) 166 (1010,0110)	492	2,2,1 2,2,1	4	GC/Ala
4	58 (0011,1010) 92 (0101,1100)	114 (0110,1010) 78 (0101,0110)	336	2,2,1 2,2,1	4	AC/Thr
5	139 (1001,0011) 153 (1001,1001)	209 (1100,1001) 195 (1100,0011)	696	2,2,2 2,2,2	4 + 2	UC/Ser
6	57 (0011,1001) 156 (1001,1100)	99 (0110,0011) 198 (1100,0110)	510	2,2,2 2,2,2	4 + 2	CG/Arg
7	46 (0011,0110) 60 (0011,1100)	116 (0110,1100) 102 (0110,0110)	324	2,2,2 2,2,2	4 + 2	CU/Leu
8	77 (0101,0101) 90 (0101,1010)	178 (1010,1010) 165 (1010,0101)	420	2,2,0 2,2,0	4	CC/Pro
9	135 (1000,0111) 225 (1110,0001)	149 (1000,1101) 169 (1011,0001)	678	1,3,1 3,1,1	2 2	GA/Asp Glu
10	27 (0001,1011) 216 (1101,1000)	83 (0100,1011) 202 (1101,0010)	528	1,3,1 3,1,1	2 2	AA/Asn Lys
11	39 (0010,0111) 228 (1110,0100)	53 (0010,1101) 172 (1011,0100)	492	1,3,1 3,1,1	2 2	UA/Tyr Stop

(continued)

Table 3.2 (continued)

1	2-3rd base	3-3rd base	4	5	6	7
ID	[R1] = [C] [R3] = [G]	[R2] = [U] [R4] = [A]	$\sum$(ID)	N1,N0,δ	y-ERG	First two bases of the codon/ amino acid
12	30 (0001,1110) 120 (0111,1000)	86 (0100,1110) 106 (0111,0010)	336	1,3,1 3,1,1	2 2	CA/His Gln
13	147 (1000,1011) 150 (1000,1110)	201 (1101,0001) 105 (0111,0001)	603	1,3,0/ 3,1,0 1,3,0/ 3,1,0	* *	Arg
14	15 (0001,0111) 240 (1110,1000	85 (0100,1101) 170 (1011,0010)	510	1,3,2 3,1,2	3 1	AU/Ile Met Ile
15	29 (0001,1101) 184 (1011,1000)	71 (0100,0111) 226 (1110,0010)	510	1,3,2 3,1,2	2 1/1	UG/Cys Trp Stop
16	51 (0010,1011) 54 (0010,1100)	204 (1101,0100) 108 (0111,0100)	417	1,3,0/ 3,1,0 1,3,0/ 3,1,0	2 *	UU/Phe Leu

$2^8 = 256$ rules for 256 primitive codons. In order to support heredity information and sustain life with wider varieties, RNA-based primitive life form got replaced with the current DNA-based life with—four bases A, G (out of purine), and C, U (out of pyrimidine) resulting in 64 codons derived out of $4^3 = 64$ base triplets. Out of the available 256 rules, we need to select 64 CA rules to represent 64 codons of current life form. In the process, our cLife model establishes a bridge between RNA-based primitive and DNA-based current life form.

Chapter 2 reports how we build up a 3NHCA (three neighbourhoods hybrid CA) employing three-neighbourhood CA (3NCA) rules expressed in 1-major/0-major format. This section reports how we get to the rule set reported in Table 3.2 out of available 256 rules.

In the 8-bit structure of 3NCA rule, there are $^8C_4 = 70$ rules referred to as Balanced Rules (BR) with four 1's and four 0's. The motivation to select BRs for our subsequent analysis stems from the observation that nature has an inherent inclination for 'symmetry'. A BR is referred to as (2, 2) or non-(2, 2) ((1, 3) or (3, 1)) rule, where a (2, 2) rule is the one having two 1's for each half byte of the rule (arranged in 1/0-major format). For example, 45 (0011 0101), 135 (1000 0111) and 169 (1011 0001) are (2, 2), (1, 3) and (3, 1) rules, respectively. Table 3.2 entries indicate how many 1's are there in a nibble (half byte). As noted in Chap. 1, we reemphasize the point that all the rules are noted in 1/0 major format in the sense, first half byte refers to next state values for RMTs (Rule Min Terms) having two 1's

(7, 6, 5, 3), while second half byte covers next state values for the 0-major RMTs (4, 2, 1, 0) with two 0's. The 8-bit pattern of each of the rules in Table 3.2 is organized in this format.

By convention, the decimal value of a binary sting is evaluated by assigning maximum/minimum weight to the leftmost/rightmost bit of a string (endianness). By reversing this convention, we define a set of transforms.

Transform 0 (Tr 0): Reversal of binary bits representing a RMT:

$$0(000), 1(001), 2(010), 3(011), 4(100), 5(101), 6(110), 7(111)$$

RMTs 7, 5, 2, 0 are palindromic as changing endianness does not change their values.

For the non-palindromic pair (6, 3), 6 gets transformed to 3 and vice versa. Similarly, 4 get transformed as 1 and vice versa. (6, 3) and (4, 1) are referred to as conjugate pairs (CoPs).

The next state bits of a 3NCA rule is denoted as $<b_7\ b_6\ b_5\ b_3\ b_4\ b_2\ b_1\ b_0>$ for the 8 RMTs $<7\ 6\ 5\ 3\ 4\ 2\ 1\ 0>$. Based on this formulation, a parameter δ is defined as follows.

$$\delta = 0, \text{if } b_6 = b_3 \text{ in 1-major half byte and } b_4 = b_1 \text{ in 0-major half byte}$$
$$\delta = 2, \text{if } b_6 \neq b_3 \text{ and } b_4 \neq b_1$$
$$\text{else } \delta = 1$$

In Table 3.2, if we interchange the conjugate pairs (CoPs) bits (b_6, b_3) and (b_4, b_1) of rule 141 $<1001\ 0101>$, it is transformed to $<1100\ 0101>$ which is Rule 197, thus we get Conjugate Rule Pair (CRP) $<141\ 197>$. If we reverse the 8-bit string of Rule 141, it generates Rule 177 $<1010\ 1001>$, so we have another CRP $<141\ 177>$. Similarly, on interchange of CoPs (b_6, b_3) and (b_4, b_1) of Rule 197 $<1100\ 1001>$, we generate Rule 163 $<1001\ 0011>$. From the above example, one can see that a rule group (RG) [141 197 177 163] with CRPs $<141\ 197>$, $<141\ 177>$, $<197\ 163>$ are generated out of Rule 141. As illustrated above, each of the 16 (sixteen) 4-rule groups (4RGs) with IDs 1–16 of Table 3.2 are generated on application of single or multiple transforms summarized next.

The Set of Transforms

The following transforms are applied on a CA rule R to derive its conjugate R' and form a conjugate rule pair CRP (R, R'). Three tables noted below reports the conjugate rule pairing (CRP) derived with these transforms. A pair of CRPs $<R, R'>$ under a transform X is represented as Tr X $(R\ R')$, where X represent a transform type Tr 1/2/3/C.

Transforms	Description
Transform 1 (Tr 1)	Interchange CoP bits $<b_6\ b_3>$ of 1-major half byte and $<b_4\ b_1>$ of 0-major half byte. For example, Rule 141 <1001 0101> becomes 197 = <1101 0101> and so Rule pairs 141 and 197 are conjugate under Tr 1 denoted as Tr 1 (141, 197)
Transform 2 (Tr 2)	Reverse the 8-bit string ($<b_7\ b_6\ b_5\ b_3>\ <b_4\ b_2\ b_1\ b_0>$) of a CA rule viewing MSB (Most Significant Bit) as LSB (Least Significant BIT) and vice versa. For example, 141 = <1001 0101> becomes 177 = <1010 1001> and vice versa. Hence, Rule pairs 141 and 177 is conjugate under Tr 2, denoted as Tr 2 (141, 177)
Transform 3 (Tr 3)	In order to derive wider variability of transformed rules, the next transform is applied for a rule R if Tr 1 (R) = Tr 2 (R), or Tr 2 (R) = R It is employed to interchange (i) Only CoP bits b_4 and b_1 of 0-major half byte with no change on 1-major half byte (ii) Only CoP bits b_6 and b_1 with no change on 0-major half byte, and (iii) Interchange both (b_6, b_3) and (b_4, b_1) For example, (i) 139 = <1001 0011> becomes <1001 1010> = 153; while 209 <1100 1001> becomes <1100 0011> = 195 under Tr3 (i). Hence <139 153> and <209 195> are conjugate pairs under Tr 3 (i). On the other hand, as per Tr3 (iii), 139 <1001 0011> becomes 209 <1100 1001> and also 153 <1001 1001> becomes 195 <1100 0011>. Thus, another 4RG [139 209 153 195] gets generated out of rule 139

The next transform realizes conjugate pair rules through complementation of bit 0 with 1 and vice versa; it is denoted as Tr C—there are four classes of such a transform.

Transforms	Description
Transform C0 (Tr C0)	A pair of rules R and R' is conjugate under Tr C0, if R can be derived out of R' through complementation of each bit of the string. There are many such pairs denoted as Tr C0($R\ R'$) such as • (57, 198) and (99, 156) in ID6 in Table 3.2 • (77, 178) and (90, 165) in ID 8 • (150, 105) in ID 13 • (15, 240) and (85, 170) in ID 14 • (29, 226) and (71, 184) in ID 15 • (51, 204) in ID 16 For each such conjugate pair, the sum of their decimal values is 255 = <1111 1111>
Transform C1 (Tr C1)	It inverts only the 0-major bits. For example, (77, 90) of ID 8 in Table 3.2 is a conjugate rule pair under Tr C1, where 77 = <0101 0101>, 90 = <0101 1010>
Transform C2 (Tr C2)	The next transform is applied, if Tr 1 (R) = R, that is, for a string $b_6 = b_3$ in 1-major and $b_4 = b_1$ in 0-major half byte It inverts non-CP bits b_2 and b_0 of 0-major half byte. Rule pair <51 54> of ID 16 is conjugate under Tr C2, where 51 = <0010 1011>, 54 = <0010 1110>. Rules 147 and 150 of ID 13 is another instance of Tr C2 (147, 150), where 147 = <1000 1011>, 150 = <1000 1110>

While all the earlier transforms are applied on a pair of rules covered by either (2, 2) or non-(2, 2) rules, the next transform specifies a Conjugate Rule Pair (CRP), where one is a (2, 2) and other one is a non-(2, 2) rule.

Transforms	Description
Transform C3 (Tr C3)	It inverts bits b_3 in 1-major and b_1 (or b_4) in 0-major half byte of a rule in the CRP. For example, non-(2, 2) rule 150 = <1000 1110> of ID 13, on inversion of b_3 and b_1, is the conjugate (2, 2) rule 156 = <1001 1100> of ID 6 and denoted as Tr C3 (150, 156). A few other such conjugate rule pairs are—Tr C3 (147, 139), Tr C3 (201, 209), Tr C3 (54, 60) and Tr C3 (108, 102), where the first rule is a non-(2, 2) rule and second one is a (2, 2) rule

Table 3.2 reports the derivation of sixteen (16) 4RGs - Decide how best to put this out of 64 codon rules by identifying conjugate rule pair on application of different transforms. Column 1 shows the Identification Number (ID) 1–16 of a 4RG with four rules ($R1$, $R2$, $R3$ and $R4$) arranged in two rows—the first row of an ID shows the details of rules $R1$ and $R3$, while second row provides the details for rules $R2$ and $R4$. Column 2 shows the rule type—whether (2, 2) or non-(2, 2). First eight (8) 4RGs (ID 1–8) cover thirty-two (32) (2, 2) rules, while ID 9-16 reports the same for non-(2, 2) 4RGs. Columns 2 and 3 show the decimal and binary pattern (in the format <$b_7 b_6 b_5 b_3 b_4 b_2 b_1 b_0$>) of the rules. Transform type, presented above, generates the conjugate rule pairs (CRPs) of each 4RG (4 rule group).

3.3.2 Generating Equivalent Rule Groups

In Table 3.2, the rules of a 4RG has four CA rules denoted <$R1$ $R2$> on first row and <$R3$ $R4$> on the second row for each of the 16 IDs. Column 6 of Table 3.2 shows the arrangement of 4RGs to Equivalent Rule Group (ERG).

The foundation of rule equivalence has been laid down with Conjugate Pairing (CRP) of rules derived with different transforms reported earlier. Two rules of a CRP derived with a transform represent a pair of Equivalent Rules (ERs). The four (2, 2) rules of ID 1–8 form 4-equivalent rule group denoted as 4ERG (4 equivalent rule group). There exists a transform for conjugation of each pair of rules in the four rules of a 4ERG. This leads to eight 4ERGs with ID 1–4 (for $\delta = 1$), ID 5–7 (for $\delta = 2$) and ID 8 (for $\delta = 0$). The parameter δ divides the eight 4RGS into three subgroups.

On the other hand for non-(2, 2) rules, two (1, 3) rules generate a 2 equivalent rule group (2ERG) and another pair of (3, 1) rules generate another 2ERG. In the process, eight 2ERGs are derived out of 4RGs with $\delta = 1$ (ID 9, 10, 11, 12) and four derived out of 4RGs with $\delta = 0$ (ID 13, 16). This leads to twelve (12) 2ERGs.

The rules $R3$ and $R4$ for all the 4RGs excepting ID 14 and 15 is a conjugate pair under a transform reported earlier. Hence, these two 4RGs (ID 14 and 15) are probed further to extract the hidden relationship of its rule pairs. The rule pair <29 71> on first row of ID 15 with two (1, 3) rules is an instance of a CRP (Tr 1 (29, 71)); consequently <29 71> forms a 2ERG. Hence, other two Rules 184 and 226 of ID 15 stand as two 1ERG (one equivalent rule group).

On the other hand, the Rule 240 <1110 1000> of ID 14 is a terminal rule having highest decimal value out of available 70 BRs. Hence, Rule 240 stands apart as a 1ERG, and it will be shown in next section that this rule represents the codon AUG corresponding to the start codon Methionine (Met). Other three rules 15, 85 and 170 of ID 14 form a 3ERG since the conjugate relationships exist as (i) Tr 1(15, 85), where 15 = <0001 0111> and 85 = <0100 1101> and (ii) Tr 2 (85, 170), where 170 = <1011 0010>. Next section establishes that these three rules <15, 85, 170>, respectively, represent the codons <AUC, AUU, AUA> corresponding to the amino acid isoleucine (Ile).

The above formulation confirms presence of eight (8) 4ERG, one (1) 3ERG, thirteen (13) 2ERG and three (3) 1ERG covering 3 Rules 240, 184 and 226. Next, in order to explore extension of 4ERG to 6ERG, we concentrate on application of transform Tr C3 on three (non-(2, 2), 0) rule pairs <147 201> and <150 105> of ID 13 and <54 108> of ID 16. Tr C3 inverts b_3 bit of 1-major and b_4 (or b_1) of 0-major half byte.

  (i)   CRP of ID 13 <147 201> = <<1000 1011> <1101 0001>> under Tr C3 becomes <<1001 0011> <1100 1001>> = <139 209> of ID 5 first row. Hence, non-(2, 2) rule pair <147, 201> of ID 13 generates a (4 + 2) = 6ERG with the (2, 2, 2) 4RG of ID 5.

 (ii) CRP of ID 13 <150 105> = <<1000 1110> <0111 0001>> under Tr C3 becomes <<1001 1100> <0110 0011>> = <156 99>, ID 6 on second and first row. So, non-(2, 2) rule pair <150, 105> of ID 13 generates a 6ERG with the (2, 2, 2) 4RG of ID 6.

(iii) CP of ID 15 <54 108> = <<0010 1110> <0111 0100>> under Tr C3 become <<0011 1100> <0110 0110>> = <60 102> of ID 7 second row. Consequently, non-(2, 2) rule pair <54, 108> generates a 6ERG with the (2, 2, 2) 4RG of ID 7.

As per the earlier formulation, the 2ERGs <147 201>, <150 105> of ID 13 and <54 108> of ID 16 generate three 6ERG with 4RGs of ID 5, 6 and 7, respectively. This leads to reduction of number of (i) 4ERG from 8 to 5 and (ii) 2ERG from 13 to 10; out of ten 2ERG, there exists nine 2ERG for amino acids and one 2ERG for two stop codons.

Following this process, the rules of sixteen (16) 4RGs of Table 3.2 leads to three (3) 6ERG, five (5) 4ERG, one (1) 3ERG (Rules 15, 85, 170 for codons AGC, AGU and AGA, respectively), ten (10) 2ERG and three (3) 1ERG. Ten 2ERG covers nine 2ERG for nine amino acids and one 2ERG for two stop codons UAG and UAA (Rules 228 and 172). Three 1ERG covers two Rules 240 and 184 assigned to amino acids (Met, Trp) and one stop codon UGA (Rule 226).

So, 20 amino acids assigned 61 rules as:

$$3 \times 6 + 5 \times 4 + 1 \times 3 + 9 \times 2 + 2 \times 1 = 18 + 20 + 3 + 18 + 2 = 61 \text{ codon rules}$$

with five equivalent rule groups denoted as y-ERGs, where $y = 6/4/3/2/1$. In addition, there are three stop codons. Column 6 of Table 3.2 shows the division of 4RGs as per the five y-ERGs (equivalent rule groups), $y = 6/4/3/2/1$.

3.3.3   Assignment of Codon Triplets A, G, C, U to the CA Rules

Codon triplets are formed by taking three neighbouring nucleotide bases at a time and marking them as first base as left, the second base in the middle and third as the right base.

Table 3.2 column 4 shows the sum of decimal values of rule number for different 4RGs with their subgroups defined as per the value of $\delta = 0, 1, 2$. The 4RGs for each subgroup of (2, 2) and non-(2, 2) groups are arranged in the table in the descending order of the sum of decimal values of rule number. Grouping of 4RGs of Table 3.2 represents the grouping of the codons derived from the analysis of the following features of 4 nucleotide bases and 16 base pairs covering first 2 bases, as shown in column 7 of Table 3.2.

Annexure 3.1 reports the analysis of the molecular structure of nucleotide bases of a codon. Based on this analysis, the nucleotide base triplets are assigned to different rule groups of Table 3.2 as detailed below.

Assignment of first group of base pairs to (2, 2, δ) 4RGS

(i)   GG > GU > GC > AC assigned to four (2, 2, 1) 4RGs (ID 1–4) having sum of decimal values 678, 528, 492, 342.

(ii)   UC > CG > CU assigned to three (2, 2, 2) 4RGs (ID 5–7) having values 696, 510, 324.

(iii)   Single base pair CC to single (2, 2, 0) 4RG (ID 8) associated with proline having value 324.

The analysis of CA rules of the 4RG of ID 7 stands apart from other 4RGs. In the physical world also, the amino acid proline displays distinct characteristics reported by many authors.

Assignment of second group of base pairs to (non-(2, 2), δ) 4RGS

(i)   The base pairs GA > AA > UA > CA are assigned to the four (non-(2, 2), 1) 4RGs (ID 9–12) having sum of decimal values as 678, 528, 492, 342.

(ii)   The base pair AG has been assigned to the 4RG (ID 13) with value 603.

(iii) Base pairs AU, UG are assigned to 4RG (ID 14 and 15) each having value 510 in view of following consideration.

Evaluation of the features of different bases of G, A and U points to the fact that the net effect of the two base pairs AU and UG are likely to be identical. Consequently, the base pairs AU and UG are assigned to two (non-(2, 2), 2) 4RGs of ID 14 and 15 having sum of decimal values as 510 and 510.

(iv) The last base pair UU has been assigned to the remaining 4RG (ID 16) having the value 417.

Assignment of third base to a CA Rule in a 4RG
Table 3.2 shows the assignment of the third base as per the following formulation displaying the transform type employed. Each 4RG covers four rules—$R1$ and $R3$ on first row, while $R2$ and $R4$ on second row. Least decimal valued rule is assigned to $R1$, followed by application of Tr1 (interchange b_6, b_3 and b_4, b_1) to derive $R2$. The rule $R3$ is derived out of $R1$ on applying Tr2 (reverse 8-bit string). If Tr1 $(R1) = R1$ (that is, $b_6 = b_3$, $b_4 = b_1$), then derive $R3$ out of $R1$ with some other transforms. Next, $R4$ is derived out of $R2$ or $R3$. Further, for non-(2, 2) rule groups, a (1, 3) least valued rule is always assigned to the base C, followed by derivation of $R2$ and $R3$.

3.3.4 Representation of x-Degeneracy Codon Groups and Design of mRNA Modelling Cellular Automata Machine (mRCAM)

The x-degeneracy codon groups are directly assigned to y-equivalent rule groups (y-ERGs), where $x = y = 6/4/3/2/1$. As a result each codon, as noted in Table 3.2, is assigned with a CA rule. A y-ERG rule group represents a degenerate codon group associated with an amino acid. There are 20 common amino acids. Once CA rules are defined for mRNA codons, an mRNA modelling CA Machine (mRCAM) is designed for an mRNA codon string by assigning the 3NCA rules for each codon of the string.

In conclusion for Sect. 3.3, a relevant observation with respect to the rule set reported in Table 3.2 is worth reporting. From study of the results presented by earlier researchers and the physical domain characteristics/information of 16×4 codon—amino acid table, we have arrived at this rules set from the bottom-up approach of designing 3NCA rule set. Top-down validation of model follows from the degeneracy results derived out of this model. Note that, this CA rule set for the 64 codons is not unique. Alternative rule design may exist to arrive at the same result of codon degeneracy. However, the process of designing the rule set for the codon set is unique. It is based on the mathematical analysis of 8-bit rules of three-neighbourhood CA on rearranging the eight RMTs (Rule Min Terms) as 1-major (7, 6, 5, 3) and 0-major groups (4, 2, 1, 0).

3.4   CA Model for Ribosomal Translation

A typical mRNA codon string derived out of pre-mRNA is shown in Fig. 3.1 on exclusion of introns in between a pair of exons, if it exists. It has 5′UTR and 3′UTR on 5′ and 3′ end, respectively, in addition to 5′ cap and poly-A tail. In the physical domain, ribosomal translation of codon string leads to protein synthesis. For building the CA model for the translation process, we proceed to design mRCAM (mRNA modelling CA Machine) to model the translation process.

3.4.1   Design of mRCAM

Once, we have developed the 64 rule set for 64 codons, we design the mRCAM by replacing each codon triplet with the corresponding 3NCA Rule.

A sample hybrid 3NHCA (three neighbourhood hybrid CA) with six 3NCA rules is noted in Fig. 3.2 for a hypothetical codon string: *<UAC AUU UUU UAU AAU UGA>*.

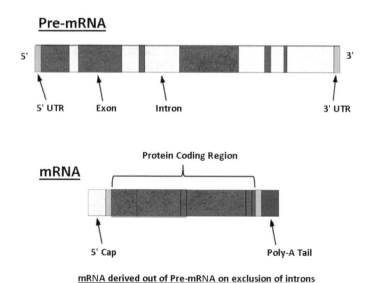

Fig. 3.1 Components of pre-mRNA and mRNA sequences

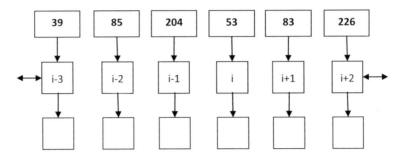

Fig. 3.2 A six-cell 3NHCA (three-neighbourhood hybrid CA)

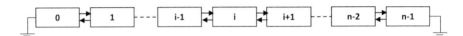

Fig. 3.3 Null boundary cellular automata

Figure 3.3 shows an n cell null boundary 3NHCA. The boundary cells 0th and
$(n - 1)$th do not have a neighbour on its left and right, respectively. For a null
boundary CA, the pseudo neighbours are assumed to be null (0). However, for our
experiment, we add a small string of dummy nucleotides at both the ends of the
mRNA string, so that the null boundary does not adversely affect the mRCAM
evolution of terminal cells. We next proceed for analysis of the evolution of
3NHCA referred to as mRCAM designed for an mRNA codon string.

As the mRCAM evolves in successive time steps, depending on the rule and its
neighbours, the next state of the cell changes. The state of ith cell is always
evaluated as a RMT value 0–7 considering the binary state of its neighbours $(i - 1)$,
i and $(i + 1)$ cells. In general, evolution of a CA cell displays two distinct regions—
transitory and cyclic (Fig. 3.4). In the transitory phase, the RMT values of a cell
changes for (say) N steps. Subsequently, a cell runs through a cycle referred to as
attractor cycle, where the RMT values keep on repeating in a cycle of (say) length
M. The parameter N is referred to as Cycle Start (CS) and M is called Cycle Length
(CL). The minimum value of CS/CL can be 1. We concentrate our analysis of these
two parameters CS and CL derived out of a mRCAM evolution.

3.4.2 CA Model for Ribosomal Translation Process

The process of ribosomal translation is presented in Sect. 1.4.4 of Chap. 1.
Figure 3.5 graphically shows the sequential steps of translation process [1]. The
translation process covers two major phases called 'Initiation' and 'Elongation'. In
the initiation phase, the 5′UTR preceding mRNA codon string is analysed for
engagement of ribosome on the start codon AUG. In the elongation phase,

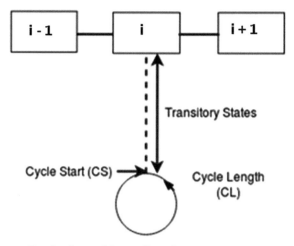

Fig. 3.4   Evolution of cellular automata through transitory and attractor cycle

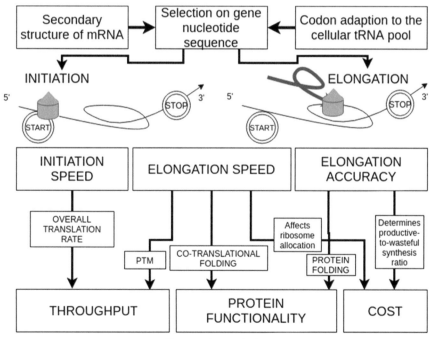

Fig. 3.5   Translation process

ribosome interrogates each codon to assign the associated amino acid to the growing nascent peptide chain. In the subsequent discussions, we report on how evolution of mRCAM models the translation process.

For a given mRNA codon string, an mRNA modelling CA Machine (mRCAM) is designed. The mRCAM is operated on initializing each CA cell with '0' for codon rules representing nine non-polar amino acids associated with 29 codons and '1' for rules representing 11 polar (6 PU + 3 PP + 2 PN) amino acids associated with 32 codons. mRCAM is operated till each cell reaches its attractor cycle. Time steps to reach the cycle are noted as CS (cycle start), while CL denotes the length of the cycle it runs through. The parameter pairs (CS, CL) are noted for each cell modelling a codon. The maximum value of CS, defined as MaxCS (Maximum CS) is derived. The MaxCS value represents the number of time steps necessary for completion of initiation phase of translation process in ribosome. Thus, the MaxCS value represents the initiation delay of the translation process. The elongation phase starts after the initiation delay, while ribosome slides over each codon to add the amino acid to the growing nascent peptide chain on large ribosomal unit.

A codon represents the information of a specific sequence of the base triplet. It does not have any co-relation with the molecular structure of nucleotide bases—C, G. U and A. In general, the CA evolution time steps do not have any co-relation with the physical system being modelled. However, for modelling codon string, we have made an exception. The CL (Cycle Length) value of each cell has been assumed to represent the time delay for the associated codon to complete its processing in the elongation phase of translation. On completion of the elongation phase for a codon string, the peptide chain leaves through the ribosomal exit port.

The acronyms used to model ribosomal translation of a wild and mutated version of codon string are noted below. The experimental set up presented subsequently refers to these acronyms. The proposed model also reports the effect of mutation on codon string. Hence, study of the mutated version of codon string is an integral part of the experimental set up to predict the effect of two types of mutations (noted as Mut1 and Mut2) on the translation time for a codon string.

- **TrD**: Translation Delay = Initiation Delay + Elongation Delay.
- **NCS**: Native Codon String transcribed out of CDS (coding DNA sequence) of a gene ID.
- **NC**: Native Codon in Native Codon String (NCS).
- **SCS**: Synthetic Codon String derived out of NCS on mutating the Native Codon (NC).
- **Mut1**: Codon (having maximum and minimum codon frequency) replaces a NC.
- **Mut2**: Codon of a polar positive residue replaces a NC. 10% additional codons for polar positive residues randomly replace NCs in a NCS.
- **CS and CL**: Cycle Start (CS) and Cycle Length (CL) illustrated in Fig. 3.4. These data is essential in CA modelling as it embeds the property of the CA rule which is used to model a codon.

- **MaxCS**: Maximum CS value for a CA cell to reach its cycle considering all the cells of RCAM modelling an mRNA codon string; it represents initiation delay.
- **SGN**: Signal graph for NCS displaying CS and CL values for each codon location. This is our standard graphical representation in different chapters for depicting RNA or DNA model, with x-axis showing serial locations of codons of a mRNA codon string (or nucleotide base sequence of a RNA/DNA strand) and y-axis showing CA parameter.
- **SGS**: Signal graph for SCS (Synthetic Codon String).
- **SGD**: Signal graph difference on subtracting CS/CL values of SGS from SGN for each codon location.
- **AR**: Active region—covers a set of codon locations where CS/CL values differ for NCS and SCS on mutating a NC.
- **CSD (Cycle start difference)**: Difference in CS value of NCS and SCS.
- **CLD (Cycle length difference)**: Difference in CL value of NCS and SCS.
- **CL Sum**: Sum of CL values = $\sum_{i=0}^{k-1} \mathrm{CL}i$ of an mRNA codon substring of length k. For a mRNA substring from xth to yth codon locations, the CL Sum is evaluated as $\sum_{i=x}^{y} \mathrm{CL}i$.
- **TrD for NCS or SCS**: Translation delay for NCS or SCS = MaxCS + sum of CLi.

Validation of CA model for ribosomal translation can be done by comparison of translation delay of native and mutated codon string for specific mutation type for which some wet lab experimental results are reported [2]. However, the review paper Diament and Tular [3] surveys all these results and pointed out the conflicting nature of reported results with respect to increase/decrease of ribosomal translation delay for the mutated version of codon string compared to that of NCS (Native Codon String). In this context, we report our results along with relevant observations with respect to the mutational effect on codon strings for 40 different genes of 4 different species.

The experimental set up for ribosomal translation is illustrated in next section as Case Study 1 for the mRCAM designed for a Homo sapiens gene. Translation delay for both for native (NCS) and its mutated versions (SCS) are compared on random insertion of two classes of mutations Mut1 and Mut2 noted earlier in the list of acronyms.

3.5   Mutational Study of mRNA Codon String—A Few Case Studies

This section illustrates the CA model for ribosomal translation reported in the earlier section. The experimental set up predicts the increase or decrease of translation delay of a Synthetic Codon String (SCS) derived with mutations introduced on NCS transcripted out of a Homo sapiens gene.

3.5.1 Case Study 1: Mutational Study on mRNA Codon String of Homo Sapiens CDS Gene ID AAB02605

This subsection presents the experimental set up, the prediction results and a few observations which have been analysed for consistency through large scale in silico experimentation with CA model reported in Sect. 3.5.2 for ten genes of four species. Prediction results refer to comparative study of translation delay for the native and mutated codon string for insertion of mutations. We report the experimental results for the NCS followed by comparison of the results for SCS (synthetic codon string) derived with mutations inserted on NCS.

Table 3.3 reports partial data retrieved out of mRCAM evolution for mRNA string of gene ID AAB02605 with start codon (AUG) as number 1 and stop codon (UGA) as number 190. The mRCAM evolution parameters model the translation of the NCS. A Codon Cluster (CC) refers to a set of adjacent codons in mRNA. While Table 3.3 shows the parameters for CC2 (codon location 100–120) and CC3 (codon location 120–140), Fig. 3.6 shows the graphical representation of parameters for CC1 (codon location 10–30), CC2 and CC3. Each row of column 1–5 of the Table 3.3 specify—RCAM cell number x modelling codon serial number x (x = 1–190), codon base triplet, rule number and rule type assigned for each codon along with CS, CL values in columns 7 and 8. Codon Frequency (CF) of NC (Native Codon) is noted in column 6 in a qualitative scale (maximum, intermediate and minimum) and these are derived from the analysis of different publications [4]. The x-axis of Fig. 3.6 shows the RCAM cell serial number along with the CF (Codon Frequency) value, rule type and encoded representation of qualitative CF values as 0/2/3, if the CF of the NC employs maximum/intermediate/minimum CF value. The CS and CL signal graphs are shown with continuous and broken line, respectively.

Figure 3.6 reports signal graph for native codon string for three codon clusters CC1, CC2 and CC3 modelled with RCAM cell clusters 10–30, 100–120 and 120–140, respectively. Cycle Start (CS) and Cycle Length (CL) signals are shown in solid and dotted line, respectively.

Table 3.3 shows partial output data derived out of mRCAM for Homo sapiens Gene ID AAB02605—CC2 (Codon Cluster 2—Codon serial No. 100–120) and CC3 (Codon Cluster 3—Codon serial No. 120–140).

Figure 3.7 displays graphical representation of results for the codon cluster CC1 for Mut1 on NC location 17. The topmost graph refers to SGN, while second and third graphs refer to SGS and SGD (displaying the difference of CS/CL values between NCS and SCS). The active region (AR, where CS/CL values differ between native and synthetic codon string) shows the region where CS/CL values differ for SGN and SGS. Cycle Start (CS) and Cycle Length (CL) signals are shown in solid and dotted line, respectively.

Results of mutation type 1 (Mut1) are derived on insertion of two codon mutations (with maximum and minimum CF—codon frequency) for a NC, one at a time. If the NC (Native Codon) has the maximum/minimum CF value, one

Table 3.3 Sample mRCAM output for two codon clusters CC2 and CC3

Codon cluster no.	Codon serial no.	Codon triplet	Rule no.	Rule type	CF			CS	CL
					Max	Inter	Min		
CC2	100	AAA	202	3:1	2.44	3.19	2.44	9	4
	101	CGG	156	2:2	1.14	1.22	0.45	3	I
	102	GTG	210	2:2	2.18	2.18	0.71	2	I
	103	AAG	216	3:1	3.19	3.19	2.44	4	1
	104	GAC	135	1:3	2.51	2.51	2.18	4	13
	105	TCG	153	2:2	0.44	1.95	0.44	4	13
	106	GAT	149	1:3	2.18	2.51	2.18	0	13
	107	GAC	135	1:3	2.51	2.51	2.18	0	13
	108	GTG	210	2:2	2.18	2.18	0.71	0	13
	109	CCC	77	2:2	1.98	1.98	0.69	0	13
	110	ATG	240	3:1	2.2	2.2	2.2	0	13
	111	GTG	210	2:2	2.18	2.18	0.71	0	13
	112	CTG	60	2:2	3.9	3.9	0.72	5	13
	113	GTG	210	2:2	2.18	2.18	0.71	6	13
	114	GGG	177	2:2	1.65	2.22	1.08	7	13
	115	AAC	27	1:3	1.91	1.91	1.7	9	13
	116	AAG	216	3:1	3.19	3.19	2.44	9	13
	117	TGT	71	1:3	1.06	1.26	1.06	9	1
	118	GAC	135	1:3	2.51	2.51	2.18	9	1
	119	CTG	60	2:2	3.9	3.9	0.72	9	2
	120	GCT	101	2:2	1.84	2.77	0.74	9	2
CC3	120	GCT	101	2:2	1.84	2.77	0.74	9	2
	121	GCA	166	2:2	1.58	2.77	0.74	'J	2
	122	CGC	57	2:2	1.04	1.22	0.45	8	2
	123	ACT	114	2:2	1.31	1.89	0.61		4
	124	GTG	210	2:2	2.18	2.18	0.71	8	4
	125	GAA	169	3:1	2.9	3.96	2.9	8	4
	126	TCT	209	2:2	1.52	1.95	0.44	22	7
	127	CGG	156	2:2	1.14	1.22	0.45	22	7
	128	CAG	120	3:1	3.42	3.42	1.23	22	7
	129	GCT	101	2:2	1.84	2.77	0.74	22	7
	130	CAG	120	3:1	3.42	3.42	1.23	23	7
	131	GAC	135	1:3	2.51	2.51	2.18	23	7
	132	CTC	46	2:2	1.96	3.9	0.72	23	7
	133	GCC	45	2:2	2.77	2.77	0.74	22	7
	134	CGA	198	2:2	0.62	1.22	0.45	22	7
	135	AGC	147	1:3	1.95	1.95	0.44	23	7
	136	TAC	39	1:3	1.53	1.53	1.22	23	7

(continued)

Table 3.3 (continued)

Codon cluster no.	Codon serial no.	Codon triplet	Rule no.	Rule type	CF			CS	CL
					Max	Inter	Min		
	137	GGC	141	2:2	2.22	2.22	1.08	23	7
	138	ATC	15	1:3	2.08	2.08	0.75	1	1
	139	CCC	77	2:2	1.98	1.98	0.69	1	1
	140	TAC	39	1:3	1.53	1.53	1.22	0	1

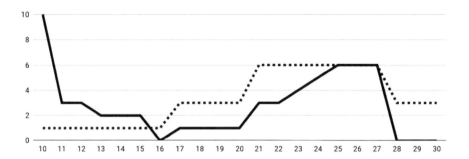

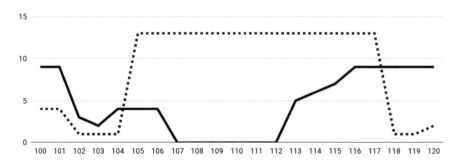

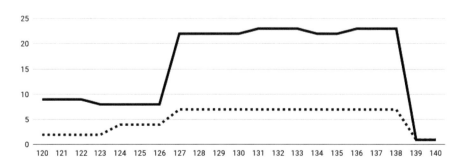

Fig. 3.6 Sample signal graphs for example codon clusters (CCs)

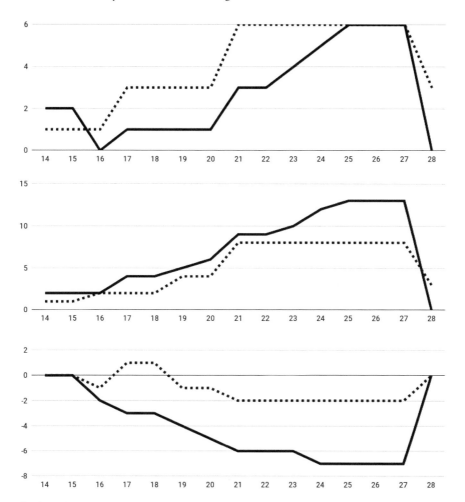

Fig. 3.7 Signal graphs for native (top) and synthetic (middle) codon strings and their difference (bottom)

mutation is inserted instead of two. That is, at least one mutation is inserted for each codon location. Figure 3.8 reports result summary for Mut1 on a few codon locations of codon clusters CC1, CC2 and CC3. The CL Difference (CLD) sum values of SCS (synthetic codon string) and NCS (native codon string) is reported in row 2 for Mut1 on the specific NC noted on row 1. The value is derived for the AR (Active Region) generated due to single point mutation on the specific NC. The negative/positive value indicates CL sum (representing the translation delay) for the SCS is higher/lesser than that of NCS. Mut1 on NC location 17 (CC1) leads to CL sum difference of (−15). While for location 108 (CC2) the CL sum for SCS is lesser (+151), for location 113 (CC2) CL sum for SCS is higher (+21). Further, mutation in location 130 of CC3 shows an instance of increase (−204) in CL sum for SCS.

Cluster No	Mut 1 on NC loc	15	16	17	18	19	20	21	22	23	24
CC1	diff in CL sum in AR	0	0	-15	-12	-16	35	-3	27	21	15
Cluster No	Mut 1 on NC loc	104	105	106	107	108	109	110	111	112	113
CC2	diff in CL sum in AR	0	-12	23	151	151	101	109	*	-41	21
Cluster No	Mut 1 on NC loc	127	128	129	130	131	132	133	134	135	136
CC3	diff in CL sum in AR	37	-10	-40	-204	-28	44	-52	49	48	16

Fig. 3.8 Result summery of Mut1 for CC1, CC2, CC3 positive/negative value indicates CL and SCS for the AR higher/lesser than NCS

In Fig. 3.8, the positive/negative value indicates CL (cycle length) sum for Synthetic Codon String (SCS) higher/lesser than that of Native Codon String (NCS) for the active region (AR).

The * for codon location 111 of CC2 indicates the fact that the codon degeneracy is 1 and so no mutation is inserted. The following observations are reported based on the detailed analysis of Mut1 on each codon location of NCS for the CDS gene ID AAB02605.

Observation 1: Insertion of codon (with maximum and minimum codon frequency) may generate zero AR (active region) or non-zero AR (where CS/CL of SCS differ from that of NCS). Instance of zero AR is reported in Fig. 3.8 for Mut1 on NC location 15, 16 of codon cluster CC1. For non-zero AR instances, the CL sum for SCS may increase or decrease compared to that of NCS.

Observation 2: The difference in CL sum is always local in the sense the AR is located around the NC location where mutation is inserted.

Observation 3: No clear pattern has emerged regarding increase/decrease of TrD (translation delay). For example, for AAB02605, CL sum for SCS is higher for 17.05% cases, and lesser for 16.12% cases, while for 66.82% cases, the CL sum difference is nil or nominal (less than 2% of translation delay (TrD)) for NCS. Increase/Decrease of CL sum depends on the context (of the residues in the AR) dynamically evaluated out of the global evolution of mRCAM for NCS and SCS. For this specific case, the value is 66.82% due to lesser length of 189 codons in the codon string for the gene AAB02605. Normally for homo sapiens genes, the percentage of nil (nominal) difference in TrD for SCS is higher in the range of 93–94%, as noted in next section with NCS having larger length of codon string.

Observation 4: The CL value (representing elongation delay) is high for codons in first few locations following start codon. This region is marked as HR (Hump Region) in recent publications. Any Mut1 in a codon of first few locations reduces the CL value significantly. For example, for CDS gene ID AAB02605, the CL value of 42 is significantly reduced in the SCS with Mut1 on first few codons of NCS.

Results Summary for Mut2 for Insertion of Codon for Polar Positive Residues for a NC of NCS

Figure 3.9 reports the result of three instances of Mut2 on random insertion of codons for polar positive residues (lys, arg and his) for NCs of NCS. For the CDS gene, AAB02605 having codons for 26 polar positive residues, 4 additional codons for polar positive residues are randomly inserted. Ten instances are analysed to evaluate the effect of Mut2 on NCS. Translation delay (TrD) = initiation delay + sum of elongation delay for each codon = 23 + 1046 = 1069 for NCS. The TrD for ten instances are computed based on mRCAM output for each Mut2 case. TrD for SCS is computed by considering the difference in CL sum values in the AR and change in initiation delay, if there is any. While for five cases TrD increased, for three cases TrD is reduced and for remaining two cases difference in TrD is nominal (less than 2% of TrD for NCS). On detailed analysis, the following observations are reported.

Observation 5: A Mut2 introduced in the initial Hump Region (HR) reduces TrD of the SCS by approximately 15% of TrD for NCS.

Observation 6: (a) Effect of Mut2 introduced in an NC (native codon) of polar positive residues in its neighbourhood—If no polar residues exist in the $(i - 2)$ to $(i + 2)$ neighbourhood, the TrD increases.

(b) On the other hand, if codon of polar positive residues exists in the neighbourhood, the change in TrD for SCS is much lesser compared to that noted under Observation 6(a).

Instance No	i^{th} loc NC replaced with Codon for residue	Presence of Codon for polar residues in the neighborhood i-2 to i+2	Active Region (AR)	Diff in TrD in AR	TrD for SCS
0	14-Arg	Yes	10-14	Nominal	1064
1	6-Lys	Yes	0-9	Large Reduction of 171	860
2	114-His	No	112-125	increase by 156	1225

Fig. 3.9 Result of three instances of Mut2 on random insertion of codons for polar positive residues (lys, arg and his)

Next subsection reports detailed study of mutational effect on large number of genes of different species.

3.5.2 Effect of Mutation on mRNA Codon String on 40 Genes of Four Species

This subsection reports the result of mutational study on 40 genes of four species—*Escherichia coli*, *Sulfolobus solfataricus*, *Homo sapiens* and *Saccharomyces cerevisiae*—listed in Table 3.4a, b. For each of the four species, experiment (as noted in earlier section) is done on ten genes.

Table 3.4b gives a summary of prediction of increase/decrease/nil (nominal change) change of translation delay for ten genes of four species for SCS (synthetic codon string due to mutation) compared to Native Codon String (NCS). Mutational study on 40 genes is conducted on random insertion of mutations, both of type Mut1 and Mut2. Average value of increase/decrease/nil or (nominal change) of translation delay is computed from this data set. The following observations have been drawn from the analysis of the data set summarized Table 3.4.

Observation 7: Compared to other species studied, the percentage of nil/nominal change under mutational pressure in Homo sapiens is significantly higher, above 93%, which means that out of 100 instances of insertion of random mutation, for more than 93 instances, no (nominal) change in translation delay has been observed.

Observation 8: Nature has designed NCS with codon frequency for NCs and codon sequence to optimize the process of translation of mRNA codon string to peptide chain of protein. Majority of mutations (synonymous or non-synonymous) inserted on NC do not change the translation delay and, so this optimized state of synthesized protein out of the translation process does not get disturbed. Hence, it can be

Table 3.4a List of four species and ten genes of each species

Species	Gene ID and number of codons
E. coli	(AAA03242–300), (AAA02601–209), (AAA03632–602), (AAA03813–643), (AAA07984–389), (AAA03999–215), (AAA03918–110), (AAA04391–485), (AAA07871–803), (AAA30530–136)
Homo sapiens	(AAA45534–3010), (AAA51729–918), (AAA52485–2350), (AAA58644–2015), (AAA62473–1158), (AAB02605–189), (AAV63975–2526), (CAA25240–1209), (CAA61107–780), (AAB02036–2870)
C. cerevisiae	(P25924–456), (P0A1S2–136), (P0A2K1–465), (P0A2T6–209), (P0DM80–486), (P02941–552)(P04787–200), (P20506–513), (P26393–291), (P58687–251)
S. salafataricus	(AAK40535–601), (AAK40574–215), (AAK40494–269), (AAK40293–312), (AAK43340–201), (CAA09918–365), (CAA11866–293), (CAA90309–344), (CAA90560–235), (CAB57529–685)

Table 3.4b   Prediction of increase/decrease/nil (nominal change) of translation delay

Experiment	Species	% of instances of translation delay		
		Increase	Decrease	Nil/nominal change
Mut1	S. solfataricus	11.35	14.04	74.60
Mut1	E. coli	9.82	9.51	80.66
Mut1	S. cerevisiae	10.01	7.22	82.76
Mut1	Homo sapiens	2.76	3.14	94.09
Mut2	S. solfataricus	18.54	14.00	67.45
Mut2	E. coli	13.10	10.00	76.10
Mut2	S. cerevisiae	12.62	16.16	71.21
Mut2	Homo sapiens	3.70	3.20	93.10

concluded that there exists an in-built tolerance to allow some mutations on codon string. Nevertheless, for certain mutations, the translation delay may change significantly resulting in the disturbance of the optimized state of synthesized protein. Deleterious effect may results for such cases. For example, as reported in Fig. 3.8, Mut1 insertion on codon location 130 of CC3 (gene ID AAB02605) resulted in high increase in translation delay by a value (-204) that may disturb the structure of the protein synthesized out of the gene).

Observation 9: The conditions under which the translation delays (TrDs) increases/decreases can be ascertained only through dynamic evaluation of the translation process. Next two subsections reports results of mutational study for two more case studies. These case studies are selected from the recent publications which report wet lab experimental results on inserting mutations on different codons of mRNA codon string.

3.5.3   Case Study 2: Study of Mutational Effect on scFv (A Fusion Protein of the Variable Heavy and Light Chains of Immunoglobulin) Codon String of Length 260

In the paper by Laurence. Guglielmi et.al. [5], the authors generated a synonymous codon library formulation study of scFV expressed in E. Coli. Majority of mutated codon string on being expressed in E. Coli displayed considerably changed solubility and antigen binding activity.

Employing the in silico experimental set up reported in Sect. 3.4.2, synonymous mutations are inserted to generate large number of MCS (Mutant Codon Strings) derived out of NCS (Native Codon String). For some codon locations, multiple mutations are inserted. The experiment is conducted on Uniprot sequence. The Cycle Length (CL) values of MCS and NCS are compared to identify the AR (active region—where CS/CL value differ for mutated version compared to that of the native). The mutational effect is marked as high, medium, low and nil, depending on the percentage of change—above 75%, in between 21 and 75%, in between 1 and 20%, and 0%, respectively.

The results, tabulated in Tables 3.5, 3.6 and 3.7 confirms large number of instances of high and medium change in CL sum values in the AR for the mutant compared to that of the native codon string of length 269 codons. The total number of mutant codon strings generated is 349. Degree of change between MCS and NCS is marked in the Tables as high/medium/low/nil. The percentage of high/medium/low/nil changes is computed as 33.81/34.01/3.72/28.30.

In Table 3.5, column 1 shows the codon location in Uniprot followed by codon triplet and amino acid in columns 2 and 3. CA rule number and rule type are noted on column 4 and 5. Column 6 shows the AR (active region where CS/CL differs between wild and mutant). Column 7 displays the CL sum for the AR for NCS (native codon string of wild) and its difference with the corresponding figure for MCS (Mutated Codon String) are reported in column 8 and 9. Finally, column 10 shows the difference in the qualitative scales (low, medium and high). The results of insertion of two instances of mutations are indicated as first instance/second instance on columns 6 to 9.

Table 3.5 Protein scFV low change mutation instances (two entries on columns 6–9 indicate results for both with'/' symbol in between)

1	2	3	4	5	6	7	8	9	10
Uniprot	CODON	AA	Rule	Rule_Type	AR	NCS CL sum	Diff in CL sum for NCS and MCS	Percent difference	Qualitative difference
141	GTG	VAL	210	2:2_1_3	135–164	327	−12	−3.67	Low
178	CCA	PRO	165	2:2_0_2	0/ 168– 182	0/ 126	0/10	0.0/7.94	Low
191	AGT	SER	201	3:1_0_2	187–200/ 187–200	65/ 62	−785/11	−1207.69/ 17.74	Low
212	ACT	THR	114	2:2_1_2	210–212/ 213–221	6/88	−6/−14	−100.0/ −15.91	Low
221	TTT	PHE	204	3:1_0_0	213–221	88	−17	−19.32	Low

Table 3.6 Protein scFV medium change mutation instances (two entries on columns 6–9 indicate results for both with '/' symbol in between)

Uniprot	CODON	AA	Rule	Rule_Type	AR	NCS CL sum	Diff in CL sum for NCS and MCS	Percent difference	Qualitative difference
5	CTG	LEU	60	2:2_2_3	6–13	91	65	71.43	Medium
8	TCT	SER	209	2:2_2_2	5–13/5–13	105/105	−273/−27	−260.0/−25.71	Medium
21	GTC	VAL	75	2:2_1_2	18–27/18–27	360/360	327/180	90.83/50.0	Medium
23	TGC	CYS	29	1:3_2_3	18–27	360	264	73.33	Medium
24	AAG	LYS	216	3:1_1_3	18–27	360	258	71.67	Medium
26	TCT	SER	209	2:2_2_2	18–29/18–27	372/360	328/258	88.17/71.67	Medium
31	ACC	THR	58	2:2_1_3	30–37	32	10	31.25	Medium
43	GGA	GLY	163	2:2_1_1	41–46/44–46	12/6	−18/3	−150.0/50.0	Medium
46	CTT	LEU	116	2:2_2_1	45–57/45–55	106/96	−150/31	−141.51/32.29	Medium

Table 3.7 Protein scFV high change mutation instances (two entries on columns 6–9 indicate results for both with '/' symbol in between)

Uniprot	CODON	AA	Rule	Rule_Type	AR	NCS CL sum	Diff in CL sum for NCS and MCS	Percent difference	Qualitative difference
6	GTG	VAL	210	2:2_1_3	5–13	105	−273	−260	High
7	CAG	GLN	120	3:1_1_3	5–13	105	−273	−260	High
9	GGA	GLY	163	2:2_1_1	9–13/5–13	49/105	29/−147	59.18/−140.0	High
10	GCT	ALA	101	2:2_1_2	5–13/5–13	105/105	−147/−525	−140.0/−500.0	High
11	GAG	GLU	225	3:1_1_3	10–13	35	29	82.86	High
12	GTG	VAL	210	2:2_1_3	5–13	105	−344	−327.62	High
16	GGG	GLY	177	2:2_1_1	17–27/14–27	362/368	209/312	57.73/84.78	High
17	GCC	ALA	45	2:2_1_3	16–27	364	311	85.44	High
18	TCA	SER	195	2:2_2_2	0/16–27	0/364	0/311	0.0/85.44	High
19	GTG	VAL	210	2:2_1_3	18–27	360	316	87.78	High
20	AAG	LYS	216	3:1_1_3	18–27	360	327	90.83	High
22	TCC	SER	139	2:2_2_2	18–27/18–27	360/360	316/327	87.78/90.83	High
25	GCT	ALA	101	2:2_1_2	18–29/18–29	372/372	331/339	88.98/91.13	High
27	GAA	GLU	169	3:1_1_0	18–29	372	310	83.33	High
28	TAC	TYR	39	1:3_1_3	18–29	372	310	83.33	High
29	ACC	THR	58	2:2_1_3	18–29	372	310	83.33	High
32	GGC	GLY	141	2:2_1_3	30–37	32	−48	−150	High
34	TAT	TYR	53	1:3_1_0	31–37	28	−28	−100	High
38	GTG	VAL	210	2:2_1_3	35–39	14	−46	−328.57	High

(continued)

Table 3.7 (continued)

Uniprot	CODON	AA	Rule	Rule_Type	AR	NCS CL sum	Diff in CL sum for NCS and MCS	Percent difference	Qualitative difference
40	CAG	GLN	120	3:1_1_3	39-42	6	−26	−433.33	High
50	GGA	GLY	163	2:2_1_1	49-57/	100/0	75/0	75.0/0.0	High
56	AGT	SER	201	3:1_0_2	52–57/52–57	90/90	72/72	80.0/80.0	High
61	TAT	TYR	53	1:3_1_0	58-62	25	20	80	High
65	TTT	PHE	204	3:1_0_0	64-67	4	−4	−100	High

3.5.4 Case Study 3: ABCC8ATP Binding Cassette Subfamily C Member 8 (Homo sapiens)

The authors of the paper 'Activating Mutations in the ABCC8 Gene in Neonatal Diabetes Mellitus' by Babenko et al. [6] reported clinical study of patients with permanent or transient mild to severe diabetes within first month after birth. Probable cause of neonatal diabetes is a form of Permanent Neonatal Diabetes Mellitus (PNDM) due to 'mutations in the KCNJ11 and ABCC8 genes'. These two genes are associated with potassium (KATP) channel expressed at the surface of the pancreatic beta cells regulating insulin release. This case study reports the mutational effect of the second gene ABACC8.

Employing the in silico experimental setup reported in Sect. 3.4.2 based on CA model, we inserted 11 non-synonymous mutations in the mRNA codon string of length 1582 transcribed out of the gene ABCC8. Seven mutations reported in the paper are covered in our study of 11 non-synonymous mutations, while the remaining 4 mutations of our study are picked up from the paper (Heterozygous ABCC8 mutations are a cause of MODY) by Bowman et al. [7]. In this paper, the authors concentrated on the clinical study of MODY (Maturity Onset Diabetes of the Young) patients displaying non-synonymous mutations (N1244D, Q485R, G214R and E100K) on the gene ABCC8. Non-synonymous mutations are noted as *XYZ*, where *X* denotes the amino acid residue associated with NC (native cdon in NCS), *Y* is the mutation site (codon location in mRNA) and *Z* refers to the residue associated with mutated codon in the MCS (Mutated Codon Strong). Residues are noted with their single letter code. The instances of PNDM and MODY, as reported in the papers [5] are detected clinically from hospital data.

For this mutational study, we operated mRCAM for codon string transcripted out of the genes. The 0/1 evolution pattern is recorded for 1000 time steps for the Native Codon String (NCS). Next, similar 0/1 evolution patterns are noted for 1000 time steps for the mutated version of NCS on insertion of each of the mutations noted earlier. The XOR operation on 0/1 evolution pattern of NCS and its mutated version is derived and reported in Fig. 3.10. XOR operation basically identifies the difference in the 0/1 evolution patterns derived out of the CA model for wild and mutant. Figure 3.10 reports the mosaic patterns for limited evolution steps. However, we confirm that the reported mosaic pattern continues for any number of evolution steps. Difference in the mosaic patterns characterizes the qualitative difference in the CA model for these two clinically identified diseases—MODY and PNDM. The mosaic patterns derived for PNDM instances differ from those of MODY. Some similarity of the mosaic pattern for MODY mutation N1244D and PNDM mutation L213R can be observed.

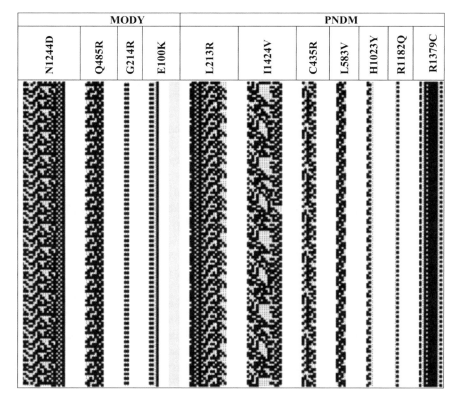

Fig. 3.10 XOR operation on 0/1 evolution pattern of NCS and its mutated version for MODY and PNDM—mutation site notes at the top of the mosaic pattern

3.6 Modelling Co-translational Folding of Nascent Peptide Chain in Ribosome

CA model for ribosomal translation is presented in Sect. 3.4. Advances in wet lab ribosomal profiling enabled researchers to predict the possibility of folding of nascent peptide chain synthesized during translation. This phenomenon of folding of nascent peptide chain is referred to as co-translational folding. The CA model designed for co-translational folding in cLife is presented in this section.

3.6.1 CA Model of Co-translational Folding in cLife

Translation of mRNA codon string transcripted out of a gene (coding DNA strand) is a universal process in all living organisms. Protein, the gene product, gets synthesized out of the translation of mRNA codon string. The elongation phase of translation for a codon is a slow process and its duration varies dynamically depending on various factors. Across different species, translation may proceed at a speed of x ($x = 5$ to 15) codons per second, adding x new peptide bonds per second to the growing nascent peptide chain. During translation, multiple ribosomes act simultaneously on the same mRNA molecule. Nature has evolved such a pipelined and parallel translation process to achieve increased efficiency of translation for an input mRNA codon string.

In last one decade, a large number of authors have confirmed co-translational folding of nascent peptide chain synthesized in ribosome [8]. A technological breakthrough of cryo-electron microscopy is expected for visualization of interplay of nascent peptide chain with the ribosomal proteins in its exit tunnel [6, 7]. In this background we present the results of co-translational folding based on CA model of ribosomal translation of mRNA codon string reported in earlier section. We note the CS (Cycle Start) and CL (Cycle Length) parameters for each cell out of mRCAM evolution. The CL value of a cell represents the translation time for the codon modelled by the CA cell. Next, difference in CL values of two adjacent cells is computed. The CA model predicts a co-translational fold if the difference in CL values is above a threshold limit.

Co-translational folding is characterized by a Folding Residue Locations (FRLs) where the nascent peptide chain folds due to variation of elongation delay associated with translation of codons. The CA model predicts a set of FRLs inserted during the translation of an mRNA codon string. The cLife models co-translational folding process based on the following assumptions:

1. Protein secondary structure covers motifs of T (Turn), S (Bend), C (Coil) in addition to Helix (H) and its variants (G/I), Beta (E) and its variant B. The region other than H and E are marked as L (unstructured region) covering T, S and C. Protein secondary structure is assumed to be generated in two phases— (i) co-translational folding during translation and (ii) post-co-translational after exit of nascent peptide chain from ribosomal exit port.

2. Co-translational folding is initiated in the elongation phase of translation while amino acid residue is added to the growing nascent peptide chain.

3. The model identifies a set of FRLs (Fold Residue Locations) at non-H/E region of a protein secondary structure.

4. The set of FRLs identified during elongation phase of translation is marked as IFRLs (Initial Fold Residue Locations). The set of IFRLs is modified or mediated by—(i) spatial constraints imposed by the ribosomal exit tunnel and (ii) as well as self-interactions of the residues of nascent peptide chain. In the process, a subset of IFRLs gets excluded from the initial list—this process excludes the set referred to as EFRL (excluded FRL). The final set of FRLs specifies the set VFRL (valid FRL) out of co-translational folding model. Thus, VFRL set = (IFRL set) − (EFRL set).

5. During post-co-translational folding, the protein folding process undergoes following two modifications—(i) some more EFRLs are identified for deletion from the VFRL and (ii) addition of some VFRLs due to folding of amino acid chain to secondary structural motifs of helix, beta and their variants. Chaperone assisted post-co-translational folding and assembly in the ER (Endoplasmic Reticulum) is currently an active research area.

As per the above assumptions, VFRL (valid FRL out of co-translation) = IFRL (initial FRL during translation) − EFRL (excluded IFRLs in ribosomal exit port).

Algorithm 1: Predict IFRL (initial folding residue location)

Input: A mRCAM for an mRNA codon string
Output: IFRL (initial fold residue location)
Steps

1. CA model for an mRNA codon string, as reported in Sect. 3.4.2, generates the parameter CL_i for ith CA cell modelling ith codon ($i = 0. 1, 2, \ldots (n - 1)$) in the elongation phase of translation.
2. Compute the difference in CL values (CLD) between a pair of adjacent $(i - 1)$th and ith cells denoted as $CLD_i = (CL_i - CL(i - 1))$.
3. Mark ith cell position as a folding residue location (FRL) if CLD_i is more than a threshold value q.
4. Generate IFRL after completion of processing of input mRNA codon string.

Sample instances of co-translational fold are graphically reported in Fig. 3.11 where difference of cycle length values is higher than a threshold value q set at 4 for the present experiment. In both the graphs, the CLD (cycle length difference) values for locations 1–24 are plotted. X-axis shows the secondary structure and Y-axis showing CL difference.

Next algorithm generates two databases marked as—VFRL (valid FRL) and EFRL (excluded FRL). Each entry in the database for VFRL covers only unstructured regions covering the secondary structural motifs T, S and C. In addition, a VFRL is assumed to be covered by the first and last two residue locations of H and E. Consequently, all the locations of folds covering residues from $(x + 2)$ to $(y - 2)$ of a H or E region (from x to y) are covered by the EFRL (excluded FRL) database.

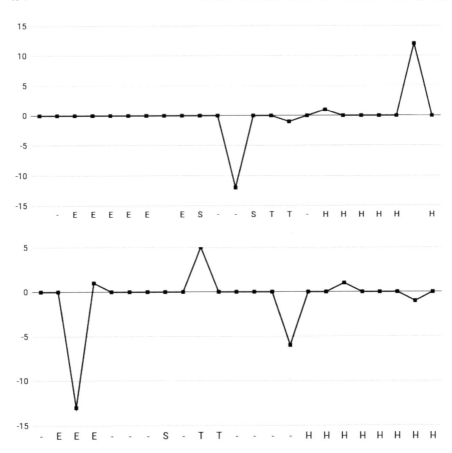

Fig. 3.11 Sample cases for co-translation folding

Algorithm 2: Generate the database for VFRL (valid FRL) and EFRL (excluded FRL)

Input: 16000 proteins selected form Dunbrack data set (dunbrack.fccc.edu).
Output: Database of VFRL and EFRL
Steps

1. Generate IFRL for each of the 16000 protein entries collected from Dunbrack data set.
2. On each entry of IFRL, generate two databases marked as

    (i)   VFRL if it covers a region for L (other than H/E), and first two and last two residues of a H and E region.
    (ii) EFRL if it covers the H/E region from $(x + 2)$ to $(y - 2)$ where a H or E region is from x to y.

Table 3.8a Sample database entries for VFRL (valid fold residue location)

Pos	Uniprot_Gene_PDB_Pos	Dihedral angles		Rule type					δ value (0/1/2)					Amino acid				
32	P0AF28_CAA31742_1A04B_5-216	0.96	3.12	2:2	1:3	1:3	2:2	2:2	0	1	2	1	1	PRO	ASP	ILE	THR	VAL
40	P0AF28_CAA31742_1A04B_5-216	0.6	−1.96	2:2	3:1	3:1	2:2	3:1	1	0	1	1	1	ALA	SER	ASN	GLY	GLU
137	P40871_AAC44632_1MDBA_1-536	−2.93	1.22	3:1	1:3	1:3	2:2	2:2	0	1	1	2	2	PHE	ASP	TYR	ARG	SER
84	Q9DF33_AAG23963_1M8TD_28-146	1.17	1.65	2:2	3:1	3:1	2:2	2:2	1	1	1	2	1	ALA	GLN	GLN	LEU	THR
465	P23847_AAA23707_1DPEA_29-535	−2.59	0.96	1:3	1:3	3:1	2:2	3:1	2	1	1	0	0	CYS	TYR	LYS	PRO	PHE

Table 3.8b Sample database entries for EFRL (excluded fold residue location)

Pos	Uniprot_Gene_PDB_Pos	Dihedral angles		Rule type					δ value (0/1/2)					Amino acid				
84	P0AF28_CAA31742_1A04B_5-216	-3.12	-2.61	2:2	1:3	2:2	2:2	1:3	2	2	1	1	0	ARG	ILE	VAL	VAL	PHE
99	O34002_AAC45662_1A59A_3-379	0.9	0.77	2:2	2:2	2:2	2:2	2:2	1	1	2	1	1	VAL	ALA	ARG	THR	ALA
453	Q9BYF1_AAF99721_1R42A_19-615	0.79	0.81	2:2	3:1	2:2	1:3	3:1	0	0	1	1	2	PRO	PHE	THR	TYR	MET
186	Q9WYS2_AAD35526_1VL8A_5-255	2.73	-2.23	1:3	2:2	1:3	2:2	3:1	0	1	1	1	2	ARG	VAL	ASN	VAL	ILE
626	P00582_CAA23607_2KFNA_324-928	0.78	0.88	2:2	3:1	2:2	1:3	2:2	0	1	2	2	2	PRO	LYS	VAL	ILE	LEU

Table 3.8a, b reports sample entries of two databases—VFRL and EFRL derived on executing the algorithms. The first column shows the codon serial number for a protein chain—its Uniprot ID, Gene ID and PDB ID chain (A/B/D residue locations) shown on column 2. For ith codon location, relevant parameters for five codon locations $((i - 2), (i - 1), i, (i + 1), (i + 2))$ are noted. Columns 4, 5 and 6 show rule types (2:2, 1:3, 3:1), δ value (0/1/2), and residue of protein chain, respectively. Column 3 reports two dihedral angles derived out of protein 3D structure—(i) between two planes with codon locations $(i - 2, i - 1, i)$ and $(i - 1, i, i + 1)$ and (ii) between planes with codon locations $(i - 1, i, i + 1)$ and $(i, i + 1, i + 2)$. These two dihedral angles are associated with the codon serial location noted on column 1. Database for valid fold residue locations stores all these parameters which are utilized for prediction of structure for a candidate protein chain out of its primary chain. We highlight the methodology of prediction of protein structure in Annexure 3.2 based on the co-translational folds VFRL identified on executing Algorithm 2 noted earlier.

Algorithm 3: Predict VFRL for an input mRNA codon string

Input: The databases for VFRL and EFRL, and the candidate mRNA codon string (for which co-translational folds are to be predicted)
Output: IFRL (initial FRL) and VFRL (valid FRL) along True/False identification if the residue location is covered by VFRL/EFRL
Steps

1. Generate IFRL (initial fold residue location) on executing Algorithm 1.
2. For each entry of IFRL identified in Step 1, check whether it is covered in the EFRL (excluded FRL) database generated by Algorithm 2.
3. Include the FRL in the VFRL (valid FRL), if it is not covered by the EFRL. Output IFRL, VFRL and True/False prediction.

3.6.2 Experimental Set Up and Sample Result Table for an Example mRNA Codon String

The experimental set up is illustrated for prediction of VFRL for the mRNA codon string of length 217 codons, transcripted out of Gene ID CAA33815; the Uniprot ID for the synthesized protein 1C1K is P13342. It is a helix (H) dominated protein. mRNA CA machine (mRCAM) for the mRNA string is designed. It is operated to identify difference in CL values between a pair of adjacent codons modelled by CA cells. The threshold value of CLD (CL Difference) for which residue location enters in VFRL list is set at value 4.

Table 3.9 reports sample prediction results derived out of mRCAM evolution for codon string derived out of gene ID CAA 33815, the Uniprot ID for the protein P13342. While Table 3.9 reports partial result table, Annexure 3.6 shows the full result table for 217 residues. Results are reported on two major columns, the left column showing relevant data for codon locations 1–109 and rest on the right column. Column 1 records the codon locations while column 2 species the fold locations identified as IFRL (Initial Fold Residue Location). Column 3 shows the fold residue location identified as VFRL (valid FRL). In order to assess the quality of prediction, we have added column 4 that shows the helix and beta regions for the protein chain. If the VFRL belongs to H (Helix) or Beta (E) except the first and last two bases (as noted on column 4 for PDB ID 1C1K), the prediction, as noted on column 5, is 'False' (e.g. residue locations 204 and 214), else the prediction is noted as 'True' as noted on column 5.

Subsequent to co-translational fold during ribosomal translation, the sample protein will undergo post-co-translational folding introducing helix and beta. The introduction of H and E during post-co-translational folding will add additional fold points and remove some of the errors introduced during co-translational folding. Scheme for deletion and addition of such fold residue locations is reported in Annexure 3.2 along with the sequential steps for prediction of protein structure out of CA-based model. Next subsection reports the results of VFRL prediction for 3000 proteins collected from Dunbrack data sets. The 16000 proteins of Dunbrack [http://dunbrack.fccc.edu] data used for design of EFRL and VFRL databases in Algorithms 2 and 3 do not cover this testing set of 3000 proteins.

Table 3.9 shows the partial table for the prediction of Fold Residue Location (FRL) for an mRNA codon strong of length 217, transcribed out of gene ID CAA33815. The Uniprot ID for the protein synthesized is P13342, PDB ID: 1C1K. Annexure 3.6 provides the entire table for the 217 locations.

3.6.3 Summary of Predicted Results for 3000 Proteins

Table 3.10 reports the percentage of error computed for a set of 3000 proteins with the CA-based model predicting VFRL (valid fold residue location) as per Algorithm 3. The list is collected from Dunbrack [http://dunbrack.fccc.edu/] data set not covered by the data used in Algorithm 2 to generate the database. The proteins are categorized as H-dominated, E-dominated, H/E-dominated and L-dominated. Depending on the number of residues, each category is further divided into large (having large no. of residues), medium (medium number of residues) and low (small number of residues). The error percentage for helix dominated proteins is higher because of false prediction of fold residue locations by our algorithm which are not covered by EFRL (Excluded Fold Residue Locations) database. As noted earlier, such erroneous predictions are handled while we predict secondary structural motifs of H and E. Majority of these errors are corrected once the secondary structural motifs H and E are correctly identified. Annexure 3.2 reports the result on inserting secondary structural motifs helix and beta derived out of post-co-translational folding.

The earlier three sections deal with (i) 3NCA rule design for codons and the CA model for codon degeneracy, (ii) ribosomal translation of codon string and mutational study of codon string along with case studies comparing wet lab results reported in

Table 3.9 Partial predicted Fold Residue Location (FRL) results for mRNA codon string

1	2	3	4	5	1	2	3	4	5
Pos	IFRL prediction	VFRL Fold points	SS motifs as noted in PDB	Comment	Pos	IFRL prediction	VFRL Fold points	SS motifs as noted in PDB	Comment
1			–		110			H	
2			E		111			H	
3	1*	1	E	True	112			H	
4			E		113			H	
5			–		114			H	
6			–		115			H	
7			–		116			H	
8			S		117			H	
9			–		118			H	
10	1*	1	T	True	119			H	
11			T		120			H	
12			–		121			H	
13			–		122			H	
14			–		123			H	
15	1*	1	–	True	124			H	
16			H		125			H	
17			H		126			H	
18			H		127			T	
19			H		128			T	
20			H		129			–	
21			H		130			S	
22			H		131			S	
23			H		132			G	
24			H		133			G	
25			H		134			G	
. . .	. . .	. . .	. . .		. . .	. . .	. . .	. . .	. . .
103			H		212			H	
104			H		213			H	
105			H		214	1*	1	H	False
106			H		215			H	
107	1*	1	H	True	216			H	
108			T		217			–	
109			H						

recent papers with cLife model results based on CA model for mRNA codon string and (iii) co-translational folding of nascent peptide chain in ribosome. In earlier discussions, we mentioned the point that the process of post-co-translational folding in ER (Endoplasmic Reticulum) is an active research area of great interest since due consideration of both co-translational and post-co-translational folding may throw new light to predict protein structure. Even after a few decades of intensive research and development including recent CASP (Critical Assessment of Protein Structure Prediction) competition, protein structure prediction seems to be an unsolved problem, specifically for a protein chain for which no partial data are available in PDB. Annexure 3.2 reports the methodology we propose to merge co-translational and post-co-translational folding to predict protein structure.

The CA model reported in earlier sections is derived out of the information of codon (nucleotide base triplet) of mRNA molecule transcribed out of coding DNA. There is another class of RNA molecules transcribed out of non-coding DNA. These are rRNA (ribosomal RNA), tRNA (translation RNA) and host of small RNAs referred to as miRNA (micro RNA), siRNA (small interfering RNA), shRNA (small hairpin RNA), etc. While rRNA and tRNA are associated with translation of codon to peptide chain of protein molecule, small RNAs plays a dominant role for controlling transcription and translation process. The next section of the current chapter address the non-coding RNA molecules.

3.7 Prediction of Secondary Structure of RNA Molecules —tRNA, miRNA, miRNA Precursor and siRNA

This section presents the CA model designed in cLife to predict secondary structure of RNA molecules. The algorithms reported to predict secondary structure of different classes of RNA molecules are based on the analysis of signal graphs derived out of ComRCAM for RNA molecule. Chapter 2 reports the design of 5NCAM, five neighbourhood uniform cellular automata referred to as BBRCAM (BackBone RNA CA Machine) and NBRCAM (Nucleotide Base RNA CA Machine). The design of ComRCAM comes out of co-evolution of BBRCAM and NBRCAM.

Annexure 3.3 reports the complete CS/CL table derived out of evolution of ComRCAM of a RNA molecule having 159 nucleotide bases (A. C, G and U). Table 3.11 shows the partial table displaying sample CS (Cycle Start)/CL (Cycle

Table 3.10 Summary of error percentage for 3000 proteins to predict VFRL

Dominance	Percentage of total database of proteins having large/medium/low number of residues	Error percentage		
		Large	Medium	Low
Helix	15.78	34.39	32.88	19.15
Beta	5.02	20.72	17.22	14.62
Helix/Beta	53.09	24.34	21.91	17.1
Unstructured	26.11	15.68	11.01	4.25

Length) values with column 2 showing the base while columns 3 and 4 displaying CS and CL parameter values for the cells modelling the bases. In general, CS values for successive base locations decreases from 5′ to 3′ end. For the example RNA, the CS value decreases from 388 to 181. For most of the ith bases (i = 1–159), the CS value decreases by 0 to (-4) compared to CS value of ($i − 1$)th base location. However, there are instances where CS value decreases by a large amount. For example, for the base locations 28 and 44, the CS value decreases by 32 and 28, respectively, compared to that of its preceding base location.

A signal graph is derived out of the CS Difference (CSD) values (shown on column 5) of the bases in adjacent locations. Figure 3.12 shows the CS Difference

Table 3.11 Sample CS/CL partial table of a RNA molecule having 159 nucleotide bases					
1	2	3	4	5	6
Pos	Nucleotide Base	CS value	CL value	CS diff	CrP
25	G	372	136	−4	
26	C	372	136	0	
27	G	372	136	0	
28	C	340	136	−32	CrP
29	C	340	136	0	
30	T	339	136	−1	
31	G	339	136	0	
32	T	338	136	−1	
33	C	338	136	0	
34	T	337	136	−1	
35	C	337	136	0	
36	G	336	136	−1	
37	T	336	136	0	
38	A	335	136	−1	
39	A	335	136	0	
40	A	334	136	−1	
41	A	330	136	−4	
42	A	330	136	0	
43	A	330	136	0	
44	T	302	136	−28	CrP
45	G	302	136	0	
46	T	302	136	0	
47	C	302	136	0	
48	A	302	136	0	
49	G	301	136	−1	
50	C	299	136	−2	

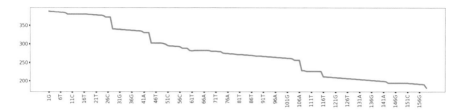

Fig. 3.12 Signal graph of CS difference values for ComRCAM evolution parameters noted in Table 3.11

signal graph for the example RNA molecule with 159 bases. While sample CS/CL table is noted in Table 3.11, the full table is reported in Annexure 3.3. The CS signal graph provides the foundation of the RNA structure prediction algorithm reported in this section. Column 2 displays the nucleotide base and CS and CL parameter values are reported on columns 3 and 4. Column 5 reports the difference in CS values of a pair of adjacent cells. Column 6 of Table 3.11 reports another parameter marked as critical pair (CRP) explained in the subsequent discussions.

Y-axis shows the Cycle Start (CS) difference parameter of Table 3.11 (full table with all 159 bases is provided in Annexure 3.3). X-axis shows the nucleotide base locations. The sharp drops in CS values in certain nucleotide locations in Fig. 3.12 graph are of interest as noted and explained in the next section.

Critical Pair (CrP) and Critical Region (CR) in Signal Graph

The CS table (e.g Table 3.11) is analysed by identifying difference in CS values of a pair of adjacent cells. The CS values, as noted earlier, decrease from 5′ to the 3′ end. For most of the pairs, CS difference value of a typical RNA molecule ranges between 0 and −4. However, for some pairs marked as critical pair (CrP), the CS difference is significantly higher, or the value for ith cell increases compared to that of $(i-1)$th cell. Based on the analysis of CS difference values for a RNA molecule, CrPs are noted for the candidate RNA molecule. CrPs identified for adjacent nucleotide base locations are marked as CR (Critical Region) that covers single or multiple CrPs in close neighbourhood; depending on the RNA molecule class, the neighbourhood may vary.

The experimental set up to predict RNA secondary structure is designed by implementing the following steps.

1. Derive BBRCAM (BackBone RNA CA Machine), NBRCAM (Nucleotide Base RNA CA Machine) and ComRCAM (Composite RNA CA Machine) for the RNA strand of nucleotide bases.
2. Note that CrP and CR out of CS table are derived for NBRCAM and ComRCAM.

3. CS graph analytics to predict secondary structure—this step differ for different RNA classes, while first two steps are common for all RNA molecules.

Signal Graph Analytics

CS signal graph analytics, as we have implemented in this book, refers to the interpretation of meaningful patterns in data, especially valuable in areas rich with recorded information available in databases. The mathematical/statistical tools, predictive models and descriptive techniques are employed to extract meaningful patterns. The process starts with the front line visualization of data/information in large biological databases to extract patterns that are relevant for top-down validation of a model designed to represent a physical system.

The focus of our signal graph analytics is to (i) analyse the parameters extracted from CA evolution subsequent to design of CA rules based on bottom-up analysis of biomolecules, followed by (ii) top-down validation of CA model parameters with reference to relevant features extracted from biological databases or wet lab experimental results.

Next few subsections of Sect. 3.7 present the background followed by the algorithmic steps for prediction of secondary structures for tRNA, miRNA precursor and miRNA. In general, a typical RNA molecule bends at the midpoint of its length with its 5′ and 3′ end coming close together. In the process, the bases on 5′ and 3′ arms may form WC (Watson–Crick) paired hydrogen bonds between the bases A, U and G, C. Such a region is referred to as stem.

3.7.1 Prediction of tRNA Secondary Structure

The nucleotide base sequences of a large number of tRNA molecules are available in tRNA database (http://GtRNAdb.ucsc.edu). A typical tRNA secondary structure, as shown in Fig. 3.13, has the following substructures in addition to the DLoop-stem, ALoop (Anticodon Loop)-stem, TLoop-stem and a variable loop in between ALoop-stem and TLoop-stem.

A more detailed structure analysis and the number of bases in each substructure of the tRNA are noted below:

- Acceptor Stem (AS) left arm bases (usually 7, for some cases 4 or 6).
- Variable number unpaired bases (usually 2, some cases may have 3 or 5) prior to DLoop-stem left arm.
- DLoop-stem (left arm—Dloop—right arm) bases (varies from 15 to 18).
- Variable number of unpaired bases (usually 0, for some cases 1 or 2), prior to ALoop-stem left arm).
- ALoop-stem (left arm—Anticodon loop—right arm) bases (variable number usually 14–17, for some cases 19 or as high as 25 if modified bases (other than normal ones) exist in the ALoop); in addition to normal bases A, C, G, T, wide variations of bases can be observed in a tRNA molecule.

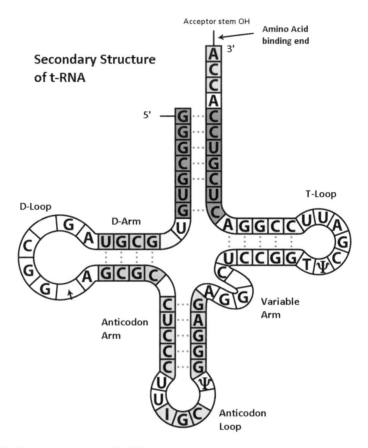

Fig. 3.13 Secondary structure of tRNA

- Variable number of unpaired bases (usually 0, for some cases non-zero), prior to VL (variable loop).
- VL with variable number of bases (usually 5, may go as high as 9, 12, 14, 15, 16, 17, …).
- Variable number of unpaired bases, prior to TLoop-stem left arm.
- TLoop-stem (left arm—TLoop—right arm) bases (usually 17, but for some cases vary from 13 to 18).
- Variable number of unpaired bases (usually 0), prior to AS right arm.
- AS (acceptor stem) right arm bases (usually 7, for some cases 4–6).
- Variable number of unpaired bases after AS right arm at 3′ ends.

Search for D, A and T stem-loop regions in a tRNA molecule proceeds within a region of minimum length 15 bases, usually bounded by a CrP (critical pair) or CR (Critical Region) derived out of CAM output for the RNA. The loop spans on either direction from the midpoint with its two stem arms on left and right. bi-directional search is initiated from the midpoint of the region towards 5′ and 3′ end such that minimum of three of WC (Watson–Crick) paired bases on left and right stem arm gets identified within the region.

Algorithm: Predict tRNA secondary structure

Input: (i) tRNA nucleotide base sequence along with its anticodon (say marked as qqq);
(ii) CS difference (CSD) table derived out of ComRCAM output for the tRNA string and the associated signal graph derived out of CSD table with X-axis displaying nucleotide base locations and Y-axis showing CSD value of each CA cell modelling a nucleotide base.

Output: Secondary structure of the tRNA molecule as expressed in the format noted in Table 3.12, each stem having two arms marked as left and right of an associated loops—D, A and T with base serial number for each stem arm and loop specified. Unpaired bases in between loop-stem structures may exist.

Analyse the ComRCAM CS table output and search for D/A/T loop-stems with two stem arms on either side of the loop. Mark a region from say r to r' base locations ($r' > r$), where ($r' - r$) is greater than 12. Algorithm initiates bidirectional search for the loop-stem from the midpoint of the region ($r' - r$)/2 towards r and r' so that maximum number of Watson–Crick (WC) paired bases (G − C) or (A − U) get identified on two stem arms of the loop.

Steps

1. Align 5′ and 3′ end bases to generate acceptor stem (AS) with maximum number of WC paired bases on 5′ and 3′ arms. Let AS be identified as $X1 - X1'$ ($X1' > X1$) at 5′ arm and $X2 - X2'$ on 3′ arm ($X2' > X2$).
2. Note the region around midpoint of tRNA string that covers anticodon qqq at location Q to ($Q + 2$). Note the CrPs from Q towards 5′ and from ($Q + 2$) towards 3′.
3. Define a region of length greater than 13 bases between a pair of CrPs marked in step 2 covering anticodon between Q to ($Q + 2$).
4. For each region noted in step 3, traverse from Q to 5′ end and ($Q + 2$) to 3′ end till maximum number of WC paired bases are identified on 5′ and 3′ arms. The region displaying maximum number of WC paired bases on two stem arms is selected as the ALoop-stem region. Let the anticodon loop-stem be identified in the region $X3$ to $X3'$.
5. Note CrP from $X1'$ (after acceptor stem 5′ end) to $X3$. Mark a region of length greater than 12 between a pair of CrPs. There may be multiple options for such a region, each of which is processed in next step.
6. For each region noted in step 5, start search for DLoop-stem from the CrP (nearest to AS 5′ arm) towards 3′ end. Region displaying maximum

Table 3.12 Predicted results of secondary structure for five tRNA molecule sequences

| Name | Accepter Stem | | Dloop | | | Aloop | | | Vloop | Tloop | | |
	5' arm	3' arm	Stem (left)	Loop	Stem (right)	Stem (left)	Loop	Stem (right)	From–To	Stem (left)	Loop	Stem (right)
chr1trna59	1–4	80–83	10–12	13–24	25–27	30–33	34–40	41–44	45–59	60–64	65–71	72–76
chr9trna1	1–7	68–74	11–13	14–20	21–23	25–28	29–45	46–49	50–54	55–58	59–63	64–67
chr6trna137	1–7	66–72	10–12	13–22	23–25	27–31	32–38	39–43	44–47	48–50	51–61	62–64
chr6trna28	1–7	87–93	10–12	13–23	24–26	41–43	44–50	51–53	54–69	70–72	73–83	84–86
chr17trna41	1–7	75–81	10–12	13–22	23–25	27–31	32–38	39–43	44–57	58–61	62–70	71–74

number of WC paired bases on two stem arms is selected as the DLoop-stem region.

7. Note the CrPs from *X2* (prior to acceptor stem (AS) arm start). Mark a region of length greater than 12 between a pair of CrPs. Multiple options for such a region may exist, each of which is processed in next step.

8. For each region noted in step 7, start search for TLoop-stem from the CrP (nearest to AS 3′ arm) towards 5′ direction. The region displaying maximum number of WC paired bases on two stem arms is selected as the TLoop-stem region.

9. Note different regions of secondary structure in the form—acceptor stem, D/A/T loop-stem, unpaired bases in between these secondary structural motifs including VL.

Results derived on execution of the algorithmic steps on a few input tRNA strands (retrieved from tRNA database) are reported in Table 3.12. The nucleotide base sequences are numbered as 1, 2, 3, …. Unpaired bases in the string including Variable Loop (VL) are implicitly indicated between loop-stem structures. The anticodon associated with the tRNA ID (e.g. Homo sapiens tRNA-Leu-CAA-6-1 for the first illustration, anticodon is CAA) is covered in the ALoop-stem. While anticodon amino acid for the tRNA guides the identification of ALoop (Anticodon Loop), the CrP location enables efficient prediction of DLoop and TLoop. The algorithm for prediction of tRNA secondary structure noted earlier is coded and executed on large number tRNA molecules retrieved from database for different species.

Table 3.13 reports the predicted secondary structure results derived on execution of the program on ten sample tRNA base sequences retrieved from the database. Annexure 3.4 reports the results of predicted secondary structure for large number of tRNA strings of different species noted in the database.

3.7.2 Two Small RNA Molecules

In recent studies, two primary classes of small RNAs referred to as microRNA (miRNA) and short interfering RNA (siRNA) have been identified to be major determinant for transcription and translation control including defense against viral invasion. Consequently, both siRNA and miRNA are used as proteomics tools to study various aspects of gene expression. Both of these small RNA molecules are derived out of longer RNA precursor by an enzyme called Dicer prior to becoming part of the protein complex called RISC (RNA-Induced Silencing Complex). Both guide the RISC to bind on the target RNA transcript. Notwithstanding such

Table 3.13 Predicted secondary structure results derived on execution of the program on ten tRNA base sequences taken from tRNA database

Name	Accepter stem		Dloop			Aloop			Vloop	Tloop		
	5' arm	3' arm	Stem (left)	Loop	Stem (right)	Stem (left)	Loop	Stem (right)	From-To	Stem (left)	Loop	Stem (right)
chr1trna30	1–7	67–73	11–13	14–22	23–25	28–32	33–39	40–44	45–49	50–54	55–61	62–66
chr1trna9	1–7	67–73	10–13	14–22	23–26	28–32	33–39	40–44	45–50	51–54	55–61	62–65
chr1trna1	1–7	67–73	11–13	14–22	23–25	28–32	33–39	40–44	45–50	51–54	53–61	62–65
chr1trna21	1–7	65–71	10–12	13–21	22–24	26–30	31–37	38–42	43–47	48–52	53–69	60–64
chr1trna37	2–7	66–71	10–12	13–23	24–26	27–32	33–39	40–45	46–48	49–53	54–60	61–65
chr1trna13	1–7	66–72	10–13	14–21	22–25	27–33	34–36	37–43	44–48	49–53	54–60	61–65
chr1trna6	1–7	64–70	10–12	13–21	22–24	27–30	31–37	38–41	42–46	47–51	52–58	59–63
chr1trna10	1–7	66–72	11–13	14–21	22–24	27–31	32–38	39–43	44–49	50–53	54–60	61–64
chr1trna2	1–5	69–73	10–13	14–22	23–26	28–32	33–39	40–44	45–50	51–54	55–61	62–65
chr1trna17	2–7	77–82	10–12	13–23	24–26	28–34	35–37	38–44	45–59	60–64	65–71	72–76

similarities, there are some major differences in the structure and function of these two small RNA molecules.

The enzyme named dicer cleaves the mature miRNA out of its precursor. A typical miRNA serves as a guide to direct the RNA-induced silencing complex (RISC) to complementary RNAs to degrade them or prevent their translation with a combination three processes—(i) site-specific cleavage, (ii) enhanced mRNA degradation and (iii) translational inhibition [9]. The referenced papers confirm the fact that rather than full complementarily of miRNA with mRNA transcript, near complementarily is more prevalent in human and plant kingdom. RISC is a ribonucleoprotein, a multi-protein complex that includes a dicer protein. One strand of miRNA (or a siRNA) is incorporated in the RISC to guide to the target site for silencing. A human dicer protein incorporating a RNA strand (with base sequence GCGAAUUCGCUU) is reported in PDB ID 4NGD (Uniprot Q9UPY3, Gene Dicer 1).

Thus, miRNA specifically regulates the synthesis of protein transcribed out of a gene, while siRNA serves dual purposes—(i) degradation of exogenous viral RNA and (ii) silencing at the level of transcripts of genome.

While miRNA has a single strand, siRNA is double stranded. Imperfect match of base sequence with the target molecule has been observed for miRNA, while there is perfect match of siRNA sequence with the target. siRNA is considered as a exogenous double-stranded RNA that is taken up externally by cells. In other words, it enters through vectors, such as viruses. Examples of vectors refer to the situation where geneticists use bits of DNA to clone a gene to create a Genetically Modified Organism (GMO). The DNA used for this process is referred to as the vector. By contrast, miRNA comes from endogenous non-coding RNA, meaning that it is made inside the cell. Such a RNA is usually found within the introns of larger RNA molecules.

Both the miRNA and siRNA, as noted earlier are derived out of miRNA precursor. In the next section, we report the CA model for miRNA precursor and algorithm for prediction of its secondary structure. Subsequent Sect. 3.7.4 discusses the prediction of miRNA secondary structure. Study of siRNA and its target gene are covered in Sect. 3.8.

3.7.3 Prediction of miRNA Precursor Secondary Structure

A single-stranded miRNA covers typically 17–24 bases. It is derived out of longer primary transcripts referred to as pri-miRNA (or miRNA precursor). The pri-miRNA display loop-stem structure (hairpin loop) in addition to other structural motifs. One of the two stem arms covers the mature miRNA. Figure 3.14 displays two instances of miRNA precursors—one refers to human and the other to elegans; precursor structures are reviewed in [10]. Mature miRNA (and its duplex miRNA) are cleaved out of pri-miRNA. The bases of the mature miRNA secondary structures are shown in red colour, while its duplex in black colour.

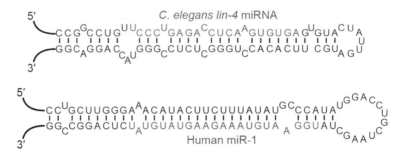

Fig. 3.14 Two examples of miRNAs derived out of miRNA precursor

The two arms from the midpoint to 5′ and 3′ end of precursor strand are denoted as 5′ and 3′ arms. Some of the observations we have drawn from study of precursors available in different publications are next presented.

Observation 1: Secondary structure of a typical precursor displays a HPL (hairpin loop) around the midpoint of the strand. The HPL lies at the centre with two arms (referred to as 5′ and 3′ arms) on either side of the loop. The bases on two arms display WC pairing—WC refers to Watson–Crick pairing of bases G and C, A and U forming hydrogen bonds. The protein named dicer cutsoff the HPL on one end and another enzyme named drosha cuts the other end of precursor to generate matured miRNA that interacts with the RNA-Induced Silencing Complex (RISC) to induce cleavage of complementary messenger RNA (mRNA) as part of the RNA interference pathway.

Observation 2: Matured miRNA and siRNA (small interfering RNA) and also other small size RNAs of usual length 20–24 bases play very important role for control (enhance and suppress) mRNA/protein production at the transcription and translation level.

Observation 3: Different structural motifs in a typical precursor are displayed in Figs. 3.15 and 3.16 noted from the paper [11]. Two arms of a precursor strand marked as 5′ arm and 3′ arm run parallel to the midpoint of the strand unless

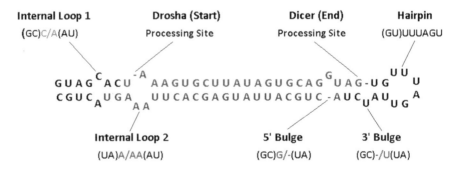

Fig. 3.15 Derivation of miRNA out of its precursor with its secondary structural motifs

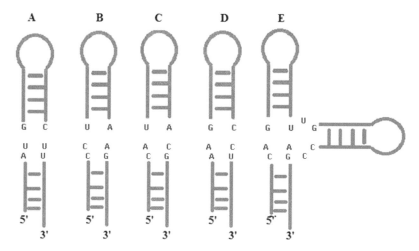

(AU)U/-(GC) (CG)-U(UA) (CG)C/A(UA) (AU)A/C(GC) (CG)A(GU)U(GC)C(GC)

Fig. 3.16 Typical secondary structural motifs of miRNA precursor molecule [11]

overridden by presence of other motifs. Stem is defined as the region where two arms run parallel with WC paired bases on them. If one section deviates from this straight line flow while other one continues in straight line flow, the deviated section is marked as 'Bulge'. If two bulges exist on the 5′ arm and its counterpart on 3′ arm, then it is also referred to as internal loop. HPL loop and its associated stem arms cover a loop with WC paired bases on the stem left and right arm. The HPL exists around the midpoint of the precursor. In addition to HPL, if another stem-loop-stem structure exists, it is referred to as branch loop.

Observation 4: Fernandez et al. [12] reported mutation of base G to A in the terminal loop of pri-mir-30c-1 precursor sequence. Such a mutation has been observed for breast and gastric cancer patients. The mutation leads to increased levels of mature miRNA. The authors have reported the predicted secondary structure for the wild version, which differs from that of the mutated strand. From this study, the authors have concluded that primary sequence determinants and RNA structure are key regulators of miRNA biogenesis.

In the above context, we report an algorithm to predict—(i) secondary structure of a miRNA precursor and (ii) a case study to predict the effect of mutation on the precursor structure. The secondary structural motifs have bulges and loops (internal loop and hairpin loop (HPL)) as note in Figs. 3.15 and 3.16 [11] displaying structural motifs marked as—(i) 5′ arm bulge, (ii) 3′ arm bulge, (iii) internal loop, (iv) multi-branch loop and (v) HPL (hairpin Loop) at the middle of a precursor with a loop and its left and right arm stem.

Algorithm to predict secondary structure accepts the ComRCAM (or NBRCAM) derived out of the nucleotide base sequence of the precursor. The signal graph for CSD values derived out of ComRCAM/NBRCAM evolution are analysed to

Table 3.14 Partial CS/CL table for pre-let-7f-2

1	G	CS	215	CL	136	CS Diff	
2	G	CS	215	CL	136	0	
3	C	CS	215	CL	136	0	
4	T	CS	214	CL	136	−1	
5	G	CS	214	CL	136	0	
6	T	CS	213	CL	136	−1	
7	G	CS	213	CL	136	0	
8	G	CS	212	CL	136	−1	
9	G	CS	212	CL	136	0	
10	A	CS	208	CL	136	−4	CrP
11	T	CS	208	CL	136	0	
12	G	CS	208	CL	136	0	
13	A	CS	208	CL	136	0	
14	G	CS	208	CL	136	0	
15	G	CS	208	CL	136	0	
16	T	CS	208	CL	136	0	
17	A	CS	207	CL	136	−1	
18	G	CS	207	CL	136	0	
19	T	CS	207	CL	136	0	
20	A	CS	206	CL	136	−1	
21	G	CS	206	CL	136	0	
22	A	CS	205	CL	136	−1	

identify secondary structural motifs—HPL (hairpin Loop), internal loop, branch loop, stem, bulge, etc. From study of secondary structure of RNA strand, we have assumed that maximizing the length of stem with WC paired bases on 5′ and 3′ arms imparts higher stability of secondary structure. In the process bulges and internal loops get inserted.

The CS signal graph analytics provides the foundation of the algorithm for prediction of secondary structure reported next. The location of CrPs on the CS signal graph guides the algorithm. CrP is defined by the CS difference between $(i - 1)$th and ith cell noted as CSD (i) that is above the threshold level.

Algorithmic steps to predict secondary structure of a precursor miRNA are noted below along with the running illustrations of an example precursor Pre-let-7f-2 out of 10 precursor structures reported in [10]. Wet lab experimental results are also reported in [13, 14]. Table 3.14 shows the partial table displaying sample CS/CL values for first 22 bases of Pre-let-7f-2 with a CrP (critical pair) at location 10 where CS value changes significantly. Any nucleotide base location on Fig. 3.17 where the CS difference values drops sharply are locations with CrPs. Y-axis displays CS value and x-axis shows the nucleotide base locations.

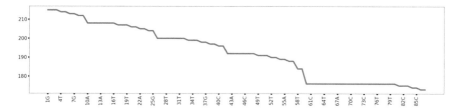

Fig. 3.17 CS difference signal graph for example, precursor pre-let-7f-2

Algorithm: Predict miRNA precursor secondary structure

Input: miRNA precursor base sequence and ComRCAM CS graph for the precursor, each CA cell representing a nucleotide base in precursor.

Output: Secondary structure of precursor displaying structural motifs—HPL (hairpin loop) around the middle of the precursor, stem, internal loop, bulge on 5′ arm and 3′ arm.

Steps

1. Note the midpoint of the precursor. Mark the two arms

   (i) 5′ arm from midpoint to the first base of 5′ arm, and

   (ii) 3′ arm from midpoint to first base of 3′ arm.

   One arm is the counterpart of the other. For Pre-let-7f-2 with 87 bases, the midpoint is cell number 43, the 5′ arm is 1–43 and 3′ arm is 43–85.

2. Derive CS difference values between the CS values of a pair of adjacent cells and note the average CS difference value for the sequence.

3. Analyse the CS difference values and set a threshold limit T to identify CrP (Critical Pair) for which CS difference is higher than the average value. Merge the adjacent CrPs to define Critical Region (CR). For precursor Pre-let-7f-2, Threshold (T) = (−4) since for most of the instances, the values are 0, (−1), (−2) and CrPs are located at cell locations on 5′ arm: 9–10, 25–26, 41–42 and on 3′ arm: 57–58, 59–60.

4. Mark the CrP nearest to the midpoint of the precursor. Mark location X nearest to midpoint and not covered by any CrP. Also note the 5′/3′ arm on which the CrP lies and adjacent CrPs if they exist. Depending on the value of T and midpoint location, multiple options may exist for X on 5′/3′ arms (for Pre-let-7f-2 the CR nearest to midpoint is at location 41–42 on 5′ arm).

5. Define two regions—(i) Region 1 (5′/3′) depending on which arm X lies: Region 1 (5′/3′) is defined as the segment from first base of 5′(3′) arm to $(X − 1)$, and its counterpart on the other arm 3′(5′), as noted in step 1. (ii) Region 2 covers cells on 5′/3′ arm not covered by Region 1. (For Pre-let-7f-2, $X = 41$ and Region 1 5′ covers cell locations 1–40 and its

counterpart on Region 1 3′ arm, while Region 2 covers all the other cells starting from location X on 5′ and 3′ arms.)

6. Traverse Region 1 (5′/3′) arm to the location $(X - 1)$ or $(X + 1)$ and its counterpart to (i) maximize stem length (with WC paired bases on either arm) on inserting bulges and loops on 5′/3′ arms wherever necessary and (ii) ensuring that a CrP lies either on stem or in loop/bulge, but not across the border; thus CrP guides the traversal for secondary structure prediction. Define a stem in a region covering k bases on 5′ and 3′ arms, if at least $k/2$ number of bases is WC paired in the region.

7. Traverse Region 2 from location X to identify stem-loop-stem of HPL (hairpin loop), while ensuring that a CrP lies either on stem or in loop, but not across the border. Presence of stem is a necessity for HPL. If stem does not exist, the next step is implemented to derive the HPL.

8. Define HPL (covering stem-loop-stem), if necessary by merging Region 1 and Region 2 to maximize WC paired bases on either arm across the boundary and ensuring presence of stem arms for a HPL.

The list of motifs identified on execution of the algorithm on precursor IDPre-let-7f-2are listed in Table 3.15. Table 3.16 reports the predicted secondary structural motifs on execution of the program for all the precursors reported by Krol [13]. The predicted secondary structural motifs for some the precursors differ from the wet lab results reported in the paper. This difference arises due to the fact that our prediction algorithm maximizes WC pairing of bases on 5′ and 3′ arms generating H-bonds in a stem. However, the wet lab experimental results display large number of H-bonds for non-WC paired bases on either arm. Future extension of the CA model covering other physical domain features associated with secondary structure formation has been planned

A bulge with large number of residues may form a HPL (hairpin loop) with hydrogen bond inserted between Watson–Crick paired bases, if they exist, around a loop at the centre. For example, the bulge at location 47–58 of precursor ID

Table 3.15 Predicted secondary structure for miRNA precursor ID pre-let-7f-2

Motif	5′ arm	Loop	3′ arm
HPL	36–39	40–42	43–46
Bulge	33–35		47–58
Stem	24–32		59–67
Internal loop	23–23		68–68
Stem	17–22		69–74
Internal loop	15–16		75–76
Stem	13–14		77–78
Bulge	11–12		79–79
Stem	5–10		80–85

Table 3.16 Prediction of secondary structure of nine pre-miRNA (other than pre-let-7f-2)[25]

(1) Pre-mir-16-1

Motif	5' arm	Loop	3' arm
HPL	36–39	40–42	43–46
Bulge	33–35		47–58
Stem	24–32		59–67
Internal loop	23–23		68–68
Stem	17–22		69–74
Internal loop	15–16		75–76
Stem	13–14		77–78
Bulge	11–12		79–79
Stem	5–10		80–85

(2) Pre-mir-17

Motif	5' arm	Loop	3' arm
HPL	42–44	45–48	49–51
Bulge	41–41		52–62
Stem	36–40		63–67
Bulge			68–68
Stem	24–35		69–76
Bulge	16–23		77–81
Stem	12–15		82–85
Bulge	6–11		86–86
Stem	1–5		87–91

(3) Pre-mir29a

Motif	5' arm	Loop	3' arm
HPL	35–37	38–49	50–52
Bulge	34–34		53–57
Stem	29–33		58–58
Internal loop	28–28		59–60
Stem	26–27		61–62
Bulge	25–25		63–63
Stem	23–24		64–69
Bulge			70–73
Stem	17–22		74–79
Bulge	14–16		80–81
Stem	8–13		82–86
Internal loop	6–7		
Stem	1–5		

(4) Pre-mir-19a

Motif	5' arm	Loop	3' arm
HPL	36–37	38–43	44–45
Bulge	31–35		46–47
Stem	29–30		48–49
Bulge			50–51
Stem	23–28		52–57
Bulge	12–22		
Stem	9–11		58–60
Internal loop	8–8		61–61

(5) Pre-mir-30a

Motif	5' arm	Loop	3' arm
HPL	31–36	37–43	44–49
Bulge			50–52
Stem	27–30		53–56
Bulge			57–58
Stem	15–26		59–70
Bulge			71–72
Stem	11–14		73–73
Bulge	9–10		

(6) Pre-mir-25

Motif	5' arm	Loop	3' arm
HPL	29–33	34–37	38–42
Bulge			43–47
Stem	26–28		48–50
Internal loop	25–25		51–51
Stem	20–24		52–56
Bulge	18–19		
Stem	10–17		57–64
Bulge			65–66

(continued)

Table 3.16 (continued)

(4) Pre-mir-19a

Motif	5' arm	Loop	3' arm
Stem	6–7		62–63
Internal loop	5–5		64–64
Stem	1–4		65–68

(5) Pre-mir-30a

Motif	5' arm	Loop	3' arm
Stem	7–8		74–75
Bulge	6–6		76–79
Stem	1–5		80–84

(6) Pre-mir-25

Motif	5' arm	Loop	3' arm
Stem	8–9		67–68
Bulge	4–7		69–70
Stem	1–3		71–73

(7) Pre-mir-15a

Motif	5' arm	Loop	3' arm
HPL	44–45	46–50	51–52
Bulge	33–43		53–57
Stem	28–32		58–61
Internal loop	24–27		62–66
Stem	19–23		67–77
Bulge	18–18		
Stem	15–17		78–80
Bulge	5–14		
Stem	1–4		81–84

(8) Pre-mir-18

Motif	5' arm	Loop	3' arm
HPL	43–44	45–52	53–54
Bulge	32–42		55–56
Stem	28–31		57–60
Internal loop	26–27		61–62
Stem	18–25		63–70
Bulge	11–17		
Stem	8–10		71–73
Bulge			74–80
Stem	1–7		81–87

(9) Pre-let-7c

Motif	5' arm	Loop	3' arm
HPL	28–29	30–34	35–36
Internal loop	25–27		37–39
Stem	20–24		40–44
Bulge			45–46
Stem	17–19		47–49
Bulge			50–67
Stem	13–16		68–71
Internal loop	9–12		72–75
Stem	6–8		76–78

Pre-let-7f-2 displays a HPL. Such HPLs are not explicitly shown in the result Table 3.16.

The case study reported next employs the same algorithm and the program code developed to predict mRNA precursor secondary structure reported earlier.

Case Study: Prediction of Secondary Structure of Precursor ID pri-miR-30c Wild and Mutant G27A

The case study covers prediction of secondary structure for the wild and mutant for the precursor pri-miR-30c reported in the paper by Fernandez [12]. G27A mutation refers to replacement of base G at location 27 of wild with base A for the mutant. The secondary structures for wild and mutant reported in the paper are reproduced in Fig. 3.18.

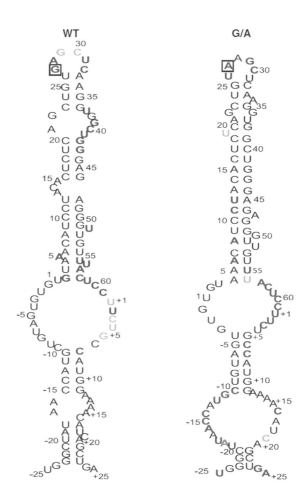

Fig. 3.18 Secondary structure for wild and mutant precursor ID Pri-mri-30c [12]

Table 3.17 ComRCAM CS/CL table for pri-miR-30c wild

1	2	3	4	5	6	7	1	2	3	4	5	6	7
Base serial no.	Base serial no. in paper	Base	CS	CL	CS Diff	CrP Loc	Base serial no.	Base serial no. in paper	Base	CS	CL	CS Diff	CrP Loc
1	−25	T	279	68	0		57	32	C	238	68	0	
2	−24	G	278	68	−1		58	33	A	236	68	−2	
3	−23	G	278	68	0		59	34	A	236	68	0	
4	−22	G	277	68	−1		60	35	G	236	68	0	
5	−21	C	277	68	0		61	36	G	236	68	0	
6	−20	T	276	68	−1		62	37	T	236	68	0	
7	−19	A	276	68	0		63	38	G	236	68	0	
8	−18	T	272	68	−4	CrP	64	39	G	236	68	0	
9	−17	A	272	68	0		65	40	C	235	68	−1	
10	−16	A	272	68	0		66	41	T	234	68	−1	
11	−15	C	272	68	0		67	42	G	234	68	0	
12	−14	C	272	68	0		68	43	G	234	68	0	
13	−13	A	272	68	0		69	44	G	228	68	−6	CrP
14	−12	T	272	68	0		70	45	A	227	68	−1	
15	−11	G	271	68	−1		71	46	G	227	68	0	
16	−10	C	271	68	0		72	47	A	226	68	−1	
17	−9	T	271	68	0		73	48	G	226	68	0	
18	−8	G	270	68	−1		74	49	G	222	68	−4	CrP
19	−7	T	270	68	0		75	50	G	222	68	0	
20	−6	A	269	68	−1		76	51	T	222	68	0	
21	−5	G	269	68	0		77	52	T	222	68	0	
22	−4	T	268	68	−1		78	53	G	222	68	0	

(continued)

Table 3.17 (continued)

1	2	3	4	5	6	7
Base serial no.	Base serial no. in paper	Base	CS	CL	CS Diff	CrP Loc
23	−3	G	267	68	−1	
24	−2	T	267	68	0	
25	−1	G	267	68	0	
26	1	T	266	68	−1	
27	2	G	266	68	0	
28	3	T	265	68	−1	
29	4	A	265	68	0	
30	5	A	264	68	−1	
31	6	A	264	68	0	
32	7	C	238	68	−26	CrP
33	8	A	237	68	−1	
34	9	T	237	68	0	
35	10	C	236	68	−1	
36	11	C	236	68	0	
37	12	T	235	68	−1	
38	13	A	235	68	0	
39	14	C	234	68	−1	
40	15	A	234	68	0	
41	16	C	234	68	0	
42	17	T	234	68	0	
43	18	C	234	68	0	
44	19	T	234	68	0	

1	2	3	4	5	6	7
Base serial no.	Base serial no. in paper	Base	CS	CL	CS Diff	CrP Loc
79	54	T	222	68	0	
80	55	T	222	68	0	
81	56	T	222	68	0	
82	57	A	218	68	−4	CrP
83	58	C	218	68	0	
84	59	T	169	68	−69	CrP
85	60	C	169	68	−1	
86	61	C	168	68	0	
87	1	T	168	68	−1	
88	2	T	167	68	0	
89	3	C	167	68	−1	
90	4	T	166	68	0	
91	5	G	166	68	−1	
92	6	C	165	68	0	
93	7	C	165	68	−1	
94	8	A	164	68	0	
95	9	T	164	68	−1	
96	10	G	163	68	0	
97	11	G	163	68	−1	
98	12	A	162	68	0	
99	13	A	162	68	−1	
100	14	A	161	68	−1	

(continued)

Table 3.17 (continued)

1 Base serial no.	2 Base serial no. in paper	3 Base	4 CS	5 CL	6 CS Diff	7 CrP Loc
45	20	C	234	68	0	
46	21	A	234	68	0	
47	22	G	234	68	0	
48	23	C	234	68	0	
49	24	T	234	68	0	
50	25	G	234	68	0	
51	26	T	238	68	4	CrP
52	27	G	238	68	0	
53	28	A	238	68	0	
54	29	G	238	68	0	
55	30	C	238	68	0	
56	31	T	238	68	0	

1 Base serial no.	2 Base serial no. in paper	3 Base	4 CS	5 CL	6 CS Diff	7 CrP Loc
101	15	A	161	68	0	CrP
102	16	C	160	68	−1	
103	17	A	160	68	0	
104	18	T	156	68	−4	CrP
105	19	C	156	68	0	
106	20	A	156	68	0	
107	21	G	156	68	0	
108	22	C	156	68	0	
109	23	T	158	68	2	CrP
110	24	G	158	68	0	
111	25	A	158	68	0	

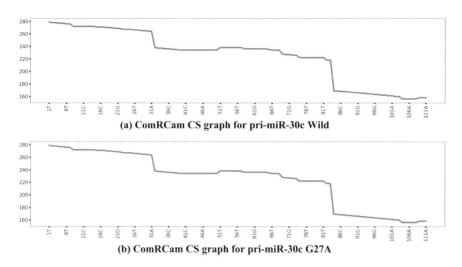

(a) ComRCam CS graph for pri-miR-30c Wild

(b) ComRCam CS graph for pri-miR-30c G27A

Fig. 3.19 CS signal graph for wild and mutant version of pri-mir-30c

Table 3.18 Secondary structure derived out of ComRCAM for the precursor pri-miR-30c wild

(1)	(2)	(3)	(4)	(5)	(6)	(7)
Structure	Location	Location	Location	Location	Location	Location
Stem	−22	−20			+20	+22
Bulge					+19	+19
Stem	−19	−18			+17	+18
Bulge	−17	−16			+12	+16
Stem	−15	−11			+7	+11
Bulge	−10	1			59	+6
Stem	2	11			49	58
Bulge	12	15				
Stem	16	20			44	48
Bulge					42	43
Stem	21	23			39	41
Bulge					36	38
HPL	24	26	27	32	33	35

Prediction of Secondary Structure for Wild from ComRCAM

The CS/CL parameter table derived out of ComRCAM output for the base sequence for the precursor is reported in three sub-tables of Table 3.17 with 111 bases divided into two major columns—left column displaying bases from 1 to 22, 22 to 44, 45 to 56 and right column showing bases 57 to 78, 79 to 100, 101 to 111. The corresponding CS signal graph for wild is reported in Fig. 3.19a, while signal graph for G27A mutant is shown in Fig. 3.19b. The serial number of 111 bases from 3′ to

5′ end of the RNA is noted in column 1 of Table 3.17, while column 2 shows the base serial number as noted in the paper [12]. Note that, there is a difference of 25 between the numbering scheme employed here versus the one used in the paper. We have numbered the base serial number as 1–111 rather than numbering the bases with positive and negative numbers. The CS difference values are noted on column 6. The CrPs (where the CS difference is outside the range 0 to −3) are reported on column 7. CrPs guide the algorithm to derive the secondary structure reported in Table 3.18. The predicted secondary structure derived out of CA model is similar to one noted in the paper [12].

Figure 3.19a shows the CS signal graph for precursor Pri-mir-30c wild and Fig. 3.19b shows the CS signal graph of the G27A mutant. Both are derived out of ComRCAM CS table, for the mutant the base G at base location 27 is replaced with A. The y-axis shows CS values and x-axis displays nucleotide bases serial number. Note that, the CS signal graphs are identical for wild and mutant G27A which demonstrates that the ComRCAM fails to detect the effect of mutation G27A. Hence, we fall back on NBRCAM to predict the mutational effect. The secondary structure predicted out of ComRCAM for wild is reported in Table 3.18 that is similar to the secondary structure reported in Fernandez paper [12].

Table 3.18 shows the secondary structure for pri-miR-30c. The HPL (hairpin loop) stem-loop-stem is shown with base serial numbers (24–26)—(27–32)—(33–35). The base serial numbers (as noted in the paper) covered by different secondary structural motifs (stem and bulge) are noted on column 2 and column 3 (on 5′ arm), column 6 and column 7 (on 3′ arm), while base serial numbers 27–32 are covered by the HPL (hairpin loop at the middle of the RNA precursor).

Prediction of Secondary Structure for Mutant G52A (G27A) from NBRCAM
ComRCAM output, as reported in Chap. 2, is derived out of co-evolution of BBRCAM (BackBone RNA CA Machine) and NBRCAM (Nucleotide Base RNA CA Machine). BBRCAM is a uniform CA designed with 5NCA rules designed for sugar–phosphate molecule. On the other hand, NBRCAM is a hybrid 5NCA designed with 5NCA rules for nucleotide bases. It has been observed that ComRCAM may fail to detect the mutational effect for some of the mutations on RNA. For example, on insertion of mutation G27A in the precursor pre-mir-30c, the ComRCAM output for the mutant displays identical output as that for the wild as reported in the signal graphs noted in Fig. 3.19a for wild and Fig. 3.19b for G27A mutant. Here, ComRCAM fails to detect this mutational effect. Consequently, we derive the secondary structure for the mutant from the CS signal graph derived out of NBRCAM. Hence, for mutational study of such RNA molecule, we execute the algorithm ('Predict: miRNA precursor secondary structure' of RNA molecule) with the NBRCAM output of the precursor rather than ComRCAM. The CS/CL parameter table derived out of NBRCAM for G27A mutation on the precursor ID pri-miR-30c is reported in Table 3.19. The CrPs are listed on column 9. The predicted secondary structure derived on execution of the program developed to predict precursor secondary structure is reported in Table 3.20. The predicted structure differ from the one reported in the paper [12]

Table 3.19 CS/CL table derived out NBRCAM for precursor pri-miR-30c mutant G27A

1 Base serial no.	2 Base serial no. as noted in paper	3 Base	5 CS	7 CL	8 CS Diff	9 CrP Loc
1	−25	T	11	1		
2	−24	G	10	1	−1	
3	−23	G	10	1	0	
4	−22	G	9	1	−1	
5	−21	C	9	1	0	
6	−20	T	7	1	−2	
7	−19	A	7	1	0	
8	−18	T	7	1	0	
9	−17	A	7	1	0	
10	−16	A	7	1	0	
11	−15	C	7	1	0	
12	−14	C	12	1	5	CrP
13	−13	A	13	1	1	CrP
14	−12	T	13	1	0	
15	−11	G	14	1	1	CrP
16	−10	C	15	1	1	CrP
17	−9	T	15	1	0	
18	−8	G	16	1	1	CrP
19	−7	T	17	1	1	CrP
20	−6	A	17	1	0	
21	−5	G	17	4	0	

1 Base serial no.	2 Base serial no. as noted in paper	3 Base	5 CS	7 CL	8 CS Diff	9 CrP Loc
57	32	C	13	1	0	
58	33	A	13	1	0	
59	34	A	13	1	0	
60	35	G	13	1	0	
61	36	G	13	1	0	
62	37	T	13	1	0	
63	38	G	9	1	−4	CrP
64	39	G	8	1	−1	
65	40	C	8	1	0	
66	41	T	7	1	−1	
67	42	G	7	1	0	
68	43	G	6	1	−1	
69	44	G	6	1	0	
70	45	A	4	1	−2	
71	46	G	4	1	0	
72	47	A	25	4	21	CrP
73	48	G	25	4	0	
74	49	G	25	4	0	
75	50	G	25	4	0	
76	51	T	25	4	0	
77	52	T	24	4	−1	

(continued)

Table 3.19 (continued)

1 Base serial no.	2 Base serial no. as noted in paper	3 Base	5 CS	7 CL	8 CS Diff	9 CrP Loc	1 Base serial no.	2 Base serial no. as noted in paper	3 Base	5 CS	7 CL	8 CS Diff	9 CrP Loc
22	−4	T	17	4	0		78	53	G	24	4	0	
23	−3	G	17	4	0		79	54	T	24	4	0	
24	−2	T	17	4	0		80	55	T	24	4	0	
25	−1	G	17	4	0		81	56	T	22	4	−2	
26	1	T	7	4	−10	CrP	82	57	A	22	4	0	
27	2	G	6	4	−1		83	58	C	22	4	0	
28	3	T	6	4	0		84	59	T	22	4	0	
29	4	A	6	4	0		85	60	C	20	4	−2	
30	5	A	6	4	0		86	61	C	20	4	0	
31	6	A	6	4	0		87	1	T	16	4	−4	CrP
32	7	C	15	1	9	CrP	88	2	T	16	4	0	
33	8	A	15	1	0		89	3	C	15	4	−1	
34	9	T	15	1	0		90	4	T	15	4	0	
35	10	C	15	1	0		91	5	G	10	4	−5	CrP
36	11	C	15	1	0		92	6	C	9	4	−1	
37	12	T	14	1	−1		93	7	C	9	4	0	
38	13	A	14	1	0		94	8	A	8	4	−1	
39	14	C	14	1	0		95	9	T	8	4	0	
40	15	A	14	1	0		96	10	G	7	4	−1	
41	16	C	14	1	0		97	11	G	6	1	−1	
42	17	T	14	1	0		98	12	A	6	1	0	

(continued)

Table 3.19 (continued)

1	2	3	5	7	8	9
Base serial no.	Base serial no. as noted in paper	Base	CS	CL	CS Diff	CrP Loc
43	18	C	11	1	−3	
44	19	T	11	1	0	
45	20	C	10	1	−1	
46	21	A	10	1	0	
47	22	G	10	1	0	
48	23	C	10	4	0	
49	24	T	10	4	0	
50	25	G	10	4	0	
51	26	T	4	4	−6	CrP
52	27	A	4	4	0	
53	28	A	5	4	1	CrP
54	29	G	5	4	0	
55	30	C	11	1	6	CrP
56	31	T	13	1	2	CP

1	2	3	5	7	8	9
Base serial no.	Base serial no. as noted in paper	Base	CS	CL	CS Diff	CrP Loc
99	13	A	6	1	0	
100	14	A	1	1	−5	CrP
101	15	A	2	1	1	CrP
102	16	C	2	1	0	
103	17	A	2	1	0	
104	18	T	2	1	0	
105	19	C	2	1	0	
106	20	A	8	1	6	CrP
107	21	G	8	6	0	
108	22	C	8	6	0	
109	23	T	8	6	0	
110	24	G	8	6	0	
111	25	A	8	6	0	

Table 3.20 Predicted secondary structure for the mutant G27A for precursor pri-miR-30C

Structure	Position	Position	Position	Position	Position	Position
Stem	−22	−20			+20	+22
Bulge	−19	−16			+12	+19
Stem	−15	−11			+7	+11
Bulge	−10	1			59	+6
Stem	2	11			49	58
Bulge	12	17			40	48
HPL	18	25	26	31	32	39

Fig. 3.20 CS signal graph derived out of NBRCAM for pri-miR-30c

derived with the package 'RNA structure software' noted under 'Method Section' [14]. The in silico prediction of the secondary structure reported in the paper for G27A mutant displays stems having hydrogen bonds with non-WC paired bases. In the CA-based prediction scheme, we have assumed that for formation of a stem with L number of bases on 5′ and 3′ arms, at least L/2 number bases should exist with proper WC paired bases (G and C, U and A) making hydrogen bonds on the base pairs on stem arms. If this condition is not satisfied, the algorithm used in this book introduces bulges rather than stem in the predicted secondary structure. Future extension of the algorithm will cover the physical domain features that affect stem formation in addition to WC pairing of bases on two stem arms.

Base serial numbers used in the Fernendez et al. paper [12] differ from the convention used in this book. Hence, both conventions are noted in Table 3.17 for the benefit of the reader.

In addition to comparing relevant physico-chemical properties for wild and mutant, one of the main focuses of the Fernendez et al. paper [12] is to study the variation of the secondary structure for the mutation G27A found to be associated with breast and gastric cancer. The secondary structures derived out of the CA model reported in Table 3.18 for wild differ considerable from the predicted structure of the mutant G27A reported in Table 3.20. Mutation study of this precursor with G27A, G27C and G27U mutants are reported in Chap. 4 under SNP (Single-Nucleotide Polymorphism) study. The results derived out these SNP (single-nucleotide polymorphism) study clearly establishes the difference of CA parameter values between the wild and G27A mutant. However, the difference of CA parameter values between wild and G27C and G27U is found to be nominal. Details are noted in 'Case Study 1' reported in Sect. 4.6 of Chap. 4.

3.7.4   Prediction of Secondary Structure of Matured miRNA

Derivation of matured miRNA out of miRNA precursor is presented in Sect. 3.7.2. Micro RNA mediated translational control process is a well-studied area [15] and in addition to the base sequence of miRNA, its secondary structure plays an important role. This section presents the CA model for prediction of secondary structure of a micro RNA (miRNA) based on the signal graph analytics of the CS signal graph derived out of ComRCAM output for miRNA molecule. For a miRNA with less than 25 bases, usually CrPs exist near the 5′ and 3′ end. CrP refers to the Critical pair where CS difference between neighbours lie outside the normal range. Another CrP at a distance of more than three bases from 5′ to 3′ end also exists that introduces a HPL (hairpin loop with stem on two arms around a loop). The CrP near the 5′ and 3′ end usually introduces open-ended bend that results in closer proximity of first and last base in miRNA secondary structure. For a miRNA with more than 25 bases, more than one CrP may exist away from 5′ to 3′ ends, each of which introduces a HPL. Further, for length of more than 25 bases, the 5′ to 3′ end bases form an acceptor stem with WC paired bases, as observed in tRNA secondary structure. For a miRNA, the algorithm defines a CR (Critical Region) between ith and $(i + 4)$th bases for a CrP between $(i − 1)$th and ith cells. In general, a miRNA or its precursor molecule displays the following structural characteristics which get reflected in the CS signal graph displaying CrPs identified out of the evolution of ComRCAM for the molecule: (i) The residues near 5′ and 3′ end make a stem (if number of bases is more than 25 bases), else it displays bends to bring the two ends closer together. (ii) Depending on the location of CrP, there exists a HPL near the central region of the molecule. Location of CrPs guides the algorithm designed to predict miRNA secondary structure.

Algorithm: Predict miRNA secondary structure

Input: miRNA nucleotide base sequence
Output: Predicted secondary structural motifs (i) HPL (hairpin loop) region, (ii) Open-ended bend region with bases on 5′/3′ arm ends, where slope continuously changes; (iii) Turn—sharp change of slope between a pair of bases in between acceptor and its adjacent stem (for length greater than 25 bases).

Steps

1. Generate ComCAM output for the input string.
2. Note the difference in CS values of each pair of adjacent cells from 5' to 3' end. For majority of cell pairs, CS difference values changes by a low negative value of 0 to −1. Identify a threshold limit that differs from most of the cell pair CS difference (CSD) values (threshold for this experimental set up is set as −3). The CS difference values outside the range of 0 to −3 are marked as CrP (Critical Pair) processed in subsequent steps of the algorithm.
3. Note the CrPs, where a CrP identifies a pair of cell location $(i − 1)$ and i where CS value changes above a threshold limit or CSD is a positive value.
4. (a) For miRNA of length less than 25: Define a CR(Critical Region) from X to $X + 4$ and Y to $Y − 4$, with (i) first CrP at loc X close to 5' (within a distance of 3 bases from 5' end) and (ii) last CrP at loc Y close to 3' end.
   (b) For miRNA of length more than 25 (artificially developed similar to an aptamer reported in Belter paper [15]: (i) generate acceptor stem out of 5' and 3' arm bases and (ii) insert a bend around 90° between a pair of bases following the acceptor stem.
5. If there is a CrP at location W (other than close to 5'/3' ends), define a CR for the region W to $W + 4$. If there are multiple CrPs in close proximity (within a distance of 3 bases), define a CR covering all the CrPs.
6. (a) For length less than 25 bases: (i) Introduce the loop of a HPL in the CR defined in Step 5 and define two stems of HPL with the bases prior and after the loop of HPL while maximizing the number of WC paired bases on two stems on either side of loop. (ii) Introduce open-ended bend with bases on 5'/3' ends not covered by the HPL.
   (b) For presence of multiple CrPs in miRNA having length greater than 25 bases: for each CR other than those processed in Step 4, execute Step 6 (a) for each CrP. (Note for the molecule TN 9.6, Step 5 identifies two CrPs at location 9 and 25, as reported in Fig. 3.25, for which two HPLs get introduced.)

The algorithm (Predict: miRNA secondary structure) has been coded and executed for some of the miRNAs listed in the database [16]. The predicted results are reported next.

Case Studies on a Family of Precursor miRNA Molecule miR-296, miR-93, miR-21 and TN-9.6

The case study focuses on a family of precursor miRNAs reported in by Belter et al. [15].

The wet lab experimental results of four miRNA precursor structure derived out of their nucleotide base sequence are reported in this paper and reproduced below as Figs. 3.21, 3.22, 3.23 and 3.24. CS signal graphs derived out of each of the four base sequences are reported in Fig. 3.25. Predicted secondary structure derived out of CA model is reported in Table 3.21. Predicted structures are similar to the wet lab results reported in the paper. The signal graph for ID miR-296 does not display any CrP in the 5′ arm base location 1–3. The first CrP is at base location 5 with value (−35). For this case, as per algorithm step 6(a), the HPL stem-loop-stem starts at location (1–4)-(5–8)-(9–12) with open-ended loop with bases at location 13–21. On the other hand for ID miR-21 and miR-93, the HPL is formed away from 5′/3′ end and hence, as per Step 6(a)(ii), open-ended loops are formed with 5′/3′ end bases not covered by HPL. For the ID TN-9.6 with 41 bases, the algorithm identifies the stem with 5′/3′ end bases and two HPLs as noted in Table 3.21. It has been

Fig. 3.21 Structure of miR-21

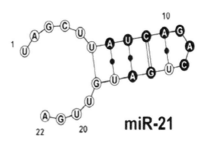

Fig. 3.22 Structure of miR-93

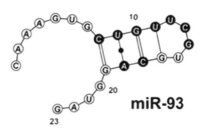

Fig. 3.23 Structure of miR-296

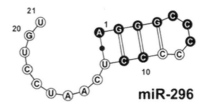

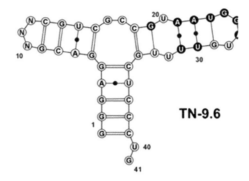

Fig. 3.24 Structure of TN-9.6

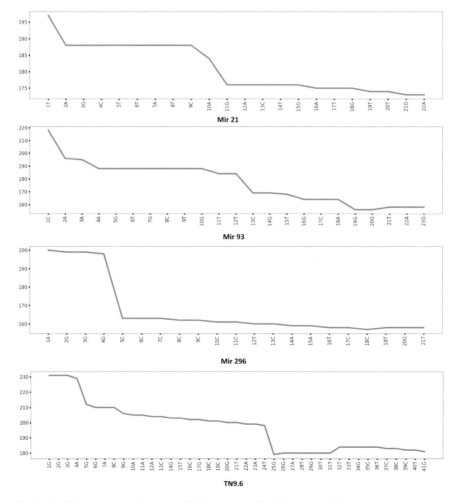

Fig. 3.25 CS signal graph for four miRNAs reported in Belter et al. [12]

Table 3.21 Predicted secondary structure of the precursor miRNAs

miRNA ID	Acceptor	Stem	Stem	Loop	Stem	Open-ended bend
miR-296			1–4	5–8	9–12	13–21
miR-93			6–8	9–15	16–18	1–5 and 19–23
miR-93			8–10	11–16	17–19	1–7 and 20–23
miR-21			7–11	12–12	13–17	1–6 and 18–22
TN-9.6	1–5	35–39	6–9	10–12	13–16	Two overhanging bases in location 40, 41
			18–23	24–27	28–33	

assumed that in order to form HPL, at least 50% of the bases in the HPL stem arms should have WC paired bases (that is GC and AU). For the second HPL noted on the last row of Table 3.21, the stem arm has 6 bases out of which 3 bases form proper WC paired bonds.

Figures 3.21, 3.22, 3.23 and 3.24 represent the secondary structures reported in Belter et al. [15].

Table 3.21 shows the predicted structures of these miRNAs. For miR-93, there are two secondary structures derived out of the algorithm which are noted on

Table 3.22a Results of four miRNA secondary structures for the miRNA sequences noted in the database (http://www.mirbase.org/)

miRNA ID	5' arm bend	Stem	Loop	Stem	3' end bend	CrP location (value)
MIMAT0001141	1–6	7–10	11–13	14–17	18–22	3 (−4), 7 (−4), 10 (−3), 18 (−10)
MIMAT0001142	1–5	6–8	9–15	16–18	19–22	6 (−17), 9 (−4), 11 (−23)
MIMAT0004109	1	2–5	6–13	14–17	18–22	7 (−5), 19 (+1)
MIMAT0010025	1–5	6–7	8–19	20–21	Nil	8 (−4), 11 (−4), 17 (−4)

Table 3.22b miRNA secondary structure for ten miRNA sequences of database (http://www.mirbase.org/)

miRNA ID	Stem	Loop	Stem	Open-ended loop
MIMAT0017553	8–10	11–15	16–18	1–7 and 19–21
MIMAT0017550	4–7	8–17	18–21	1–3 and 22–21
MIMAT0017551	1–4	5–8	9–12	13–21
MIMAT0017551	9–11	12–14	15–17	1–8 and 18–21
MIMAT0006732	14–16	17–16	17–19	1–13 and 20–23
MIMAT0006736	8–10	11–17	18–20	1–7 and 21–21
MIMAT0006736	10–12	13–18	19–21	1–9 and 22–21
MIMAT0014618	4–7	8–10	11–14	1–3 and 15–22
MIMAT0030605	3–5	6–15	16–18	1–2 and 19–22
MIMAT0030604	3–6	7–15	16–19	1–2 and 20–26

second and third row. TN-9.6 has 41 bases and its structure is reported on fourth and fifth rows with two overhanging bases 40 and 41. Predicted secondary structure of these precursor miRNAs displaying stem-loop-stem and open-ended bends on 5′ and 3′ ends are similar to those noted in the paper Belter et al. [12].

Subsequent to the earlier case study, we now present the prediction results in Table 3.22a, b for some of the miRNA ID noted in miRNA database http://www.mirbase.org [16]. Results noted in Table 3.22a are derived on execution of the algorithm. Serial numbers of bases are reported for each region—5′ arm bend, HPL (stem-loop-stem), 3′ end bend, critical pair location and value. In addition, we have predicted a large number of miRNA sequences reported in the database [16]. A result summary of ten more cases displaying HPL (hairpin loop: stem-loop-stem) and open-ended loop base locations are reported in Table 3.22b.

In addition to miRNA, another small length non-coding RNA named siRNA plays an important role for gene regulation [17]. Although there are certain similarities between miRNA and siRNA, as pointed in Sect. 3.6.2, their interaction for gene regulation differs. Next section concentrates on analysis of siRNA molecule.

3.8 Analysis of Signal Graph of siRNA and its Target Gene

The siRNA is derived out of a duplex RNA (formed by sense and antisense strands), each having 21–25 nucleotide bases. The precursor siRNA covers the interconnected sense and antisense strands, as illustrated below for a typical siRNA with two overhanging bases on either end.

Sense strand: 5′ C A G C T G G G T C C C T C T T C A A T T
Anti-sense strand: 3′ T G G T C G A C C C A G G G A G A A G T T

The enzyme dicer trims the double-stranded siRNA precursor to generate siRNA sense and antisense strands. These processed RNAs are incorporated into the RNA-induced silencing complex (RISC) which is guided by siRNA strand to the target gene mRNA transcript to control the biological processes of transcription. The NCBI probe database stores the wet lab results with respect to binding of a siRNA on its target gene mRNA transcript. The database stores—(i) the siRNA ID along with its base sequence and (ii) the base sequence of the target gene for each siRNA along with the region where siRNA sequence match fully or partially.

The siRNA, as reported in the paper Rohloff et al. [18], has higher specificity of gene regulation compared to that of miRNA. Both miRNA and siRNA based therapeutic agents are being investigated for treatment of wide varieties of diseases. For last one decade gene regulation study [19] has received significantly higher emphasis in view of large increase of patients attacked with deadly disease like cancer. Different experimental results have confirmed that small RNAs like miRNA, siRNA and shRNA (short hairpin RNA), etc., play a key role for gene regulation.

In last one decade, a large number of researchers [20] have concentrated on efficient design of siRNA for a given target gene. The main focus of all these studies is to analyse the sequence of nucleotide bases in siRNA that targets a specific gene. In their paper Riba et al. [21] authors reported an in silico experimental set up that analyses both off and on-target genes for different siRNAs.

A significant observation is noted in the Belter et al. [19]—"We think that the knowledge of the miRNA structure may give a new insight into miRNA—dependent gene regulation mechanism and be a step forward in the understanding their function and involvement in cancerogenesis". This observation motivated us to undertake the signal graph analytics of miRNA to predict its secondary structure. We established the point that the secondary structure of miRNA precursor and mature miRNA can be derived out of signal graphs generated out of RCAM for the molecules. We continue with the same approach of signal graph analytics to predict binding of siRNA on an mRNA transcript derived out of the target gene.

Our experimental setup is designed with siRNAs reported in NCBI probe database [22] along with the target genes. The CS signal graphs for siRNA strand and mRNA transcript are derived out of ComRCAM designed for siRNA sequence and the mRNA transcript sequence. In the CA-based model reported in Sect. 3.7.2, we established the fact that the secondary structure of miRNA including its HPL (hairpin loop) can be predicted from the CS signal graph derived out of CS Difference (CSD) values between a pair of CA cells, each modelling a nucleotide base. The CS signal graph displays CS values derived out of ComRCAM evolution for a RNA molecule. In this background, we proceed to analyse the signal graphs derived out of ComRCAM for a siRNA molecule and also for RNA transcript derived out of target gene reported in NCBI Probe database. Two sample signal graphs for a siRNA ID 8817008b (base serial number 1–19) and RNA transcript (base serial number 1–3756) derived out of target gene ID ADCY8 noted in NCBI Probe database [22] are reported in Fig. 3.26. Partial signal graph for RNA transcript (from serial number 3225–3243) is noted in Fig. 3.26. For subsequent discussions, we refer to the target gene by the RNA transcripted out of it. The x-axis of Fig. 3.26 shows the bases and their serial number, while y-axis reports the CS (Cycle Start) values associate with a CA cell. CSD (Cycle Start Difference) between $(i - 1)$th and ith locations is noted as CSD (i). For this case, the siRNA base sequence match fully with the base locations 3225–3243 (between a pair of dotted vertical lines) of target gene for the siRNA. For subsequent discussions, we shall refer to the target gene by the RNA transcript derived out of it. Signal graph analytics on signal graphs of siRNA and RNA transcript is reported in the rest of this section. The state-of-the-art analysis of binding a siRNA on its target gene is determined by the similarity of base sequences along with relevant parameters extracted from the physical domain features through wet lab experiment. The focus of our signal graph analytics is to utilize these parameters along with the additional parameter derived out of signal graphs of siRNA and mRNA transcript of target gene. We report the algorithm for reduction of off target binding of a siRNA on other genes. By convention, this condition is referred to as 'False Positive' while on-target binding is noted as 'True Positive'. In addition we also minimize 'True Negative' cases where a siRNA fails to detect its on-target gene.

The focus of the signal graph analytics is to identify the parameter out of the pair of signal graphs (siRNA and RNA transcript) that represents the physical domain condition for binding of siRNA with the target gene. Based on the study of wet lab results of NCBI with the results derived out of our CA-based model, we proceed to design the algorithm. We retrieve NCBI data and derive a pair of CS Signal graph for siRNA and RNA transcript having a region (say M) where siRNA sequence match fully or partially. So length of M may vary from (say) 5/6–19 (number of bases in siRNA).

The variables and acronyms used in the algorithmic steps are defined below. Variable X refers to the onserial locations are associated with signal graph of RNA transcript of the gene, while X' refers to that associated with signal graph for the siRNA.

The siRNA signal graph is denoted as M', while M refers to the signal graph region having the same base sequence (full or partial match). This identical base sequence for siRNA and a region of gene is a prerequisite for binding of siRNA on the target gene. We define two regions adjacent to M referred to as LoM (Left of M) and RoM (Right of M), each of length 20 bases.

PS refers to Primary Signal that has the maximum negative value PSVal (greater than −9) at the location PSLoc. For M' these are marked as PSVal', PSLoc', and PSVal, PSLoc for transcript region of the (LoM-M-RoM). If a secondary signal (SS) value less than PS) exists at ith location, while a PS exists at $(i − 1)$ or $(i + 1)$, then the SS value is added to the PS value.

PSLoc' is aligned with PSLoc to select 19 bases out of (LoM-M-RoM) on either end of PSLoc (marked as 5' and 3' end), that is aligning PS of siRNA signal graph with that of transcript signal graph. The Secondary Signals (SSs) outside the range 0 to −3 and less than PS value are marked in M' and (LoM-M-ROM).

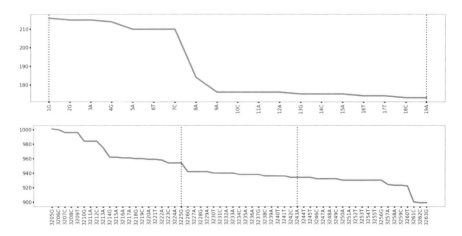

Fig. 3.26 8817008b/ADCY8

Secondary signals (SSs) in the signal graph are noted as two vectors on 5′/3′ end of PSLoc: 5′-SSV and 3′-SSV within a length of 19 bases around Primary Signal location (PSLoc) identified on alignment of PSLoc′ with PSLoc.

The parameters for M′ are noted as PSVal′, PSLoc′, 5′-SSV′ and 3′-SSV′, while for the signal graph for the transcript region covering (LoM-M-RoM), these variables are noted as PSVal, PSLoc, 5′-SSV and 3′-SSV.

Cardinality of SSV's for M′ are noted as 5′-CSSV′, 3′-CSSV′, while for the transcript region, these are noted as 5′-CSSV, 3′-CSSV.

The algorithm aligns M′ PSLoc′ with transcript (LoM-M-RoM) primary signal location PSLoc and computes the cardinality of secondary signal vectors both for M′ and transcript (LoM-M-RoM). If the cardinality of 5′/3′ secondary signal vectors for M′ and gene (LoM-M-RoM) signal graphs are identical, we predict it as a case of true positive.

Degree of Match (DoM) of M′ signal graph with that of gene region is defined with a score of 0–2, 0 signifying best match and 2 as worst. If DOM is more than 2, it is marked as a case of true negative instance.

Algorithm: Predict true positive and true negative instances of binding a candidate siRNA on mRNA transcript of target gene

Input: Signal graph for the candidate siRNA M′ of 19 bases and the signal graph of mRNA transcript for the region (LoM-M-RoM) of the target gene, where siRNA base sequence match with the base sequence of M region.
Output: True Positive/True Negative instances with respect to binding candidate sgRNA on-target gene.
Steps

1. Analyse siRNA signal graph M′ to mark—PSVal′, PSLoc′.
2. Analyse mRNA transcript graph for (LoM-M-RoM) to mark PSVal and PSLoc. In case, left end of LoM or right end of RoM has the PS signal, consider the next PS signal close to M region that displays base sequence match with M′.
3. Align M′ PSLoc′ with PSLoc of transcript (LoM-M-RoM) to select 19 bases out of (LoM-M-RoM) on 5′ and 3′ end of PSLoc.
4. Note the SS vectors on 5′ and 3′ end of PSLoc′ marked as 5′ SSV′ and 3′ SSV′ and the corresponding parameters for (LoM-M-RoM) signal graph marked as 5′-SSV, 3′-SSV.
5. Compute the cardinality of 5′ and 3′ SSV's of M′ marked as 5′ CSSV′ and 3′ CSSV with the corresponding values of cardinality of transcript (LoM-M-RoM) marked as 5′ CSSV and 3′ SSV. Compare the SS vector cardinality of M′ and (LoM-M-RoM).
6. (a) For the cases of best match, assign the score 0 and worst match as 2, while assigning the score as 3 for no match. The cases having score above 2 are marked as True Negative instances.

(b) Instances of exception in the event of no PS/SS (primary/ secondary signal in signal graph) are treated as true positive cases, if the following conditions are valid—(i) signal graph of siRNA M' has no PS/SS, while mRNA Transcript M region has a very high valued PS (highest or close to highest) (ii) signal graph for LoM-M-RoM has no PS/SS, while siRNA signal graph M' has a very high signal value.

Table 3.23 shows the data retrieved from the NCBI Probe database: siRNA ID, the RNA transcript derived out of its target gene along with matched regions. There are instances of full (or partial) match of siRNA base sequence with that of target gene. Table 3.23a, b show five instances of full and partial sequence match, respectively. The regions of matched base sequence are noted on column 2 and 4. The corresponding signal graph pairs are reported sequentially in Figs. 3.26, 3.27, 3.28, 3.29 and 3.30 for Table 3.23a cases. Figures 3.31, 3.32, 3.33, 3.34 and 3.35 sequentially report the signal graph pairs for the cases of Table 3.23b. The base type and serial locations are displayed on the X-axis, while Y-axis shows the CS Difference (CSD) values. On execution of the algorithm, signal graph similarity has been identified for each pair of signal graphs as per the criteria set in Steps 3–6 of the algorithm. So, each of these ten cases are marked as true positive cases.

Code has been developed for the algorithm for generation of large-scale prediction results reported in Annexure 3.5 with the input data for siRNA and target gene retrieved from NCBI Probe database [22].

Table 3.24 reports sample prediction results (DOM on column 8) for ten cases derived on execution of the program code for the algorithm. The siRNA/gene ID is noted on column 1. The location of PS/SS (Primary and secondary signals) for M',

Table 3.23 Gene ID and siRNA ID for full and partial match of bases sequence

(a) Table with full matched sequence of siRNA with M region of target gene			
Gene	Match region M of transcript	siRNA	Match Pos of siRNA
ADCY8	3225–3243	8817008b	1–19
ADK	288–306	8817009b	1–19
AGXT2L2	1244–1262	8817057b	1–19
AKR1C2	536–554	8817035b	1–19
ANGPTL2	1209–1227	8817020b	1–19
(b) Table with partial matched sequence of siRNA with M region of target gene			
Gene	Match region M of transcript	siRNA	Match Pos of siRNA
ATPBD1B	104–109	8817287b	5–10
ADIPOR1	463–468	817306b	1–6
ATPBD1C	92–96	8817304b	15–19
ANKRD13A	350–357	8817296b	8–15
BTBD1	970–976	8817324b	4–10

Table 3.24 True positive instances with respect to binding of siRNA on mRNA transcript

1	2	3	4	5	6	7	8
siRNA ID/Gene ID	M′	LOM	M	ROM	LOM′/ROM′	LOM/ROM	DOM
8817008b/ADCY8	8	3207–3213	3214	3215–3225	0/0	1/0	1
8817009b/ADK	9	271–278	279	280–289	0/1	1/1	1
8817057b/AGXT2L2	19	1248–1265	1266		0/0	0/0	0
8817035b/AKR1C2	13	542–553	554	555–560	0/0	0/0	0
8817020b/ANGPTL2	*	1208–1221	1222	1223–1226	0/0	0/0	0
8817287b/ATPBD1B	15	107–120	121	122–125	0/0	2/0	1
8817306b/ADIPOR1	9	438–445	446	447–456	0/0	0/0	0
8817304b/ATPBD1C	*	92–92	93	94–110	0/0	0/0	0
8817296b/ANKRD13A	15*	318–331	332	333–336	0/0	0/0	0
8817324b/BTBD1	9	957–964	965	966–975	0/0	0/0	0

LoM, M and RoM are, respectively, shown on columns 2–5. For seven cases, the score is 0—that is, best match of M′ with the specified region of LoM-M-RoM specified on column 7 as per the criteria noted in steps 3–6 of the algorithm. Same cardinality of secondary signals on 5′ and 3′ ends of primary signal leads to best match score of 0. For three cases, as noted on row 1, 2, and 6, the DOM (Degree of Match) is 1 indicating that binding of siRNA is inferior compared to the cases showing DOM value 0.

For the exception cases (Algorithm Step 6(b))—serial numbers 5, 8, 9 (* marked on column 2 in Table 3.24) signal graph display very high signal value either for M′ or M in the corresponding signal graphs noted in Figs. 3.30, 3.33 and 3.34, respectively. While for ID8817020b/ANGPTL2, 8817296b/ANKRD13A, the siRNA do not have any PS (primary signal), the M region of transcript displays

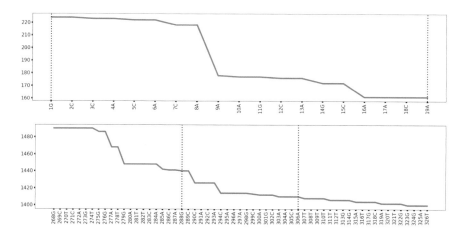

Fig. 3.27 8817009b/ADK

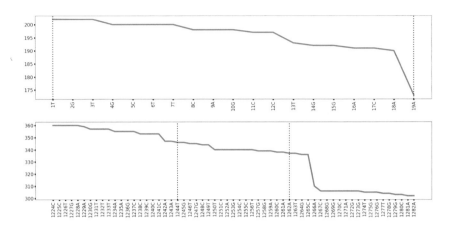

Fig. 3.28 8817057b/AGXT2L2

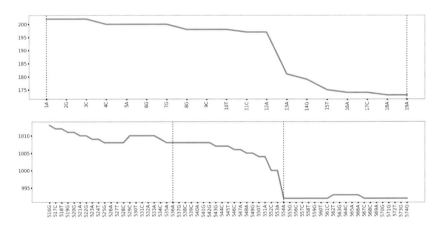

Fig. 3.29 8817035b/AKR1C2

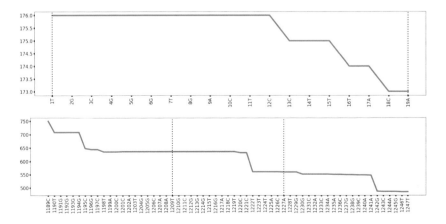

Fig. 3.30 8817020b/ANGPTL2

very high signal of valued PS (−71 and −91, respectively). On the other hand for the ID 8817296b/ANKRD13A, the transcript full region of LoM-M-RoM displays no PS, but the siRNA display very high signal valued PS.

In Table 3.24, serial numbers 1–5 refer to cases of full match of siRNA bases with M, while partial match cases are noted under serial numbers 6–10.

Signal graph pairs for full match cases are shown in Figs. 3.26, 3.27, 3.28, 3.29 and 3.30, while Figs. 3.31, 3.32, 3.33, 3.34 and 3.35 display the signal graph pairs for partial match instances (Serial numbers 6–10 of Table 3.24).

The siRNA Primary Signal (PS) between base location 7 and 8 (Fig. 3.26) is aligned with the PS of target gene Primary Signal (PS) between base location 3213 and 3214 with bases A and G, respectively. Subsequent to PS alignment, the number of secondary signals on 5′ and 3′ end of siRNA and gene signal graph are (0 and 0) and (1 and 0), respectively. Hence, it is a true positive case.

The siRNA Primary Signal (PS) between base base locations 8 and 9 (Fig. 3.27) is aligned with the PS of Target gene Primary Signal (PS) between base locations 278 and 279 with bases T and G, respectively. Subsequent to PS alignment, the number of secondary signals on 5′ and 3′ end of siRNA and gene signal graph are (0 and 1) and (1 and 1), respectively. Hence, it is a true positive case.

The siRNA Primary Signal (PS) between base locations 18 and 19 (Fig. 3.28) is aligned with the PS of Target gene Primary Signal (PS) between base locations 1265 and 1266 with bases C and A, respectively. Subsequent to PS alignment, the number of secondary signals on 5′ and 3′ end of both signal graphs are 0 and 0. Hence, it is a true positive case.

The siRNA Primary Signal (PS) between base locations 12 and 13 (Fig. 3.29) is aligned with the PS of target gene between base locations 553 and 554 with bases A and A, respectively. Subsequent to PS alignment, the number of secondary signals on 5′ and 3′ end of both signal graphs are 0 and 0. Hence, it is a true positive case.

Figure 3.30 displays an exception case with no PS for siRNA while the target gene Primary Signal (PS) shows a very high PS between base locations 1221 and 1222 with bases C and T, respectively. This is an Exception Case as per the Algorithm Step 6b(i). Hence, it is a true positive case.

For the next few figures (Figs. 3.31, 3.32, 3.33, 3.34 and 3.35), it is to be noted that the matching region is not the entire sequence but a partial sequence. But the graphs demonstrate that they are still true positive cases based on the algorithm reported earlier to predict True Positive and True Negative cases. The signal graphs for these partial match instances are explained in subsequent paragraphs for Figs 3.31 to Fig. 3.35 with column 8 of Fig 3.24 displaying prediction results.

The siRNA Primary Signal (PS) between base locations 14 and 15 is aligned with the PS of Target gene Primary Signal (PS) between base locations 120 and 121 (Fig. 3.31) with bases G and G, respectively. Subsequent to PS alignment, the number of secondary signals on 5′ and 3′ end of siRNA and gene signal graphs are (0 and 0) and (2 and 0), respectively. Hence, it is a true positive case but a partial match based on the sequence.

The siRNA Primary Signal (PS) between base locations 8 and 9 is aligned with the PS of target gene Primary Signal (PS) between base locations 445 and 446 (Fig 3.32) with bases C and T, respectively. Subsequent to PS alignment, the number of

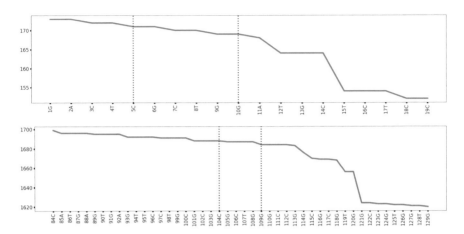

Fig. 3.31   8817287b/ATPBD1B

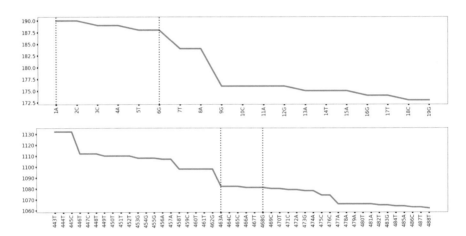

Fig. 3.32   8817306b/ADIPOR1

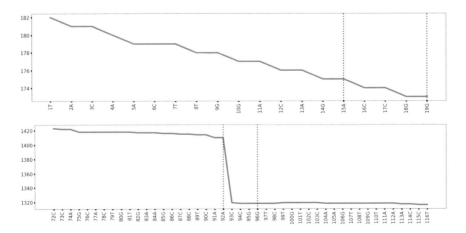

Fig. 3.33   8817304b/ATPBD1C

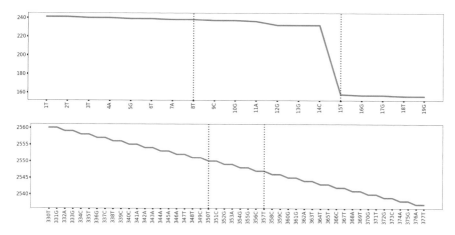

Fig. 3.34   8817296b/ANKRD13A

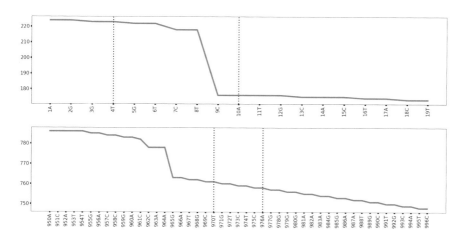

Fig. 3.35   8817324b/BTBD1

secondary signals on 5′ and 3′ end of both signal graphs. Hence, it is a true positive case but a partial match based on the sequence.

The signal graph for siRNA does not have PS/SS signal. This is an instance of exception handling noted in Step 5 of the algorithm. Number of SS signals for siRNA is 0 and 0 with very high signal value of (−91) in location 92–93 (3.33) of target gene signal graph. Hence, it is a true positive case but a partial match based on the sequence.

The signal graph for gene does not have PS/SS signal. This is an instance of exception handling noted in Step 5 of the algorithm. Number of SS signals for gene is 0 and 0 with very high signal value of (−74) in location 14–15 of siRNA (Fig. 3.34). Hence, it is a true positive case but a partial match based on the sequence.

The siRNA Primary Signal (PS) between base locations 8 and 9 is aligned with the PS of target gene Primary Signal (PS) between base locations 964 and 965 (Fig. 3.35) with bases A and G, respectively. Subsequent to PS alignment, the number of secondary signals on 5′ and 3′ end of both signal graphs are 0 and 0. Hence, it is a true positive case but a partial match based on the sequence.

The algorithm (predict true positive and true negative instances of binding a candidate siRNA on mRNA transcript of target gene) has been coded in Python and operated on the data retrieved for NCBI Probe database. Annexure 3.5 presents the results derived on execution of the program on data retrieved from NCBI Probe Database.

We close this chapter on RNA with a brief introduction to RNA-binding proteins.

3.9 RNA-Binding Protein (RBP)

RNA-binding proteins are a large and varied group of proteins that are associated with post-transcriptional gene regulation. Figure 3.36 displays various regulatory regions of DNA that control the transcription process. RNA-binding proteins (RBPs) bind to various regions either upstream or downstream from the coding region of the RNA to control five major processes in mRNA metabolism—splicing, polyadenylation, export, translation and decay. A small domain of around 80 amino acids of RBP displays a motif referred to as RRM (RNA Recognition Motif) with four beta strands and two alpha helixes. The RRM mainly controls various biological functions. As summarized in reference [21], a RBP also binds either to the

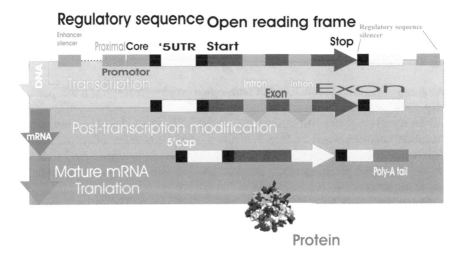

Fig. 3.36 Translation process

double or single-stranded miRNA/siRNA to derive a multi-protein complex referred to as RISC (RNA-Induced Silencing Complex). The single strand of miRNA/siRNA acts as a template for RISC to recognize the specific region of mRNA transcript to initiate gene silencing and also for defense against viral infection.

The review paper 'Origins and Mechanisms of miRNAs and siRNAs' by Carthew et al. [23] reports the prevailing status of research with respect to interaction of miRNA and siRNA with different classes of RBPs. The major interaction site of RBP with miRNA and siRNA is their HPL structure with stem-loop-stem which gets identified from the analysis CS signal graph presented in earlier sections.

3.10 Summary

We started this section on Ribonucleic acid (RNA) with the statement—'origin of life' on earth remains one of the deepest mysteries in modern science. For building cLife (CA-based Life) model, we have assumed that primitive life was initiated on earth with RNA-based codon string with triplet of two primitive molecules—purine and pyrimidine. Such a primitive codon can be modelled with three-neighbourhood two-state per cell CA. There are 256 CA rules out of which we selected 64 rules to model 64 codons of current life form with information flow from DNA to RNA followed by protein synthesis from mRNA. Mathematical model for codon degeneracy has been established based on CA rule analysis. Subsequent model for ribosomal translation has been designed to study mutational effect on codon string. One of the major emphases of the model is to investigate co-translational folding since cLife assumes that the major determinants of protein secondary structure are based on co-translational and post-co-translational folding. We have not presented any methodology for post-co-translational folding in the main text of this chapter. We have projected our methodology to model post-co-translational folding along with the prediction of 3D structure of a protein in Annexure 3.2.

The rest of this chapter covers models for predicting structure of different classes of RNA molecules—tRNA, miRNA precursor and miRNA. cLife assumes that RNA secondary structure plays a leading role for different biological functions in living cells of different species. It has been reported in recent publications that small RNA molecules (miRNA, siRNA, etc.) exert significant influence on transcription and translation processes. In this context in Sect. 3.8 of this chapter, we report the CA model for interaction of siRNA and the RNA transcript derived out of its target gene noted in NCBI Probe database.

The CA model is designed based on the signal graph analytics of the CS/CL signal graphs derived out of mRCAM (for codon string) and ComRCAM (for RNA molecules) designed with CA rules reported in Chap. 2. This methodology enables us to implement bottom-up design model from detailed analysis of biomolecules.

The signal graph analytics, as we have implemented in different section of this chapter, refer to the interpretation of meaningful patterns in the graphs, especially valuable in areas rich with recorded information available in biological databases.

The mathematical/statistical tools, predictive models and descriptive techniques are employed to extract meaningful patterns. The process starts with the front line visualization of data/information in large biological databases to extract patterns from the signal graphs that are relevant for top-down validation of the model designed to represent structure and function of biomolecules. This methodology of signal graph analytics has been carried forward in subsequent Chap. 4 dealing with DNA.

Questions

1. How does a living cell utilize the information content of a codon in a mRNA string? Why the code is degenerate?
2. Illustrate equivalent CA rules under a transform on the 8 bit structure of a CA rule modelling a codon. How CA rule equivalencies represent codon degeneracy?
3. Note the parameters and associated methodology employed for design of equivalent groups of 3NCA rules.
4. What is the importance of 5′ and 3′ UTR on either end of mRNA codon string? Note the sequential steps of codon translation process. Explain the terms 'elongation delay', 'co-translational folding' and 'ribosomal exit port'?
5. 'Synonymous codon mutation does not have any effect on protein product'— True or False? If 'False', explain the possible reasons from the study of published literature on the effect of deleterious effect of synonymous mutations.
6. 'All non-synonymous codon mutation will affect biological function'—True or False? Justify your answer from the study of published literature on the effect of deleterious effect of non-synonymous mutations.
7. What are the different 'motifs' observed in the secondary structure of a RNA strand? Explain the importance of the three sites of Dloop, anticodon loop and Tloop observed in the secondary structure of a tRNA molecule.
8. Name a few small size RNAs with 20–24 bases. 'Small size RNA molecules play important role for biological function'—True or False? Justify your answer from the study of the results reported in recent publications with respect to control of Transcription process by small length RNAs.
9. Explain the acronym 'RISC' and its importance to control translation process. What are the roles of siRNA in this biological function?
10. What are the major differences between siRNA and miRNA? From the study of recent publications explain their role as possible therapeutic agents for different diseases.
11. Explain the terminologies—dicer enzyme, siRNA precursor, aminoacyl-tRNA, argonaute protein and signal graph analytics methodology of CA model.
12. Note the sequential steps proposed in the CA model for co-translational folding.
13. 'In addition to the sequence of nucleotide bases of a RNA strand, its secondary structure plays a determinant role for its biological function'—True or False? Justify your answer from study of published literature in this field.

14. 'Critical Pair (CrP) plays important role in the CA model'—True or False? Justify your answer explaining how this parameter is employed in CA model building.

15. (a) Write a short note on two primitive biomolecules purine and pyrimidine out of which nucleic acid molecules are formed—explaining their similarity and differences. (b) Describe the biological functions performed by the biomolecules 'Riboswitch' and 'Ribozyme'. (c) Write a short note on the essential biological functions a RNA molecule is capable of performing.

Annexure

Annexure 3.1: Analysis of Nucleotide Bases to Assign Cellular Automata (CA) Rules

Section 3.3.3 reports the assignment of Cellular Automata (CA) rules to the codon triplets. Such an assignment demands detailed analysis of the atoms of different nucleotide bases and the associated sequence in different codons, as presented in this annexure.

The relevant features of the four nucleotide bases are retrieved from PubChem database (https://pubchem.ncbi.nlm.nih.gov/) which are analyzed in this annexure. The volume surface of Nitrogen (N), Oxygen (O), and Carbon (C) atom are reported as 15.3, 18.1 and 9.9, respectively as noted in the paper [24]. The polar positive, polar negative and non-polar surface areas for four bases are computed for each of the bases. The value of XLogP3 for G, A, U and C are -1.0, -0.1, -1.1 and -1.7, respectively. This value specifies how hydrophilic or hydrophobic a molecule is. Higher hydrophobicity is indicated by lesser negative value in the comparative scale for four bases.

Based on the analysis of these parameters, the bases are arranged in the following descending order: G > A > U > C with G and C, assigned maximum and minimum weight respectively. Among the bases, the atomic structure of base **A** stands apart from others since it does not have any oxygen atom in its atomic structure and so it does not have any negatively charged surface area. As per XLogP3 values, base **A** can be viewed as most hydrophobic, while **C** is least hydrophobic among the four bases.

For comparative analysis of 16 base pairs with left and middle base (first two bases), it has been assumed that the parameters of the left base play more dominant role. For example, the base **A** is assigned higher weight than the base **C** for the pair AC; on the other hand higher weight is assigned to C than A for the pair CA. Considering the special characteristics of base A, and the XLogP3 values, the 16 base pairs are divided into the following two groups.

(i) The first group consists of the pairs GG, GU, GC, UC, CG, CU, CC, none of which covers the base A. In a comparative scale, these pairs are assumed to be neither hydrophilic nor hydrophobic. The base pair **A** followed by **C** is added in this list since **A** is least hydrophilic, while **C** is most hydrophilic.

(ii) The second group covers the six base pairs involving base A that is least hydrophilic—GA, AA, UA, CA, AG and AU. These pairs are assumed to be either hydrophilic or hydrophilic. Two other base pairs UG and UU are added to this list since both U and G have similar XLogP3 values -1.1 and -1.0. Further, the base U displays polar positive surface area (30.6) and polar negative surface area (36.2).

The first group of base pairs are assigned to (2, 2) 4RGs having balanced rules subdivided as per the parameter value δ. The descending order of base pairs is derived based on higher weight to the left base and sum of polar surface area for each pair. Each pair is associated with an amino acid as per the parameter value δ elaborated next.

- $\delta = 1$ GG > GU > GC > AC
- $\delta = 2$ UC > CG > CU
- $\delta = 0$ CC

The second group of base pairs are assigned to non-(2, 2) 4RGs (ID 9–16) with its subgroups detailed next. Each pair is associated with two or three amino acids including stop codons.

- $\delta = 1$ GA > AA > UA > CA
- $\delta = 0$ AG
- $\delta = 2$ AU > UG
- $\delta = 0$ UU

The base pairs are next assigned to different 4RGs, as reported in main text Sect. 3.3.3, based on the sum of decimal values of CA rules associated with 4RGs and parameter δ.

Annexure 3.2: Outline of CA Model for Post co-Translational Folding and Prediction of Protein Structure

CA model for co-translational folding is explained in Sect. 3.6. The algorithm designed to predict the fold points (of nascent peptide chain translated out of codon string) is reported along with the results derived on executing the software code developed for the algorithm.

We have experimented with the mRNA codon string of 16000 proteins retrieved from Dunbrack data set (http://dunbrack.fccc.edu) and designed two databases named as EFRL (Excluded Fold Residue Location) and VFRL (Valid Fold Residue Location). We predict true/false cases of prediction of fold residue location (FRL) for an input codon string based on these EFRL and VFRL databases. Based on the foundation laid down in Sect. 3.6 with the CA model for co-translational folding, we take the next step to model post-co-translational folding.

This annexure highlights the methodology we are implementing for post-co-translational folding and prediction of protein structure. We have assumed that during post-co-translational folding, secondary structures of helix and beta get introduced on protein chain. Each of helix and beta introduces folds at the start and end on the protein chain. In addition to the folds detected during co-translational folding, we add these new folds due to helix and beta. Fold Segments (FSs) are next derived with the amino acid residues between each pair of folds identified during co-translational and post-co-translational folding. On either end of such a segment dihedral angles of first and last two residues are noted. The Fold Segment Database (FSDB) designed out of large number of proteins lays the foundation for prediction of protein structure. The sequential steps of this methodology are noted below which are being finalized prior to development of the program.

Step 1: Develop the fold segment databases (FSDB0, FSDB1, FSDB2 and FSDB3) with the results of true VFRL, and FRL specified on first and last base of secondary structures of helix and beta noted in PDB for 16000 Dunbrack proteins.

– While FSDB0 covers, each of the segments derived out of 16000 proteins, the next three databases merge adjacent fold segments as follows.
– FSDB1 and FSDB2 merges each FSDB0 entry with the left and right neighbouring segments, respectively.
– The FSDB3 merges each of FSDB0 with its left and right neighbours. These databases have been designed.
– Each entry in FSDB records two dihedral angles of the first and last residue locations as explained in Sect. 3.6 while introducing two databases for EFRL (excluded FRL) and VFRL (valid FRL).

Step 2: Select three public domain packages for prediction of helix and beta in a protein chain.

Step 3: In order to predict three-dimensional structure of a candidate protein chain, we execute the Steps 4–7.

Step 4: Derive co-translational fold list out of VFRL (valid fold residue location) noted in Sect. 3.6.

Step 5: Run the three packages (noted in Step 2) to predict helix and beta structure in the candidate input chain. Identify helix and beta locations in the candidate chain based on majority voting of the results derived out of three packages. Based on location of helix and beta in the candidate chain, next step modifies the VFRL list.

Step 6: Modification of VFRL (Valid Fold Residue Location) list and derivation of composite VFRL for the candidate chain—mark a VFRL entry as 'False' if it is covered within a helix and beta excepting first and last two bases. Remove such 'False' entries. Subsequent to exclusion of such 'False' FRLs from VFRL, add additional folds at the start and end of location of each helix and beta. This is marked as composite VFRL List.

Step 7: Derive fold segments between each pair of folds noted in the composite VFRL list generated in Step 6.

Step 8: For each fold segment identified in Step 7, search FSDB (fold segment data base) designed in Step 1.

Step 9: Stitch the fold segments identified in Step 8 to predict the three-dimensional structure of the input candidate protein chain while taking into consideration the dihedral angle pairs noted for start and end residues of a segment. Stitching of a pair of adjacent segments extracted from FSDB demand search of correct options out of available multiple options.

Annexure 3.3: CS/CL Signal Graph of an Example RNA Molecule and Derivation of CrP (Critical Pair)

CS signal graph is derived out of Cycle Start (CS)/Cycle Length (CL) parameter table for an example RNA molecule having 159 nucleotide bases reported below. The column 1 and column 2 of the table below report the serial number of the nucleotide bases of the RNA. The CS and CL parameters derived out of ComRCAM are noted on column 3, 4 and on column 9, 10 on two major columns 1 to 6 and 7 to 12 respectively. The CS signal graph is derived by subtracting CS value of $(i - 1)$th cell from that of ith cell modelling respectively the $(i - 1)$th and ith location bases. The CS Difference (CSD) values are noted on column 5 and 11. Most of the CS difference values lie in the range 0 to -4, hence the threshold limit is set as -4. The CS difference value that is outside this range is marked as CrP (critical pair) reported on Col 6 and 12. For example, the first CrP in the list at location 28 is -32 since CS values for base location 27 is 372, while the value for location 28 is 340. A positive CS difference value outside the range of 0 to -4 is also marked as CrP as noted for location 62. The CrP values are analysed and processed in different algorithms reported in the main body of this chapter. Location of CrPs derived out of CS signal graphs provides the foundation of signal graphs analytics projected for different classes of RNA molecules.

The CS signal graph derived out of CS difference values is reported below for the RNA molecule. Signal graph analytics of such graphs represents various physical domain features of the RNA molecule as reported in different section of this chapter (Fig. 3.37).

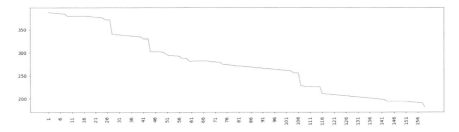

Fig. 3.37 Cycle Start (CS) signal graph of RNA molecule with COMRCAM evolution

Pos	AA	CS Val	CL Val	CSD	CrP	Pos	AA	CS Val	CL Val	CSD	CrP
1	G	388	136			81	T	271	68	0	
2	G	387	136	−1		82	T	271	68	0	
3	C	387	136	0		83	T	270	68	−1	
4	T	386	136	−1		84	A	270	68	0	
5	C	386	136	0		85	A	269	68	−1	
6	T	385	136	−1		86	T	269	68	0	
7	A	385	136	0		87	T	268	68	−1	
8	T	384	136	−1		88	A	267	68	−1	
9	G	380	136	−4		89	G	267	68	0	
10	G	380	136	0		90	C	267	68	0	
11	C	380	136	0		91	T	266	68	−1	
12	T	380	136	0		92	G	266	68	0	
13	T	380	136	0		93	G	265	68	−1	
14	A	380	136	0		94	G	265	68	0	
15	G	380	136	0		95	A	264	68	−1	
16	T	380	136	0		96	A	264	68	0	
17	T	380	136	0		97	T	263	68	−1	
18	G	379	136	−1		98	G	263	68	0	
19	G	379	136	0		99	G	262	68	−1	
20	T	378	136	−1		100	T	262	68	0	
21	T	378	136	0		101	G	261	68	−1	
22	A	377	136	−1		102	G	261	68	0	
23	A	377	136	0		103	C	260	68	−1	
24	A	376	136	−1		104	A	256	68	−4	
25	G	372	136	−4		105	C	256	68	0	
26	C	372	136	0		106	A	256	68	0	
27	G	372	136	0		107	C	228	68	−28	CrP
28	C	340	136	−32	CrP	108	T	228	68	0	
29	C	340	136	0		109	C	226	68	−2	
30	T	339	136	−1		110	C	226	68	0	
31	G	339	136	0		111	T	226	68	0	
32	T	338	136	−1		112	G	226	68	0	
33	C	338	136	0		113	T	226	68	0	
34	T	337	136	−1		114	A	226	68	0	
35	C	337	136	0		115	G	226	68	0	
36	G	336	136	−1		116	T	211	68	−15	CrP
37	T	336	136	0		117	C	211	68	0	
38	A	335	136	−1		118	C	210	68	−1	
39	A	335	136	0		119	C	210	68	0	
40	A	334	136	−1		120	A	209	68	−1	
41	A	330	136	−4		121	G	209	68	0	
42	A	330	136	0		122	C	208	68	−1	
43	A	330	136	0		123	T	208	68	0	

(continued)

(continued)

Pos	AA	CS Val	CL Val	CSD	CrP	Pos	AA	CS Val	CL Val	CSD	CrP
44	T	302	136	−28	CrP	124	A	207	68	−1	
45	G	302	136	0		125	C	207	68	0	
46	T	302	136	0		126	T	206	68	−1	
47	C	302	136	0		127	C	206	68	0	
48	A	302	136	0		128	A	205	68	−1	
49	G	301	136	−1		129	G	205	68	0	
50	C	299	136	−2		130	G	204	68	−1	
51	C	295	136	−4		131	A	204	68	0	
52	T	294	136	−1		132	G	203	68	−1	
53	G	294	136	0		133	A	203	68	0	
54	A	293	136	−1		134	C	202	68	−1	
55	G	293	136	0		135	T	202	68	0	
56	C	292	136	−1		136	G	201	68	−1	
57	A	288	136	−4		137	A	201	68	0	
58	A	288	136	0		138	A	200	68	−1	
59	C	288	136	0		139	G	200	68	0	
60	A	282	136	−6	CrP	140	C	199	68	−1	
61	T	281	136	−1		141	A	199	68	0	
62	T	282	136	1	CrP	142	G	198	68	−1	
63	T	282	136	0		143	G	194	68	−4	
64	C	282	136	0		144	A	194	68	0	
65	T	282	136	0		145	G	194	68	0	
66	A	282	136	0		146	G	194	68	0	
67	C	282	136	0		147	A	194	68	0	
68	A	282	136	0		148	T	194	68	0	
69	A	280	136	−2		149	C	194	68	0	
70	A	280	136	0		150	G	194	68	0	
71	T	280	136	0		151	C	194	68	0	
72	T	279	136	−1		152	T	193	68	−1	
73	A	279	136	0		153	T	193	68	0	
74	T	275	136	−4		154	G	192	68	−1	
75	T	274	68	−1		155	A	192	68	0	
76	A	274	68	0		156	G	191	68	−1	
77	T	273	68	−1		157	C	191	68	0	
78	T	273	68	0		158	C	190	68	−1	
79	T	272	68	−1		159	C	181	68	−9	CrP
80	T	271	68	−1							

Annexure 3.4: Predicted tRNA Secondary Structures for tRNA Sequences Retrieved from the Database - GtRNAdb. ucsc.edu

The algorithim to predict secondary structure of a tRNA molecule is reported in Sect. 3.7.1. Representative results are reported in Figs. 3.12 and 3.13. The algorithm has been coded and executed for large number base sequences of tRNA molecules of different species. The predicted results are tabulated below. The basic structure of tRNA molecule for wide variety of species is identical with three loops D, T, anticodon loops and variable length V loop.

Name	Accepter	D Stem	D Loop	D Stem	A Stem	A Loop	A Stem	V Loop	T Stem	T Loop	T Stem	Stem
chr19trna13	1–7	10–12	13–22	23–25	27–31	32–38	39–43	44–48	49–51	52–62	63–65	66–72
chr1trna50	1–5	10–12	13–22	23–25	27–31	32–38	39–43	44–47	48–52	53–60	61–65	68–72
chr6trna125	2–7	10–12	13–22	23–25	27–31	32–38	39–43	44–49	50–52	53–61	62–64	66–71
chr6trna126	1–3	5–7	8–12	13–15	28–32	33–39	40–44	45–49	50–52	53–63	64–66	71–73
chr6trna127	1–3	6–8	9–24	25–27	28–32	33–39	40–44	45–59	60–62	63–69	70–72	73–75
chr6trna122	1–3	10–12	13–23	24–26	28–32	33–39	40–44	45–49	50–52	53–63	64–66	71–73
chr2trna13	1–7	15–17	18–22	23–25	27–31	32–42	43–47	48–68	69–73	74–80	81–85	86–92
chr2trna12	1–7	10–12	13–17	18–20	22–26	27–33	34–38	39–42	43–47	48–53	54–58	59–65
chr2trna17	1–7	11–13	14–20	21–23	28–34	35–33	34–40	41–46	47–49	50–60	61–63	64–70
chr19trna14	1–7	10–12	13–22	23–25	27–31	32–38	39–43	44–47	48–50	51–62	63–65	66–72
chr6trna9	1–7	10–12	13–22	23–25	27–31	32–38	39–43	44–48	49–51	52–62	63–65	66–72
chr1trna78	1–7	10–12	13–22	23–25	27–31	32–38	39–43	44–48	49–51	52–62	63–65	66–72
chr8trna5	4–7	9–11	12–18	19–21	23–27	28–35	36–40	41–44	45–49	50–56	57–61	62–65
chr8trna2	1–7	10–12	13–22	23–25	27–31	32–42	43–47	48–66	67–69	70–75	76–78	86–92
chr8trna1	1–3	11–13	14–18	19–21	28–30	31–34	35–37	38–46	47–49	50–64	65–67	68–70
chr8trna9	1–7	10–12	13–22	23–25	27–31	32–38	39–43	44–57	58–60	61–71	72–74	75–81
chr8trna8	1–3	10–12	13–22	23–25	27–30	31–39	40–43	44–48	49–52	53–61	62–65	70–72
chr6trna95	1–4	10–12	13–22	23–25	27–31	32–38	39–43	44–48	49–51	52–58	59–61	62–65
chr6trna94	4–6	10–12	13–22	23–25	29–33	34–36	37–41	42–48	49–51	52–62	63–65	70–72
chr6trna97	1–3	5–7	8–16	17–19	27–30	31–39	40–43	44–47	48–50	51–65	66–68	70–72
chr6trna96	1–7	10–12	13–22	23–25	27–31	32–38	39–43	44–47	48–50	51–62	63–65	66–72
chr6trna91	1–7	11–13	14–18	19–21	27–30	31–39	40–43	44–48	49–51	52–62	63–65	66–72
chr6trna93	1–3	11–13	14–21	22–24	27–31	32–38	39–43	44–48	49–53	54–60	61–65	67–69
chr6trna99	1–7	10–12	13–22	23–25	27–31	32–38	39–43	44–47	48–50	51–61	62–64	65–71
chr6trna98	1–3	10–12	13–23	24–26	28–32	33–39	40–44	45–59	60–62	63–69	70–72	73–75

(continued)

(continued)

Name	Accepter	D Stem	D Loop	D Stem	A Stem	A Loop	A Stem	V Loop	T Stem	T Loop	T Stem	Stem
chr16trna8	1–4	6–8	9–22	23–25	27–31	32–38	39–43	44–47	48–52	53–61	62–66	69–72
chr16trna6	1–3	6–8	9–22	23–25	27–31	32–38	39–43	44–47	48–52	53–61	62–66	70–72
chr16trna1	1–3	7–9	10–16	17–19	27–31	32–38	39–43	44–48	49–52	53–61	62–65	69–71
chr16trna3	2–7	10–12	13–22	23–25	28–31	32–38	39–42	43–48	49–53	54–60	61–65	66–71
chr1trna98	1–3	10–12	13–23	24–26	28–32	33–39	40–44	45–52	53–55	56–63	64–66	71–73
chr1trna99	1–5	12–14	15–19	20–22	28–31	32–35	36–39	40–48	49–52	53–59	60–63	67–71
chr12trna15	1–5	7–9	10–24	25–27	29–34	35–39	40–45	46–57	58–60	61–66	67–69	70–74
chr1trna92	1–3	12–14	15–19	20–22	28–31	32–38	39–42	43–44	45–47	48–61	62–64	70–72
chr12trna10	1–3	12–14	15–22	23–25	27–31	32–38	39–43	44–53	54–56	57–62	63–65	69–71
chr12trna12	1–3	12–14	15–22	23–25	27–31	32–38	39–43	44–53	54–56	57–62	63–65	69–71
chr2trna18	1–7	12–14	15–19	20–22	30–32	33–36	37–39	40–47	48–51	52–60	61–64	65–71
chrXtrna10	1–7	10–12	13–23	24–26	28–32	33–39	40–44	45–49	50–52	53–63	64–66	67–73
chr6trna128	1–7	10–12	13–23	24–26	28–32	33–39	40–44	45–49	50–52	53–63	64–66	67–73
chr7trna29	1–5	10–12	13–21	22–24	27–30	31–37	38–41	42–47	48–51	52–60	61–64	67–71
chr15trna8	1–7	10–12	13–22	23–25	27–31	32–38	39–43	44–47	48–50	51–61	62–64	65–71
chr15trna9	1–7	10–12	13–22	23–25	27–31	32–38	39–43	44–47	48–51	52–60	61–64	65–71
chr6trna50	1–7	10–12	13–22	23–25	27–31	32–38	39–43	44–57	58–60	61–71	72–74	75–81
chr7trna21	1–5	10–12	13–21	22–24	27–30	31–37	38–41	42–47	48–51	52–60	61–64	67–71
chr7trna20	1–4	10–12	13–21	22–24	26–30	31–37	38–42	43–47	48–50	51–61	62–64	67–70
chr7trna23	1–5	10–12	13–21	22–24	26–30	31–37	38–42	43–47	48–51	52–60	61–64	67–71
chr7trna22	1–3	10–12	13–21	22–24	27–30	31–37	38–41	42–48	49–51	52–60	61–63	69–71
chr15trna1	1–7	10–12	13–22	23–25	27–31	32–38	39–43	44–47	48–51	52–60	61–64	65–71
chr15trna3	1–5	10–12	13–22	23–25	28–31	32–38	39–42	43–48	49–52	53–61	62–65	68–72
chr19trna3	1–3	11–13	14–22	23–25	30–36	37–35	36–42	43–47	48–53	54–62	63–68	71–73

(continued)

(continued)

Name	Accepter	D Stem	D Loop	D Stem	A Stem	A Loop	A Stem	V Loop	T Stem	T Loop	T Stem	Stem
chr19trna4	1–3	5–7	8–15	16–18	26–31	32–38	39–44	45–49	50–52	53–57	58–60	65–67
chr6trna54	1–7	9–11	12–25	26–28	29–33	34–36	37–41	42–49	50–54	55–61	62–66	67–73
chr19trna6	2–7	10–12	13–22	23–25	29–33	34–36	37–41	42–49	50–52	53–61	62–64	66–71
chr6trna56	2–4	6–8	9–14	15–17	29–31	32–38	39–41	42–33	34–36	37–61	62–64	65–67
chr6trna59	1–3	10–12	13–23	24–26	28–32	33–39	40–44	45–55	56–58	59–66	67–69	71–73
chr6trna168	1–3	10–12	13–23	24–26	28–32	33–39	40–44	45–49	50–52	53–63	64–66	71–73
chr5trna5	1–7	10–12	13–22	23–25	27–31	32–38	39–43	44–48	49–51	52–62	63–65	66–72
chr6trna159	1–4	10–12	13–23	24–26	31–30	31–40	41–40	41–50	51–53	54–62	63–65	70–73
chr6trna158	1–7	10–12	13–22	23–25	27–31	32–38	39–43	44–47	48–54	55–59	60–66	66–72
chr6trna155	1–4	10–12	13–23	24–26	31–30	31–40	41–40	41–50	51–53	54–62	63–65	70–73
chr6trna154	1–3	10–12	13–23	24–26	31–30	31–40	41–40	41–50	51–53	54–62	63–65	71–73
chr6trna157	3–7	10–12	13–22	23–25	27–31	32–38	39–43	44–48	49–53	54–60	61–65	66–70
chr6trna151	1–7	11–13	14–20	21–23	26–30	31–37	38–42	43–51	52–54	55–61	62–64	65–71
chr6trna150	1–6	10–12	13–22	23–25	27–31	32–38	39–43	44–48	49–51	52–62	63–65	67–72
chr6trna153	1–7	10–12	13–22	23–25	27–31	32–38	39–43	44–48	49–51	52–62	63–65	66–72
chr6trna51	1–7	10–12	13–22	23–25	27–31	32–38	39–43	44–57	58–60	61–71	72–74	75–81
chr19trna1	1–3	10–12	13–23	24–26	28–32	33–39	40–44	45–52	53–55	56–63	64–66	71–73
chr19trna2	1–7	11–13	14–23	24–26	28–31	32–38	39–42	43–47	48–52	53–59	60–64	65–71
chr1trna59	1–4	10–12	13–24	25–27	30–33	34–40	41–44	45–59	60–64	65–71	72–76	80–83
chr9trna1	1–7	9–11	12–24	25–27	30–33	34–40	41–44	45–54	55–58	59–63	64–67	68–74
chr9trna1	1–7	11–13	14–20	21–23	25–28	29–45	46–49	50–54	55–58	59–63	64–67	68–74
chr6trna137	1–7	10–12	13–22	23–25	27–31	32–38	39–43	44–47	48–50	51–61	62–64	66–72
chr1trna118	1–5	10–13	14–22	23–26	29–33	34–44	45–49	50–136	137–140	141–149	150–153	154–158
chrXtrna1	1–3	6–8	9–23	24–26	27–32	33–37	38–43	44–47	48–51	52–60	61–64	69–71

(continued)

(continued)

Name	Accepter	D Stem	D Loop	D Stem	A Stem	A Loop	A Stem	V Loop	T Stem	T Loop	T Stem	Stem
chr6trna28	1–7	10–12	13–23	24–26	41–43	44–50	51–53	54–69	70–72	73–83	84–86	87–93
chr6trna44	1–7	10–12	13–22	23–25	27–31	32–38	39–43	44–57	58–60	61–71	72–74	75–81
chr14trna19	1–7	10–12	13–22	23–25	33–35	36–46	47–49	50–69	70–72	73–83	84–86	87–93
chr17trna41	1–7	10–12	13–22	23–25	27–31	32–38	39–43	44–57	58–61	62–70	71–74	75–81

Annexure 3.5: Predicted Results for Binding of siRNA Molecules on RNA Transcript of Target Gene

The algorithm to predict binding of siRNA with target gene is reported in Sect. 3.8.

Table 3.24 reports sample prediction results for ten cases derived on execution of the program code for the algorithm. The siRNA/gene ID is noted on column 1. The next table reports a number of prediction results (DOM) on last column for siRNA and target gene retrieved from NCBI Probe database [22].

siRNA ID/Gene ID	M'	LOM	M	ROM	LOM'/ROM'	LOM/ROM	DOM
8811589b/UBE2T	19	300–317	318		0/0	0/0	0
8812102b/UBE2E3	8	518–524	525	526–536	0/0	0/1	1
8812967b/P2RX3	–	586–590	591	592–604	0/0	0/1	1
8811547b/UFC1	19	114–131	132		1/0	0/0	1
8811418b/UBE2L6	15	437–450	451	452–455	1/0	0/0	1
8811391b/UBE2S	19	160–177	178		0/0	1/0	0
8811310b/UBE2I	8	155–161	162	163–173	0/0	0/0	0
8812922b/P2RX3	5	326–329	330	331–344	0/0	0/1	1
8817061b/PHF19	19	1565–1582	1583		0/0	0/0	0
8810803b/UBE2 M	4	414–416	417	418–432	0/1	0/2	1
8810630b/UBE2D3	12	332–342	343	344–350	0/0	0/0	0
8811746b/UBE2J1	19	489–506	507		1/0	1/0	0
8812175b/UBE2V1	19	444–461	462		0/0	3/0	1
8811576b/UBE2T	15	472–485	486	487–490	1/0	0/0	1
8811073b/RAB6IP1	19	3672–3689	3690	–	0/0	1/0	1
8811496b/UFC1	4	–	–	–	0/0	0/0	0
8812990b/P2RX3	19	108–125	126		2/0	0/0	1
8811406b/UBE2S	10	88–96	97	98–106	1/1	1/0	1
8812149b/UBE2E3	–	227–227	228	229–245	0/0	0/2	1
8811687b/UBE2 W	–	258–262	263	264–276	0/0	0/0	0
8811792b/UBE2 K	19	330–347	348		1/0	0/0	1
8811775b/UBE2 K	19	382–399	400		0/0	0/0	0
8817034b/CYP4Z1	15	1314–1327	1328	1329–1332	1/0	0/1	2
8817011b/OR12D3	19	–	–	–	0/0	0/0	0
8811564b/UBE2T	19	537–554	555		0/0	2/0	1

(continued)

(continued)

siRNA ID/Gene ID	M'	LOM	M	ROM	LOM'/ROM'	LOM/ROM	DOM
8811442b/UBE2L6	–	262–266	267	268–280	0/0	0/0	0
8812959b/P2RX3	19	670–687	688		0/0	0/0	0
8810822b/UBE2 M	15	279–292	293	294–297	1/0	2/0	1
8811779b/UBE2 K	13	388–399	400	401–406	0/0	0/0	0
8812723b/SOST	–	212–226	227	228–230	0/0	0/0	0
8812136b/UBE2E3	8	366–372	373	374–384	0/0	0/1	1
8810843b/UBE2 M	11	118–127	128	129–136	0/1	0/1	0
8810821b/UBE2 M	–	291–292	293	294–309	0/0	0/0	0
8812738b/SOST	–	109–109	110	111–127	0/0	0/1	1
8811513b/UFC1	15	310–323	324	325–328	1/0	0/0	1
8811561b/UBE2T	19	537–554	555		0/0	2/0	1
8810642b/UBE2D3	9	284–291	292	293–302	0/0	1/0	1
8812124b/UBE2E3	19	383–400	401		0/0	0/0	0
8811205b/UBE2H	–	452–455	456	457–470	0/0	0/1	1
8811017b/RAB6IP1	7	3046–3051	3052	3053–3064	0/1	0/0	1
8810855b/UBE2 M	5	69–72	73	74–87	0/1	0/1	0
8812094b/UBE2E3	19	563–580	581		0/0	2/0	1
8812027b/UBE2G1	–	425–435	436	437–443	0/0	0/0	0
8811703b/UBE2 W	–	221–226	227	228–239	0/0	0/0	0
8817021b/KCNT1	19	1336–1353	1354		0/0	1/0	1
8811319b/UBE2I	19	116–133	134		0/0	0/0	0
8811622b/UBE2T	8	184–190	191	192–202	0/0	1/0	0
8817088b/DVL2	15	785–798	799	800–803	0/0	0/0	0
8810760b/UBE2B	–	187–188	189	190–205	0/0	0/1	1
8811283b/UBE2I	19	313–330	331		0/0	0/0	0

Annexure 3.6: Fold Residue Prediction

Section 3.6.2 showed a partial table in Table 3.9 with a partial number of residues. The next table shows the complete results for prediction of Fold Residue Location (FRL) for mRNA codon string of length 217, transcribed out of Gene ID CAA33815, Uniprot ID for the protein synthesized is P13342, PDB ID 1C1K.

(1) Pos	(2) IFRL prediction	(3) VFRL fold points	(4) SS motifs as noted in PDB	(5) Comment	(1) Pos	(2) IFRL prediction	(3) VFRL fold points	(4) SS motifs as noted in PDB	(5) Comment
1			–		110			H	
2			E		111			H	
3	1*	1	E	True	112			H	
4			E		113			H	
5			–		114			H	
6			–		115			H	
7			–		116			H	
8			S		117			H	
9			–		118			H	
10	1*	1	T	True	119			H	
11			T		120			H	
12			–		121			H	
13			–		122			H	
14			–		123			H	
15	1*	1	–	True	124			H	
16			H		125			H	
17			H		126			H	
18			H		127			T	
19			H		128			T	
20			H		129			–	
21			H		130			S	
22			H		131			S	
23			H		132			G	
24			H		133			G	
25			H		134			G	
26			H		135			G	
27			H		136			T	
28			H		137			S	
29			H		138			–	
30	1*	1	H	True	139			B	
31			T		140			T	
32			T		141			T	

(continued)

(continued)

(1) Pos	(2) IFRL prediction	(3) VFRL fold points	(4) SS motifs as noted in PDB	(5) Comment	(1) Pos	(2) IFRL prediction	(3) VFRL fold points	(4) SS motifs as noted in PDB	(5) Comment
33	1*	1	S	True	142			T	
34			–		143			T	
35			–		144			B	
36			T		145			–	
37			T		146			H	
38			T		147			H	
39			T		148			H	
40			T		149			H	
41			T		150			H	
42			–		151	1*	1	H	True
43			–		152			H	
44			–		153			T	
45	1*	1	–	True	154	1*	1	T	True
46			–		155			S	
47			H		156	1*	1	S	True
48			H		157			–	
49			H		158			H	
50			H		159			H	
51			H		160			H	
52			H		161			H	
53			–		162			H	
54			S		163			H	
55			–		164			H	
56			H		165			H	
57			H		166			H	
58			H		167			H	
59			H		168			H	
60			H		169	1*	1	–	True
61			H		170			H	
62			H		171			H	
63			H		172			H	
64			H		173			H	
65			H		174			H	
66			–		175			H	
67			–		176			H	
68			H		177			H	
69			H		178			–	
70			H		179			–	

(continued)

(continued)

(1) Pos	(2) IFRL prediction	(3) VFRL fold points	(4) SS motifs as noted in PDB	(5) Comment	(1) Pos	(2) IFRL prediction	(3) VFRL fold points	(4) SS motifs as noted in PDB	(5) Comment
71			H		180			–	
72			H		181			H	
73			H		182	1*	1	H	True
74			H		183			H	
75	1*		H	False	184	1*		H	False
76			H		185			H	
77			H		186			H	
78			H		187			H	
79			H		188			H	
80			H		189			H	
81			H		190			H	
82			–		191			H	
83			H		192			H	
84			H		193			H	
85			H		194			H	
86			H		195			H	
87	1*	1	S	True	196			H	
88			S		197			E	
89			–		198			E	
90			G		199			E	
91			G		200			–	
92			G		201	1*	1	H	True
93			T		202			H	
94			T		203			H	
95			H		204	1*		H	False
96			H		205			H	
97			H		206			H	
98			H		207			H	
99			H		208			H	
100			H		209			H	
101			H		210			H	
102			H		211			H	
103			H		212			H	
104			H		213			H	
105			H		214	1*	1	H	False
106			H		215			H	
107	1*	1	H	True	216			H	
108			T		217			–	
109			H						

References

1. Clancy, S., Brown, W.: Translation: DNA to mRNA to protein. Nat. Educ. **1**(1), 101 (2008)
2. Tuller, T.: Selected Publications. Tel Aviv University, www.cs.tau.ac.il/~tamirtul/Sublinkes/Tuller_Publications.html
3. Diament, A., Tuller, T.: Estimation of ribosome profiling performance and reproducibility at various levels of resolution. Biol. Direct **11**(1), 24 (2016)
4. Stothard, P.: The sequence manipulation suite: JavaScript programs for analyzing and formatting protein and DNA sequences (2000)
5. Guglielmi, L., et. al.: Expression of single-chain Fv fragments in E. coli cytoplasm. Methods Mol Biol. 215–224 (2009)
6. Babenko, A.P., Polak, M., Cavé, H., Busiah, K., Czernichow, P., Scharfmann, R., Bryan, J., Aguilar-Bryan, L., Vaxillaire, M., Froguel, P.: Activating mutations in the ABCC8 gene in neonatal diabetes mellitus. N. Engl. J. Med. **355**(5), 456–466 (2006)
7. Bowman, P., et al.: Heterozygous ABCC8 mutations are a cause of MODY. Diabetologia **55**(1), 123–127 (2012)
8. Gingold, H., Pilpel, Y.: Determinants of translation efficiency and accuracy. Mol. Syst. Biol. **7**(1), 481 (2011)
9. Iwakawa, H., Tomari, Y.: The functions of microRNAs: mRNA decay and translational repression. Trends Cell Biol. **25**(11), 651–665 (2015)
10. Krol, J., et al.: Structural features of microRNA (miRNA) precursors and their relevance to miRNA biogenesis and small interfering RNA/short hairpin RNA design. J. Biol. Chem. **279**(40), 42230–42239 (2004)
11. Biao, Liu et. al.: Analysis of secondary structural elements in human microRNA hairpin precursor. BMC Bioinform. **17**, 112 (2016)
12. Fernandez, N., et al.: Genetic variation and RNA structure regulate microRNA biogenesis. Nat. Commun. **8**, 15114 (2017)
13. Lambe, A.T., et al.: Transitions from functionalization to fragmentation reactions of laboratory secondary organic aerosol (SOA) generated from the OH oxidation of alkane precursors. Environ. Sci. Technol. **46**(10), 5430–5437 (2012)
14. Deigan, K.E., et al.: Accurate SHAPE-directed RNA structure determination. Proc. Natl. Acad. Sci. **106**(1), 97–102 (2009)
15. Belter, A., et al.: Mature miRNAs form secondary structure, which suggests their function beyond RISC. PLoS ONE **9**(11), e113848 (2014)
16. Griffiths-Jones, S., et al.: miRBase: tools for microRNA genomics. Nucleic Acids Res. **36**(suppl_1), D154–D158 (2007)
17. Gibb, E.A., et al.: Human cancer long non-coding RNA transcriptomes. PLoS ONE **6**(10), e25915 (2011)
18. Rohloff, J.C., et al.: Nucleic acid ligands with protein-like side chains: modified aptamers and their use as diagnostic and therapeutic agents. Mol. Ther.-Nucleic Acids **3**, e201 (2014)
19. Keeler, A.M., ElMallah, M.K., Flotte, T.R.: Gene therapy 2017: progress and future directions. Clin. Transl. Sci. **10**(4), 242–248 (2017)
20. Reynolds, A., et al.: Rational siRNA design for RNA interference. Nat. Biotechnol. **22**(3), 326 (2004)
21. Riba, A., et al.: Explicit modeling of siRNA-dependent on-and off-target repression improves the interpretation of screening results. Cell Syst. **4**(2), 182–193 (2017)
22. Chalk, A.M., et al.: siRNAdb: a database of siRNA sequences. Nucleic Acids Res. **33**(suppl_1), D131–D134 (2005)
23. Carthew, R.W., Sontheimer, E.J.: Origins and mechanisms of miRNAs and siRNAs. Cell **136**(4), 642–655 (2009)
24. Gerstein, M., Tsai, J., Levitt, M.: the volume of atoms on the protein surface: calculated from simulation, using Voronoi polyhedra. J. Mol. Biol. **249**, 955–966 (1995). https://doi.org/10.1006/jmbi.1995.0351
25. Loughlin, F.E., et al.: Structural basis of pre-let-7 miRNA recognition by the zinc knuckles of pluripotency factor Lin28. Nat. Struct. Mol. Biol. **19**(1), 84 (2012)

Chapter 4
Cellular Automata Model
for Deoxyribonucleic Acid (DNA)

"The body itself is an information processor. Memory resides not just in brains but in every cell. No wonder genetics bloomed along with information theory. DNA is the quintessential information molecule, the most advanced message processor at the cellular level—an alphabet and a code, 6 billion bits to form a human being".

—James Gleick

4.1 Introduction

On 25 April 1953, Nature magazine published the paper entitled 'Molecular Structure of Nucleic Acids: A Structure of Deoxyribose Nucleic Acid' by Watson and Crick [1]. Since that time, DNA (Deoxyribose Nucleic Acid) is the most popular acronym in Biology. Perhaps its popularity in general media is not only due to its meticulous and simultaneous competitive discovery (in early 1950s) but the symmetry and beauty of its glorious double helix structural form. Yet this magnificent structure that reads like a book probably depends on its precursor, the RNA, as proposed in 'RNA world hypothesis' referred to in Chap. 3. The three-dimensional symmetrical structure of DNA stores hereditary information in its 'data memory' and has a 'functional memory' to store code for biological functions. In this chapter, we present the modelling of DNA using Cellular Automata (CA). Signal graph derived out of DCAM displays symmetric and regular pattern which is possibly the reflection of DNA's beautiful 3D structure. We will take more time to explore this phenomenon. This chapter has five sections:

- DNA helical structure representation in cLife (CA-based Life)
- Functionally important regions of DNA string
- Exon–intron boundary of mRNA transcript
- CA model for representation of Single Nucleotide Polymorphism (SNP)
- cLife model for sgRNA binding on target gene editing site for CRISPR/Cas 9

© Springer Nature Singapore Pte Ltd. 2018
P. P. Chaudhuri et al., *A New Kind of Computational Biology*,
https://doi.org/10.1007/978-981-13-1639-5_4

4.2 DNA Helical Structure Representation in cLife

Let us reintroduce the structure of DNA. We can see from Fig. 4.1 how the two sugar–phosphate backbones hold the nucleotides (ATGC). Corresponding CAM should be designed to model this physical domain feature. Figure 4.2 shows the molecular structure of double strand of DNA, and Fig. 4.3 shows the molecular structure of sugar–phosphate backbone of DNA with attached nucleotide base.

What is cLife Depicting?

When Cellular Automata Machine (CAM) runs, we obtain a series of parameters which we have introduced in Chap. 2 and derived for RNA in Chap. 3. These CA parameters hold good for all macromolecules (DNA, RNA and protein). The graph itself has no physical resemblance with the actual molecule; however, our target is to find a certain change of CA parameter Cycle Start (CS) value at a particular nucleotide position, which forms Critical Pairs (CrPs) between the values of CS (Y-axis) against nucleotide position which has correspondence or correlation in wet lab modelling (known verified data). Thus, a signal graph represents a structural property of the macromolecule being modelled.

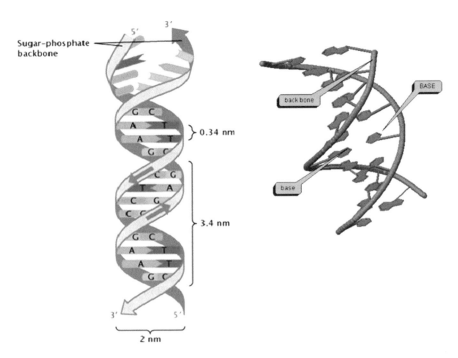

Fig. 4.1 Helical structure of DNA

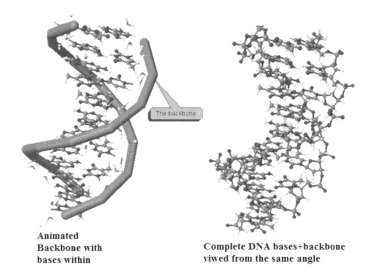

Fig. 4.2   Molecular structure of double strand of DNA

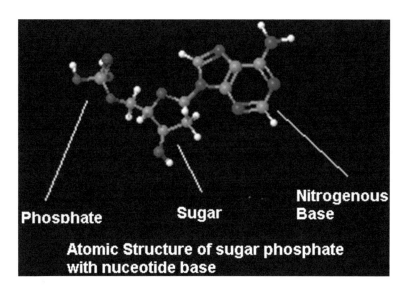

Fig. 4.3   Molecular structure of sugar–phosphate backbone of DNA with attached nucleotide base

Section 2.6 of Chap. 2 reports the design of 5NCA rule for sugar–phosphate molecule and nucleotide bases of DNA. This design conforms to the basic principle of bottom-up approach for designing CA rules for the building blocks of DNA molecule (sugar–phosphate and nucleotide bases A, C, G, T) from the information content of their molecular structure.

As described in Chap. 2, three CAMs are designed with five-neighbourhood CA (5NCA) rules:

- Backbone DNA CA Machine (BBDCAM)—it is a uniform 5NCA representing CA model for DNA sugar–phosphate backbone.
- Nucleotide Base DNA CA Machine (NBDCAM)—it is a hybrid 5NCA representing CA model for nucleotide base sequence in a DNA strand.
- Composite DNA CA Machine (ComDCAM) derived out of co-evolution of BBDCAM and NBDCAM to represent a DNA strand covering sugar–phosphate backbone and bases covalently bonded to backbone sugar molecule.

DNA molecule is modelled in cLife with these three CAMs. From the evolution of CAMs, we derived Cycle Start (CS)/Cycle Length (CL) parameter values, CS signal graph and 0/1 evolution pattern. The output of different CAMs for Gene ID J00272 with 607 bases is illustrated for its full length in Annexure 4.1 for both sense and antisense strands. We represent partial tables and graphs in the main text. Table 4.1a–c displays the CS/CL parameter values and sum of Rule Min Terms (RMTs) in a cycle for each CA cell for the serial base locations 300–350 of three CAMs. Table 4.2a–c shows the corresponding values of the antisense strand of the Gene ID J00272.

Cycle Start (CS) signal graph and 0/1 evolution pattern for a hypothetical DNA strand are reported in Sect. 2.6 of Chap. 2. Figure 4.4a–c shows the partial CS signal graph derived out of the evolution of BBDCAM, NBDCAM and ComDCAM for the DNA sense strand (CS parameter values tabulated in Table 4.1). The 0/1 evolution pattern derived out of these three CAMs are reported in Fig. 4.5a–c, respectively, for sense strand. The corresponding figures for antisense strand are reported in Fig. 4.6a–c showing CS signal graphs (CS parameter values tabulated in Table 4.2), and 0/1 evolution patterns shown in Fig. 4.7a–c. The binary pixels shown in 0/1 evolution is noted with black for binary value '1' and white for '0'.

BackBone DNA CA Machine (BBDCAM) is a uniform 5NCA with same rule for each cell representing a sugar–phosphate molecule linked to a nucleotide base. The seed for the cells for all the CAMs are set as '**1**' if the cell represents a purine base (A or G) and '**0**' for pyrimidine base (C or T). The Nucleotide Base DNA CA Machine (NBDCAM) is a hybrid CA with a CA cell representing a base (A, C, G, T) with its 5NCA rule as noted in Table 2.5 of Chap. 2. While designing CA rules for different building blocks for biomolecules in Chap. 2, we pointed out the fact that both CA rule and seed affect the CA evolution. The evolution of CA cells in transient states, in general, are influenced by the CA rule assigned to the cell, while seed determines the switching point from transient to cyclic states in a cell (that is Cycle Start (CS) parameter value). Consequently, evolution pattern for sense and antisense strands are identical with CS values steadily decreasing by 2 after a group

Table 4.1 Partial sense strand CS/CL table for Gene ID J00272

Position	Nucleotide base	(a) BBDCAM			(b) NBDCAM			(c) ComDCAM		
		CS	CL	Cycle total	CS	CL	Cycle total	CS	CL	Cycle total
300	G	201	34	533	0	3	37	320	68	1520
301	G	199	34	539	0	3	74	320	68	1524
302	G	199	34	550	0	1	28	319	68	1529
303	C	199	34	540	0	1	25	319	68	1509
304	T	199	34	521	0	1	18	319	68	1501
305	G	197	34	515	4	1	4	319	68	1580
306	C	197	34	504	5	1	9	319	68	1547
307	A	197	34	514	5	1	18	318	68	1580
308	T	197	34	533	10	1	4	319	68	1576
309	C	195	34	539	11	1	9	319	68	1572
310	T	195	34	550	11	1	18	319	68	1599
311	G	195	34	540	11	1	4	319	68	1620
312	C	195	34	521	12	1	9	319	68	1568
313	A	193	34	515	12	1	18	315	68	1648
314	A	193	34	504	12	1	4	306	68	1685
315	A	193	34	514	12	1	9	306	68	1657
316	G	193	34	533	12	1	18	305	68	1639
317	G	191	34	539	9	1	4	304	68	1530
318	G	191	34	550	10	1	9	304	68	1572
319	G	191	34	540	10	1	18	303	68	1496
320	C	191	34	521	13	1	4	303	68	1532
321	G	189	34	515	13	1	8	302	68	1417
322	T	189	34	504	13	1	16	302	68	1477
323	C	189	34	514	13	1	0	301	68	1468
324	G	189	34	533	13	1	0	301	68	1455
325	G	187	34	539	12	1	0	300	68	1553
326	A	187	34	550	12	1	1	300	68	1587
327	C	187	34	540	12	1	2	300	68	1595
328	A	187	34	521	12	1	4	299	68	1639
329	A	185	34	515	12	1	9	292	68	1566
330	G	185	34	504	12	1	18	292	68	1546
331	T	185	34	514	11	1	4	292	68	1575
332	G	185	34	533	11	1	8	292	68	1505
333	C	183	34	539	11	1	16	292	68	1494
334	A	183	34	550	11	1	1	292	68	1503
335	G	183	34	540	11	1	2	291	68	1580
336	C	183	34	521	10	1	4	291	68	1578
337	T	181	34	515	11	1	9	291	68	1578

(continued)

Table 4.1 (continued)

Position	Nucleotide base	(a) BBDCAM			(b) NBDCAM			(c) ComDCAM		
		CS	CL	Cycle total	CS	CL	Cycle total	CS	CL	Cycle total
338	G	181	34	504	11	1	18	290	68	1539
339	C	181	34	514	12	1	4	290	68	1501
340	T	181	34	533	13	1	9	289	68	1579
341	G	179	34	539	13	1	18	289	68	1606
342	T	179	34	550	20	1	4	288	68	1536
343	G	179	34	540	21	1	9	288	68	1617
344	C	179	34	521	21	1	18	282	68	1519
345	C	177	34	515	22	1	4	282	68	1517
346	T	177	34	504	22	1	8	277	68	1547
347	G	177	34	514	22	1	16	277	68	1477
348	A	177	34	533	22	1	1	277	68	1434
349	T	175	34	539	22	1	2	276	68	1482
350	G	175	34	550	25	1	4	276	68	1505

Table 4.2 Antisense strand CS/CL table (partial) for Gene ID J00272

Position	Nucleotide base	(a) BBDCAM			(b) NBDCAM			(c) COMDCAM		
		CS	CL	Cycle total	CS	CL	Cycle total	CS	CL	Cycle total
300	C	201	34	533	3	1	25	326	68	1501
301	C	199	34	539	3	1	18	326	68	1548
302	C	199	34	550	4	1	4	325	68	1611
303	G	199	34	540	5	1	9	325	68	1611
304	A	199	34	521	5	1	18	324	68	1607
305	C	197	34	515	7	1	4	324	68	1566
306	G	197	34	504	8	1	9	320	68	1584
307	T	197	34	514	8	1	18	320	68	1585
308	A	197	34	533	11	1	4	308	68	1527
309	G	195	34	539	12	1	9	308	68	1535
310	A	195	34	550	12	1	18	308	68	1583
311	C	195	34	540	13	1	4	306	68	1549
312	G	195	34	521	16	1	9	306	68	1579
313	T	193	34	515	16	1	18	306	68	1510
314	T	193	34	504	17	1	4	306	68	1508
315	T	193	34	514	17	1	8	306	68	1498
316	C	193	34	533	17	1	16	305	68	1544
317	C	191	34	539	17	1	0	305	68	1569
318	C	191	34	550	17	1	0	304	68	1653

(continued)

Table 4.2 (continued)

Position	Nucleotide base	(a) BBDCAM			(b) NBDCAM			(c) COMDCAM		
		CS	CL	Cycle total	CS	CL	Cycle total	CS	CL	Cycle total
319	C	191	34	540	10	1	0	304	68	1566
320	G	191	34	521	9	1	0	304	68	1581
321	C	189	34	515	9	1	0	304	68	1543
322	A	189	34	504	8	1	1	290	68	1568
323	G	189	34	514	8	1	2	286	68	1554
324	C	189	34	533	10	1	4	286	68	1492
325	C	187	34	539	11	1	9	280	68	1469
326	T	187	34	550	11	1	18	279	68	1545
327	G	187	34	540	12	1	4	279	68	1535
328	T	187	34	521	13	1	9	278	68	1522
329	T	185	34	515	13	1	18	278	68	1556
330	C	185	34	504	13	4	18	266	68	1467
331	A	185	34	514	13	4	36	234	68	1480
332	C	185	34	533	13	4	74	234	68	1568
333	G	183	34	539	10	4	20	233	68	1520
334	T	183	34	550	10	4	40	233	68	1556
335	C	183	34	540	10	4	16	232	68	1531
336	G	183	34	521	10	4	32	232	68	1510
337	A	181	34	515	5	1	1	231	68	1534
338	C	181	34	504	5	1	3	231	68	1582
339	G	181	34	514	4	1	7	230	68	1554
340	A	181	34	533	6	1	15	230	68	1527
341	C	179	34	539	6	1	30	229	68	1572
342	A	179	34	550	6	1	28	229	68	1595
343	C	179	34	540	6	1	25	228	68	1639
344	G	179	34	521	6	1	18	228	68	1600
345	G	177	34	515	7	1	4	222	68	1616
346	A	177	34	504	10	1	9	222	68	1547
347	C	177	34	514	10	1	18	222	68	1510
348	T	177	34	533	14	1	4	220	68	1503
349	A	175	34	539	14	1	8	220	68	1457
350	C	175	34	550	14	1	16	220	68	1425

of four bases displaying same CS value. This pattern is clearly visible in Tables 4.1a and 4.2a and the corresponding CS signal graphs of BBDCAM reported in Figs. 4.4a and 4.6a. Section 4.2 reports further analysis of this pattern and its relationship with physical domain structural features. Patterns/parameters

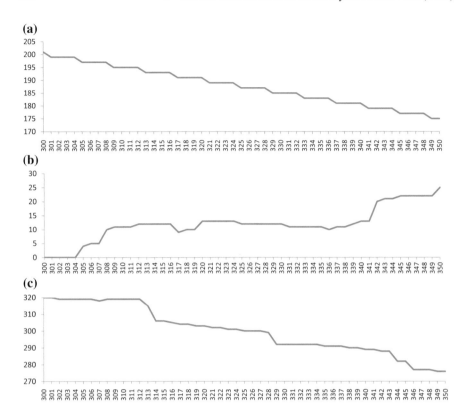

Fig. 4.4 **a** Partial BBDCAM CS signal graph for sense strand (*Y*-axis: CS; *X*-axis: nucleotide base location), **b** partial NBDCAM CS signal graph for sense strand (*Y*-axis: CS value; *X*-axis: nucleotide base location), and **c** partial ComDCAM CS signal graph for sense strand (*Y*-axis: CS; *X*-axis: nucleotide base location)

derived out of ComDCAM (or NBDCAM) evolution are analysed for top-down validation of CA model in subsequent sections of this chapter.

Evolution of Backbone DNA CA Machine (BBDCAM) models the helical structure of DNA molecule. A few observations are noted below with respect to this model from the study of CS/CL graph for any region of a gene, as illustrated for sense and antisense strands of Gene ID J00271 in Figs. 4.4, 4.6 and Tables 4.1, 4.2.

Observation 1: The BBDCAM output displays decreasing Cycle Start (CS) values and constant Cycle Length (CL) values. Note the change in CS values for successive locations; if four successive cell locations i, $(i + 1)$, $(i + 2)$, $(i + 3)$ have the same CS value (say x), then next four location $(i + 4)$, $(i + 5)$, $(i + 6)$,

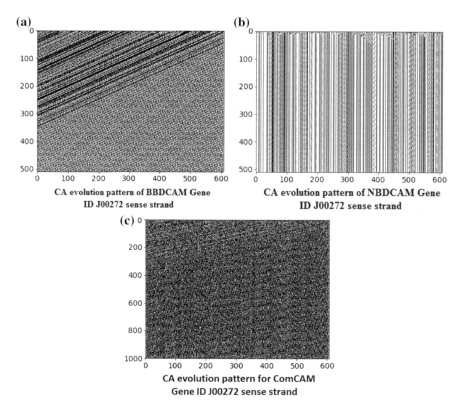

Fig. 4.5 a BBDCAM 0/1 evolution pattern for sense strand (Y-axis: time step of evolution; X-axis: nucleotide base location), **b** NBDCAM 0/1 evolution pattern for sense strand (Y-axis: time step of evolution; X-axis: nucleotide base location), and **c** ComDCAM 0/1 evolution pattern for sense strand (Y-axis: time step of evolution; X-axis: nucleotide base location)

$(i + 7)$ have CS value $(x - 2)$. This condition, as noted in Fig. 4.1a and Table 4.1a, is valid for any successive eight locations across the gene length.

Observation 2: Although RMTs for each CA cell differ, the sum of RMT values denoted as CycTot in the cycles for eight successive cells also repeat after eight time steps. For the example gene (NCBI ID J00271 Table 4.1a), the CS values for locations 305–308 are 197, while for 309–312, the CS value is 195. The CycTot values for the corresponding eight cells are ((515, 504, 514, 533), (539, 550, 540, 521)). The repetition of CycToT values after every eight cell blocks, as reported in Fig. 4.4a and Table 4.1a, is valid throughout the length of the BBDCAM representing the gene.

Observation 3: CycTot values repeating after every eight cell locations can be represented as sinusoidal wave of period 8. However, on considering the spatial scale of cell locations, it can be observed that after every four cell locations, the CS

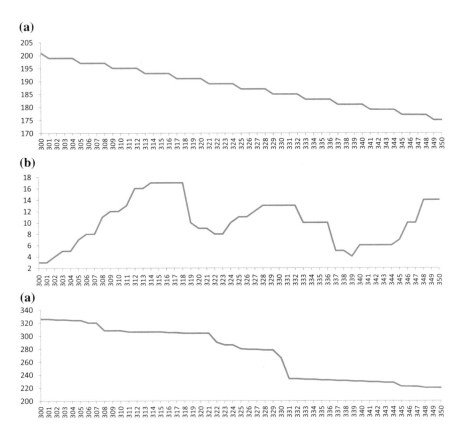

Fig. 4.6 a BBDCAM CS signal graph for antisense strand (*Y*-axis: CS value; *X*-axis: nucleotide base location), **b** NBDCAM CS signal graph for antisense strand (*Y*-axis: CS value; *X*-axis: nucleotide base location), and **c** ComDCAM CS signal graph for antisense strand of Gene ID J00272 (*Y*-axis: CS value; *X*-axis: nucleotide base location)

value gets reduced by a value 2. This condition is valid for any successive eight locations.

Observation 4: In a temporal scale, the CycTot graph is synchronized with a time period of 8. Further in the spatial domain, after every four cells, the Cycle Start (CS) value decreases by 2. As a result, the CS value between a pair of adjacent blocks of 4 CA cells modeling a pair of blocks of 4 adjacent nucleotide bases decreases by value 2. Hence, it is observed that in a spatio-temporal scale BBDCAM output gets synchronized with a time period of 8 + 2 = 10.

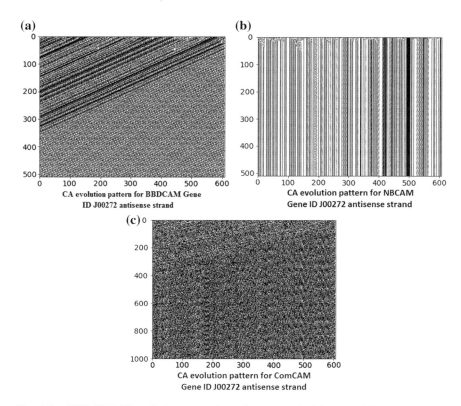

Fig. 4.7   a BBDCAM 0/1 evolution pattern for antisense strand of Gene ID J00272 (*Y*-axis: time step of evolution; *X*-axis: nucleotide base location), **b** NBDCAM 0/1 evolution pattern for antisense strand (*Y*-axis: time step of evolution; *X*-axis: nucleotide base location), and **c** ComDCAM 0/1 evolution pattern of antisense strand of Gene ID J00272 (*Y*-axis: time step of evolution; *X*-axis: nucleotide base location)

Observation 5: The four observations noted above refer to sense strand of gene. These are also valid, as noted in Table 4.2a and Fig. 4.6a, for the antisense strand of the gene.

Observation 6: Figure 4.1 shows the typical double helix structure of a DNA strand. It shows the physical distance between a pair of bases as 0.34 nm. After every four bases, both the sense–antisense strands take helical turn associated with two base locations. Consequently, after a spatial displacement of 10 base locations the helical structure, as shown in Fig. 4.2, it repeats at a distance of 10×0.34 nm = 3.4 nm.

Insights from the CA Model of DNA

Figure 4.1 shows the schematic structure of the DNA double helix. Figure 4.2 shows the molecular structure on the left side where we have shown the sugar–phosphate backbone coloured in orange, emphasizing its contribution in forming the helical structure.

A complete helix is formed with eight nucleotide pairs. In this case, it is equivalent to the two-dimensional representation of one wavelength (sinusoidal). The 'wavelength' itself is 3.4 nm (nanometer). It was calculated by Crick and subsequently validated from X-ray crystallography and other methods. The distance between two nucleotide bases is 1/10th, i.e., 0.34 nm.

We have mentioned that the Cellular Automata Machine (CAM) when targeted for a particular molecule is a representation that is correlated with real-world measurements. Once we have a high confidence level of correlation, we apply for blind testing (unknown) and validate that CAM. The maximum Cycle Start (CS) value is basically the number of steps taken by the full CAM to complete the equivalent structure. However, if we view an enlarged portion of the CS graph of BBDCAM, we seem to have a spatial representation of the double helix backbone.

The CS pattern of the Backbone DNA Cellular Automata Machine (BBDCAM) graph shows clearly the inherent regular structure of the DNA backbone structure.

In the above context, we make our final observation as follows. An alternative representation of 3D structure of DNA helical structure can be observed in the spatio-temporal evolution of BBDCAM for any DNA strand (sense and antisense) across the length of the strand. The meaning of 'spatio-temporal synchronization', as noted earlier, needs some elaboration. So far as CA evolution is concerned, we have a dimensional reduction (three dimensions to two dimensions) in our CS graph. The CS (cycle Start) parameter does not represent any physical domain time; at best one can conjecture that CA evolution captures some structural features of a biomolecule. These characteristics of our CA model can be observed in Chaps. 3, 4 and 5.

4.3   Functionally Important Regions of DNA

The experimental methods for identifying gene location are not based on direct examination of DNA molecules but instead rely on detection of the RNA molecules that are transcribed from genes. All genes are transcribed into RNA [2].

In CA studies, we fall back into mRNA transcript to study the DNA behaviour and associated properties. This is because RNA is the dynamics biology dictated by information coded on the DNA. We could compare the DNA as 'instruction set' in a not so volatile memory; DNA is less fragile, which it should be as it keeps the information of hereditary for future generation, but the external action can only be seen through the mRNA that is created. That is why present biology is seen as a cycle DNA → RNA → Protein. But, as mentioned before, any cycle must start somewhere. Authors of this book point to the fact that this cycle started with RNA, the RNA → Protein experiment went on through unknown millions of years till the chemical combination dropped an oxygen atom in its sugar–phosphate molecule and found a stable formation of chemicals that will enable inheritance. Before the advent of DNA, the combination of RNA and other organic complexes should have been far more dynamic but there was little control. Any cyclic process is a stable system. DNA → RNA → Protein would thus be a stable system which would replicate and also have both positive and negative feedback and as well as enhancing and suppressing mechanism. Perhaps it was not yet life, but it was a chemical soup which could recombine and thus replicate. A small mutation in the memory (DNA) would give varieties; some varieties through millions of years of collision dynamics may have given rise to what is now known as the simplest single cell organism.

In the case where the gene is discontinuous, the primary transcript is processed to remove the introns and link the exons. Techniques that map the positions of transcribed sequences in a DNA fragment could be used to locate exons and entire gene. The transcript is longer than the coding part of the gene. Usually, it begins several tens of nucleotides upstream of the initiation codon and continues several tens or hundreds of nucleotides downstream of the termination codon. Consequently, transcript analysis does not provide a precise definition of the start and end of the coding region of a gene. However, it does point to the fact that a gene is present in a particular region, and it can locate the exon–intron boundaries.

A gene is the basic physical and functional unit in a DNA strand. It stores instructions to make molecules called proteins. The discipline of genetics covers genes, its variations and heredity information of living organisms. Subsequent to Gregor Mandel's work in late nineteenth century [3], genetics study received the attention of a large number of researchers. There exist large volumes of published literature and databases that cover a rich repertoire of results on different areas of genetics. The human genome project has further given a big leap forward in this field to study the structural and functional characteristics of genes across different species. In this context, we reproduce below different modules of NCBI gene prediction tool Genomon [4]. Figure 4.8 shows different modules of this tool. NCBI

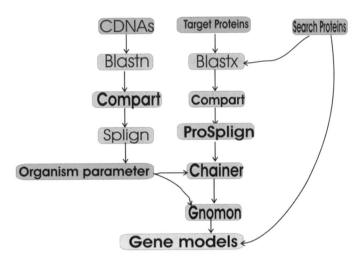

Fig. 4.8 NCBI gene prediction

gene prediction employs both homology searching and ab initio modelling from first principle.

Gene transcription is a highly complex biological process involving a large number of proteins and also small length siRNA and miRNA presented in Chap. 3. Figure 4.9 displays different functionally important regions of an RNA transcript (simply noted as transcript in subsequent discussions) derived out of a gene in a DNA strand. It covers enhancer/silencer and promoter regions which are the binding sites for different protein complexes. Pre-mRNA and mRNA protein-coding regions of Fig. 4.9 have been detailed in Fig. 4.10. Derivation of matured mRNA out of mRNA transcript is reported in Sect. 4.4. The term

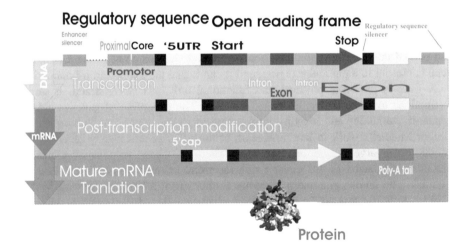

Fig. 4.9 Functionally important regions of an RNA transcript

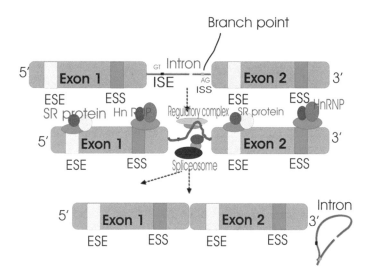

Fig. 4.10 Intron splicing out of pre-mRNA (ESE and ESS are the exon splicing enhancer and silencer; ISE and ISS are the intron splicing enhancer and silencer)

'Transcriptome' refers to the set of all RNA molecules in one cell or a population of cells. Most important components of an RNA transcript are—(i) mRNA codon string derived out of Coding DNA Sequence (CDS) that is translated to a protein chain, (ii) non-coding region out of which essential RNA molecules (tRNA, ribosome, miRNA precursor, etc.) are formed, and (iii) small molecules such as miRNA, siRNA derived out of miRNA precursor molecule. CA model for all these components is reported in Chap. 3.

Figure 4.9 shows the functionally important regions of a DNA [5]. It covers the promoter, enhancer, silencer regions associated with protein binding for control of transcription process. Post-transcriptional modifications generating mRNA transcript displaying intron–exon boundary have been detailed in Fig. 4.10. Finally, with the exclusion of introns, the matured mRNA is formed along with 5' UTR, 5' cap on 5' end and 3' UTR, poly-A tail at 3' end. Each step of the transcription process and post-transcriptional modifications to generate matured RNA is a complex process involving a large number of RNA binding proteins (RBPs).

The concept of bottom-up model building approach for cLife followed by top-down validation has been proposed in Chap. 1. The design of CA rules to represent information content of sugar–phosphate molecules and nucleotide bases (reported in Chap. 2) conforms to our basic modelling approach of bottom-up design of CAMs for DNA molecule. This is followed by top-down modelling of DNA characteristic features with patterns derived out of DNA CAMs.

The functionally important regions of a transcript covering transcription control regions are summarized below. Relevant materials noted in Fig. 4.10 are also covered in this summary.

1. The 5′ UTR begins at the Transcription Start Site (TSS) and ends one nucleotide base prior to start codon (mostly AUG). While in prokaryote the length of the 5′ UTR is usually three to ten bases long, while for eukaryotes, the length may vary from one hundred to several thousand bases.

2. In addition to the number of bases covered, the elements of a eukaryotic and prokaryotic 5′ UTR differ significantly. While the eukaryotic 5′ UTR contains a sequence referred to as KCS [6] (Kozak consensus sequence—ACCAUGG) covering a start codon, the prokaryotic 5′ UTR contains a Ribosome Binding Site (RBS) covering a base sequence referred to as the SDS [7] (Shine Dalgarno sequence—AGGAGGU) of length three to ten base pairs from the start codon.

3. In order to regulate translation process, the eukaryotic 5′ UTR contains regulatory element called upstream Open Reading Frames (uORFs) covering start and stop codons (Fig. 4.11(b)). Unlike prokaryotes, a large percentage (around 35%) of human eukaryotic 5′ UTRs displays introns in between a pair of exons.

4. Figure 4.10 displays the activation of enzyme complex 'spliceosome' for splicing of introns out of mRNA to derive matured mRNA codon string ready for translation. As detailed in Fig. 4.10 and reported in different publications [8], most of the introns, in general, start with base pair TG and end with AG. The algorithm we report next utilizes this information from physical domain along with the information derived out of signal graph generated by the CA model.

5. Splice site location in mRNA points to exon–intron boundary.

The next section reports the CA model for mRNA transcript from which matured RNA is generated on exclusion of introns in between exons. In addition to prediction of exon–intron boundary, the CS signal graph derived out of CAM evolution lays the foundation for analysis of the splicing process, including differential/alternative splicing. The mutational study of DNA viewed as its counterpart on RNA transcript is reported in Sect. 4.5.

4.4   Exon–Intron Boundary of mRNA Transcript

By convention, a sequence of nucleotide bases in a transcript is divided into six reading frames, three from 5′ end and three from 3′ end. Each reading frame covers a sequence of consecutive non-overlapping base triplets. The base triplet, out of which an amino acid is synthesized, as presented in Chap. 3, is referred to as codon. The Coding DNA Sequence (CDS) covers a codon string that gets translated to a protein chain. A CDS starts with a start codon (ATG) and stops with a stop codon (TAG/TAA/TGA). The start codon and stop codon of a CDS is covered by the same reading frame.

The matured mRNA ready for translation is derived on splicing of introns by an enzyme referred to as 'spliceosome'—a large and complex molecular machine found in cell nucleus of eukaryotic cell. There are prokaryotic mRNA transcripts and also some eukaryotes transcripts where introns are absent. However, in general, eukaryotic transcripts display presence of introns. Through the presence of introns and their alternative splicing, nature has realized wider biodiversity in the evolution of different species. Further, even for the same species, alternative splicing allows synthesis of multiple protein products out of the same gene, each having different amino acid sequences. The regulated process of differential splicing allows inclusion or exclusion of a particular exon in the synthesized protein in different tissues or under some specific physical condition in living cells. Such a phenomenon allows nature to efficiently encrypt information. However, decryption of such encoded information, specifically in the context of prevailing physical condition in living cells is a challenging task. Identification of genetic variants affecting splicing in RNA sequencing population studies is still in its infancy [9].

With the advances in the experimental analysis of gene databases derived out of human genome project, identification of exon–intron boundary has become a routine process. However, synthesis of multiple proteins out of same gene across the genome has attracted the attention of many researchers [10] who reported the study of different aspects of differential splicing including characterization of introns for specific organism. A recent survey [11] addressed the issue of evolutionary pressure across genome from the study of 72 eukaryotic organisms.

At present exon–intron boundary prediction is more of a 'hit and miss' process [12]. Based on the CA model, this section reports an in silico framework alternative to the current approaches published in the literature. However, rather than exon–intron boundary prediction, the focus of this section is to present a framework of signal graph analytics of CA-based model that provides a foundation to study different aspects of differential splicing. We hope this approach will reduce cost and time for wet lab experiment to identify alternative splicing (AS).

Figure 4.11 displays a pre-mRNA and the matured mRNA with splicing of introns. Prerequisite for design of algorithm to predict exon–intron boundary is to study published paper and wet lab results including the information available in NCBI/HGNC databases [13]. The algorithm has been designed by collecting all the relevant information from the physical domain along with the analysis of the parameters derived out of signal graph of ComRCAM modelling an mRNA transcript. Experience gained from the analysis of signal graphs in Chap. 3 has enabled us to design the algorithm for prediction of exon–intron boundary.

In the review article by Patricia et al. [14], the authors reported important aspects of 5′ UTR-mediated regulation site noted in Fig. 4.11a, while Fig. 4.11b shows the layout of mRNA protein-coding region derived out of Pre-mRNA. The mRNA coding region covers 5′ cap, 5′ UTR (UnTranslated Region) on 5′ end and 3′ UTR and poly-A tail on 3′ end (Fig. 4.11b). A few relevant information of 5′ UTR is retrieved from published literature [15] and shown in Fig. 4.11. This information and analysis of CS/CL signal graphs for mRNA nucleotide base sequence guide the design of the algorithmic steps for prediction of boundary between exon and intron.

The variables and acronyms used in the algorithm are presented next. We have considered genes where an intron in a transcript is bounded with GT and AG on 5′ and 3′ ends with standard start, stop codons.

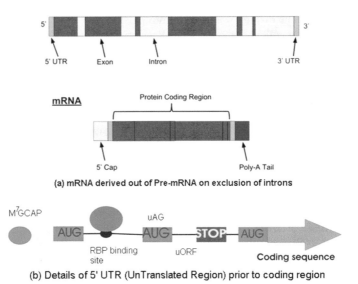

(a) mRNA derived out of Pre-mRNA on exclusion of introns

(b) Details of 5' UTR (UnTranslated Region) prior to coding region

Fig. 4.11 Pre-mRNA and coding region of matured mRNA (*Note*—base T of gene is replaced by U in mRNA transcript)

1. A start/stop codon triplet or GT/AG base pair (on the 5'/3' end of an intron) in a transcript is denoted by the sequence number of its leftmost base noted in the transcript sequence. While base triplet ATG denotes the start codon, three stop codons are TAA, TAG and TGA.

2. The last base of an exon (say x = A/T/C/G) is added to GT as xGT for characterizing differentiation of end of different classes of exons. Similarly, the first base (say y) of an exon is added at the end of AG as AGy (y = A/T/C/G).

3. The variable W (W = ATG/xGT/AGy/stop codon) is used to denote any one of these three-letter codes that are represented as a triplet (say) <v1 v2 v3>, where v1, v2 and v3 represent a numerical value of cycle start difference between a pair of adjacent CA cells (CSD) derived out of CS/CL table of ComRCAM for the transcript sequence.

4. The stop codon in a transcript is a valid one if the total number of bases in the Coding DNA Sequence (CDS) (after exclusion of bases covered by introns) is divisible by 3 in order to ensure that start and stop codon are in the same reading frame. For example, for NCBI Gene (Nucleotide) ID J00277, the start and stop codons are at base locations 1664 and 3350 covering 1687 bases, while the number of bases covered by three introns is (267 + 153 + 697) = 1117. Hence, the number bases in CDS is (1687 − 1117) = 570 divisible by 3. Hence, stop codon at location 3350 is a valid one.

5. A gene transcript displays multiple ATG and stop codons. It also shows multiple GT and AG pair that defines an intron boundary. The algorithm identifies valid W by employing the Splicing Rule (SR) set characterized by a set of parameters reported in Table 4.3.

Table 4.3 Splicing Rule (SR) set—valid and invalid Splicing Rule (SR) set for start codon ATG

1	2	3	4		5		6				7	8	9	10
ID	ATG	Location of ATG	CSDF start	CSDF end	5' end val	3' end val	Pattern 1	Pattern 2	Pattern 3	Pattern 4	CL value	Loc of CL	ATG Pos in CSDF	DF CSD sign
(a) Valid ATG SR set. Valid W (W = ATG) table														
J00271	−1:0:−1	838	826	842	−4	−4	0	47	47	6	136	CLF0	3'	−ive −ive
J00277	0:0:0	1664	1615	1672	−4	1	16	45	33	7	272	CLF1	3'	−ive +ive
X00806	0:−1:0	1086	1054	1089	15	3	0	83	11	6	136	CLF0	3'	+ive +ive
X03821	−1:0:−1	279	258	284	7	−4	0	67	22	11	272	CLF1	3'	+ive −ive
X15227	−1:0:−1	505	462	508	3	−4	0	68	26	6	272	CLF2	3'	+ive −ive
X15863	0:0:−1	1	0	12	0	−21	23	15	38	23	272	CLF0	5'	+ive +ive
X54518	0:0:0	1744	1740	1747	−110	2	0	0	75	25	272	CLF3	3'	−ive −ive
X65606	0:0:−6	514	513	516	−4	−6	0	0	50	50	272	CLF0	3'	−ive −ive
X69215	0:−1:0	443	439	453	−4	−4	0	80	0	20	136	CLF1	5'	−ive −ive
Z12021	−2:0:0	1857	1827	1876	−16	−8	48	4	30	18	544	CLF3	5'/3'	−ive −ive
AF038869	0:−1:0	73	67	85	−14	−4	0	84	0	16	136	CLF0	5'	−ive −ive
AF053233	−3:0:0	25	18	30	−14	−17	23	15	38	23	272	CLF1	3'	−ive −ive
AF092923	−1:0:−1	328	315	386	−23	−12	0	97	0	3	68	CLF0	5'	−ive −ive
BC093995	0:−1:0	42	0	77	0	−4	0	95	3	3	68	CLF0	5'/3'	+ive +ive
DQ168992	−1:0:−1	504	490	509	−18	−65	0	40	40	20	136	CLF1	3'	−ive −ive
J00272	0:0:−8	163	162	165	−4	−8	0	0	50	50	136	CLF1	3'	−ive −ive
(b) Partial invalid ATG SR set similar to (a)														
J00271	−1:0:−1	45	28	49	8	−4	0	73	18	9	272	CLF0	3'	+ive +ive
J00271	−1:0:−1	166	155	175	−6	−14	0	86	0	14	272	CLF0	5'/3'	+ive +ive
J00271	0:−4:0	193	189	196	−8	−8	0	0	50	50	272	CLF1	3'	+ive +ive
J00271	0:−1:−3	317	307	323	−12	−8	18	24	41	18	272	CLF1	3'	+ive +ive
J00277	−1:0:−1	154	151	166	−4	−35	0	88	0	13	272	CLF0	5'	+ive +ive

(continued)

Table 4.3 (continued)

1	2	3	4		5		6				7	8	9	10
ID	ATG	Location of ATG	CSDF start	CSDF end	5' end val	3' end val	Pattern 1	Pattern 2	Pattern 3	Pattern 4	CL value	Loc of CL	ATG Pos in CSDF	DF CSD sign
J00277	−8:0:0	353	353	381	−8	1	0	34	66	0	272	CLF0	5'	+ive +ive
J00277	0:0:0	367	353	381	−8	1	0	34	66	0	272	CLF0	5'/3'	+ive +ive
J00277	0:−1:0	488	482	512	−23	−4	0	90	0	10	272	CLF0	5'	+ive +ive
J00277	−1:0:−1	1082	1073	1093	−6	−9	0	86	0	14	272	CLF0	5'/3'	+ive +ive
J00277	−2:0:0	1536	1527	1545	−4	−15	47	11	21	21	272	CLF0	3'	+ive +ive
X00806	0:−4:0	49	34	52	−4	−29	0	42	42	16	272	CLF0	3'	+ive +ive
X00806	−7:−8:0	132	132	138	−7	−4	0	0	57	43	272	CLF0	5'	+ive +ive
X00806	0:0:0	187	185	201	−4	−4	0	47	47	6	272	CLF0	5'	+ive +ive
X00806	−1:0:−4	199	185	201	−4	−4	0	47	47	6	272	CLF0	3'	+ive +ive
X00806	−4:0:0	228	228	243	−4	−11	0	13	63	25	272	CLF0	5'	+ive +ive
X00806	0:0:0	269	262	285	−10	−11	25	25	50	0	272	CLF1	5'	+ive +ive
X00806	0:−1:0	355	349	378	−6	−4	0	93	0	7	272	CLF1	5'	+ive +ive
X00806	−1:0:−1	441	428	449	9	−4	0	73	18	9	272	CLF1	3'	+ive +ive
X00806	1:0:0	488	488	493	1	1	0	0	67	33	272	CLF2	5'	+ive +ive
X00806	0:−1:0	619	593	627	−4	1	0	86	11	3	272	CLF2	3'	+ive +ive
X00806	−1:0:0	643	635	654	−10	−11	30	20	45	5	272	CLF2	5'/3'	+ive +ive
X00806	−1:0:−1	709	706	713	−4	−4	0	75	0	25	272	CLF3	5'	+ive +ive
X00806	−1:0:−1	747	729	761	−4	−4	0	67	30	3	272	CLF3	5'/3'	+ive +ive

6. Cycle Length (CL) value for the ComRCAM output of a gene varies. CL value is used as one of the CA parameters to predict intron–exon boundary. A region covering base locations (say) x to y having same CL value is referred to as CL frame (CLF); it is divided into four equal subregions CLF0, CLF1, CLF2, CLF3.
7. The Splicing Rule (SR) set specifies the condition for which a W in a transcript is a valid one.
8. A 'skip' tag is used to implement alternative splicing. The tag represents the physical domain condition under which a specific exon is skipped in an mRNA. Prior to processing an exon, if this tag is 'on', then the exon is skipped in the mRNA.

Algorithm: Predict exon–intron boundary

Input: (i) CS signal graph derived out of CS/CL table of mRNA Transcript of the gene; (ii) The Splicing Rule (SR) set characterized with six parameters illustrated in Table 4.3 and explained below; (iii) 'skip' tag on/off—the tag is set 'on' to skip ith exon (i = 1, 2, –) prior to its processing.
Output: exon–intron–exon boundary identified in Steps 3 and 4, with first exon starting with ATG, and last exon ending with a stop codon (TAA, TAG, TGA) along with location of xGT, AGy on intron boundary.
Steps

1. Find the Cycle Start Difference Frame (CSDF) with Cycle Start Difference (CSD) outside the range of 0 to −3 around each W = {ATG, GT, AG, stop codon}
2. Apply the Splicing Rule (SR) set to locate valid ATG at the start of first exon.
3. Search for valid xGT (to locate the end of last exon search initiated in earlier step/iteration, where x = G/A/T/C is the last base of exon). In parallel search for valid stop codon, if identified, mark all GT as invalid, mark this as the last codon of last exon of matured mRNA and stop, else go to next step.
4. Search for current valid AGy (end of the intron search initiated in last step, y = G/A/T/C is the first base of next exon start loc). If 'skip' tag is on, skip the next exon by searching the next AGy instead of current AGy.
5. Go back to step 3.

   Splicing Rule (SR) set is to Identify Valid W (W ∈ {ATG, xGT, AGy, stop codon}). Different rules of the SR set are illustrated with reference to valid W in NCBI Genes noted in column 1 of Table 4.3a—valid ATG list. The variables used in different columns of the table are explained below. The table specifies six parameters employed in the splicing rule set. Table 4.3a notes the parameters for valid W (W = ATG)—each of the parameters is satisfied for a valid W. Table 4.3b reports a partial list of ATGs derived out of signal graph for transcripts of the genes. The full table is shown in Annexure 4.2. Each entry of Table 4.3b gets invalidated since it fails to satisfy the range of at least one of the six parameters for valid ATG.

Variables and six parameters used for splicing rule set are noted in Table 4.3a along with corresponding column numbers.

- Column 1: NCBI gene ID.
- Column 2: Parameter 1 – W triplet value <v1 v2 v3>, where v1, v2, v3 is an integer value with positive or negative sign; the integer values are mostly positive or negative values with the integer ranging from 0 to 4.
- Column 3: Location of first base of W (for W = ATG, column 3 shows the serial base location of A).
- Column 4: Cycle Start Difference Frame (CSDF) start and end base location.
- Column 5: CSDF is bounded by Critical Pair (CrP) on either end; a Cycle Start Difference (CSD) value outside the region of 0 to −3 is marked as CrP.
- Column 6: Parameter 2—Four different patterns observed in a CSDF (Pattern 1: −q followed by 0s, where q = 2–4), Pattern 2: <−1 0>; Pattern 3: sequence of 0s; and Pattern 4: any pattern other than Pattern 1, or 2, or 3).
- Column 7: Parameter 3—Cycle Length (CL) value of cycle length frame (CLF) covering W.
- Column 8: Parameter 4—Location of W in the subregion of cycle length frame (CLF)—four subregions are marked as CLF0, CLF1, CLF2, CLF3.
- Column 9: Parameter 5—Location of W is marked as 5′ (or 3′) end of CSDF if W is closer to 5′ (3′) end. For large size CSDF, W may be marked as 5′/3′ indicating that W is not close to either 5′ or 3′ end.
- Column 10: Parameter 6—Sign of Cycle Start Difference (CSD) values on the CSDF (four different combinations: (+ive +ive), (+ive −ive), (−ive +ive), (−ive −ive)) where +ive is Positive and −ive is Negative.

Similar to valid/invalid ATG list, we identify valid/invalid list for W = xGT, AGY and stop codon. For GT, we look for xGT (x = A/C/G/T, x is the last base in an exon), and in the process, we search for four different triplets for four different values of x. Same is the approach for four different triplets AGy (y = A/C/G/T, y is the first base in an exon). Minimum length of an exon and intron is assumed to cover at least 12 bases.

Based on the splicing rule set derived for W (W ∈ {ATG, xGT, AGy, stop codon}), the algorithm predicts exon–intron–exon boundary. Manual implementation of the algorithm is illustrated in Table 4.4 along with explanatory note for each of the illustrations. The experimental set up noted above is based on ComRCAM. The set up highlights the approach undertaken in cLife to predict intron-exon boundary. The experiment with NBRCAM is in progress to develop a more refined/efficient scheme.

The definitions of the various columns in Table 4.4 are as follows:

- Column 0: Serial number of illustrations.
- Column 1: Gene ID—ComRCAM operated with the transcript derived out of the gene generating CS/CL table, location of W in the table and associated CA parameters noted in columns 2–9.
- Column 2: W = ATG location predicted as per the six parameters illustrated in Table 4.3a.
- Column (3, 4), (5, 6), (7, 8): Specifies exon–intron and intron–exon boundary with columns 3 and 5 showing the locations of W (xGT and AGy) on the first row for each of the illustrations. Row 2 shows the numerical values of the triplet as noted in ComRCAM CS/CL table. The third row reports the location of Cycle Start Difference Frame (CSDF) covering the triplet, while row 4 specifies the values of Cycle Start Difference (CSD) values on the 5′ and 3′ boundary of the CSDF.
- Column 7: Location of W = stop codon (TAA or TAG or TGA).
- Column 8: Comments are added for better readability of table entries.

Alternative Splicing Rule (ASR) for an Exon–Intron Sequence
Let a sequence be represented as

$$\textbf{exon1---intron1---intron}(i-1)\textbf{---exon}(i)\textbf{---intron}(i)\textbf{---}$$
$$\textbf{exon}(i+1)\textbf{---intron}(i+1)\textbf{--exon}(n)$$

In order to skip ith exon in the matured mRNA, prior to identifying AG to locate end of $(i-1)$th intron, put a 'skip' tag on. If skip tag is on, then step 4 of the algorithm skips the search for valid AG for $(i-1)$th intron; instead it searches for valid AG for the ith intron. In the process, ith exon is skipped. The 'on' status of skip tag setting represents the physical domain condition that demands production of a protein without ith exon. This basic process can be tuned and utilized to implement alternative splicing.

We close this section with a few observations in respect of 'alternative splicing' whereby multiple protein products are synthesized out of the same gene (Fig. 4.12). There are an estimated 20,000–25,000 human protein-coding genes. The estimate of the number of human genes has been repeatedly revised down from initial prediction of 100,000 or more as genome sequence quality and gene finding methods have improved. The HUGO Gene Nomenclature Committee (HGNC) is the only internationally recognized authority for assigning standardized nomenclature to human genes. Currently, the HGNC database contains almost 40,000 approved gene symbols, over 19,000 of which represent protein-coding genes [16]. No doubt, in the years to come, more number of protein-coding genes will be identified. Higher percentage alternative splicing can be observed in higher eukaryotes. This phenomenon establishes the fact that for higher level of species, nature has coded the information in a much more compact form to meet the demand of various physical conditions.

Table 4.4 Predicted exon-intron boundary

S. No.	Gene ID	ATG loc	First xGT location	First Agy location	Second xGT location	Second Agy location	Stop location	Comments
1	X15963	24–26					TAA 459–461	Gene has no intron—hence gene covers one exon at nucleotide bases location 24–461. All the stop codons TAG (at 296, 317, 408), TGA (155, 164, 178, 203), TAA (211, 304, 344, 365) are invalid
2	X03819	46–48	AGT 452–454	AGA 1115–1117			TAA 1216–1218	NCBI reports non-standard start codon CGC (location 13–15) and stop codon ACA (location 1211–1213)—it has one intron at nucleotide base location 453–1116
			Triplet <–9 0 –2>	Triplet <–3 0 0>				
			CSDF (452 459)	CSDF (1109 1124)				
			Value (–9 –7)	Value (–26 –21)				
3	J00271	838–840	GGT 865–867	AGG 1165–1167	AGT 1232–1234	AGG 1434–1436	TGA 1525–1427	Gene has first intron (866–1166) and second intron (1233–1435); second row shows the triplet, and third row shows the CSDF covering the triplet; fourth row shows the CSD value
			Triplet <–1 0 0>	Triplet <0 0 0>	Triplet <0 –9 0>	Triplet <0 –1 0>		
			CSDF (858 874)	CSDF (1161 1185)	CSDF (1229 1236)	CSDF (1420 1438)		
			Value (–4 –4)	Value (–4 –4)	Value (–20 –9)	Value (1 1)		
4	X03821 ComRCAM output	279–281	GGT 455–457	AGG 671–673	GGT 807–809	AGC 956–958	TAG 1188–1190	Gene has first intron (456–672) and second intron (808–957), second row shows the triplet, third row shows the CSDF covering the triplet, and fourth row shows the CSD value
			Triplet <0 –1 1>	Triplet <–12 0 0>	Triplet <–1 0 –1>	Triplet <0 0 –18>		
			CSDF (451 457)	CSDF (671 676)	CSDF (787 820)	CSDF (955 958)		
			Value (–27 1)	Value (–12 –4)	Value (–18 –4)	Value (–4 –18)		

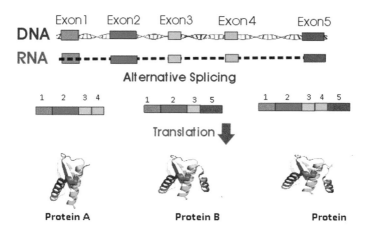

Fig. 4.12 Typical instance of alternative splicing

Alternative splicing is a highly regulated complex process involving a large number of proteins and small size RNA molecules under different prevailing conditions such as developmental and tissue-specific, transduction of foreign genetic materials, cellular stress conditions including temperature, pH values, etc. Production of alternatively spliced proteins is a naturally occurring phenomenon. As a result, any major variations in alternative splicing may lead to different diseases including cancer. In this background, a large number of research groups have started investigating this area [17].

4.5 Representation of Single Nucleotide Polymorphism (SNP) in cLife

This section concentrates on evaluation of mutational effect in a gene due to SNP and its modelling by cellular automata.

Single Nucleotide Polymorphism (SNP)

Subsequent to completion of human genome project two acronyms which have attracted major attention are Single Nucleotide Polymorphism (SNP) and Genome-Wide Association Study (GWAS). SNP refers to a single point variation of nucleotide base in a specific location of genome. SNPs can be observed throughout a person's genome due to various reasons. On an average, one SNP can be observed in every 300 bases of a human genome. That is, around 10 million SNPs may exist in the human genome. Hence, occurrence of the variation for more than 1% of the population is considered as an acceptable SNP. It may appear in a coding or non-coding region of a

DNA. While most of the SNPs do not have any effect, some of these have been observed to be associated with diseases. Our major focus in this chapter is to develop CA model to study disease-related SNPs.

4.5.1   Defining the Problem

SNP study can be employed as a molecular marker to investigate a specific trait/ disease inheritance in a family tree. Further, variation of individual's response to a drug type (for example, sulfa allergy), susceptibility to an environmental factor, risk of developing a particular disease, etc. are likely to be linked to SNP map of an individual. Consequently, the current emphasis on 'personalized medicine' heavily depends on the SNP map in the individual's genome. As a result, genome-wide association study has become an important tool to track genetic variation leading to specific trait not only for humans but across all living organisms of interest. For example, Genome-Wide Association Study (GWAS) is commonly employed to achieve higher yield of different crops.

cLife approach to SNP study

For the study of a specific SNP on a gene, we derive its mRNA transcript. Both ComRCAM and NBRCAM are designed for the nucleotide base sequence of the transcript. CAM outputs are analysed both for Wild (W) and mutated (M) version of the transcript. The CS signal graph is next derived out of Cycle Start (CS)/Cycle Length (CL) output of CAM evolution. The analysis of the signal graph enables identification of CA parameters which can be mapped to physical domain features. These features are co-related to the deviation of mutant from that of wild gene identified either through wet lab study or clinical analysis. Signal graph analytics methodology has been designed out of the analysis of published wet lab/clinical analysis results on disease-related SNP study. On establishing the experimental setup based on such published results for a set of mutants, prediction results are next derived for any arbitrary SNP mutant in the gene. Validation of prediction results may proceed based on reported wet lab results or clinical study not taken into consideration while designing the in silico experimental set for CA model of cLife. The methodology has been implemented in the algorithm reported in Sect. 4.5.2.

CA Model for SNP

In this section, we concentrate on building CA model for SNP in DNA. We have reported in Sect. 4.3 that we study different aspects of DNA from its RNA transcript. We develop the model with base T of DNA replaced with the base U for transcript. So mutation on a gene is mapped on to its mRNA transcript and we study the effect of mutation based on the evolution of CAMs designed for wild and mutated versions of the transcript base sequence. The foundation for such a study has been laid down in the case study reported in Sect. 3.7.2 of Chap. 3 for prediction of difference of the mRNA precursor ID pre-miR-30c [18] wild and its mutant version. The secondary structure of precursor is reported in Sect. 3.6.3 of Chap. 3 for wild and mutant derived out of ComCAM and NBRCAM evolution of the mRNA precursor of the transcript. While NBRCAM, as reported in Chap. 2, is designed out of nucleotide base sequence, ComRCAM is derived out of co-evolution of backbone RCAM (BBRCAM) and NBRCAM. BBRCAM is designed as a uniform CAM with rules designed out of sugar–phosphate molecule.

Although we analyse the evolution of both ComRCAM and NBRCAM to evaluate the mutational effect on a gene, we first evaluate the effect of mutation out of NBRCAM designed out of the base sequence of a transcript. If NBCAM fails to capture the mutational effect, we fall back on ComRCAM output to evaluate the mutational effect. The reverse situation may also exist, that is, ComRCAM may fail to capture the mutational effect. For example, in Sect. 3.7.3 of Chap. 3, we reported that the ComRCAM output for wild and mutant (mutation G27A [18]) are identical (shown in Fig. 4.13). That is, ComRCAM failed to capture the mutational effect. For this case, we evaluated the effect of mutation from NBRCAM.

We evaluate SNP effect from NBRCAM. CS signal graph derived out of NBRCAM evolution for an example transcript sequence (ID pre-mir-30C) wild and mutant version is reported in Fig. 4.14. The signal graphs differ in the region covering base locations 32–54. This deviation of mutant from wild is analysed to derive CA model parameter mapped to physical domain features. If both the CAMs (NBRCAM and ComRCAM) fail to capture the mutational effect we conclude that the given mutation does not affect the gene.

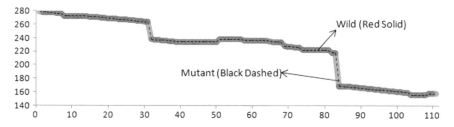

Fig. 4.13 Identical CS signal graph of ComRCAM for wild and mutant (G27A) for precursor ID pre-miR-30c (*Y*-axis: CS value; *X*-axis: nucleotide base location)

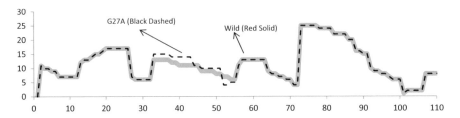

Fig. 4.14 Different CS signal graphs of NBRCAM for wild and mutant (G27A) for precursor ID pre-mir-30c (*Y*-axis: CS value; *X*-axis: nucleotide base location)

4.5.2 Design of Algorithm for Study of Mutational Effect of SNP in a Gene

The algorithm identifies the CA model parameters to be mapped to the physical domain features reported in wet lab analysis or clinical study. Acronyms and variables used in the algorithm are explained next.

- **Wild (W)**: Wild version of mRNA transcript derived out of input gene coding or non-coding region.
- **Mutant (M)**: Mutant M derived out of Wild (W) on replacing a base in W. For coding region, the mutation is synonymous if there is no change in amino acid residue derived out of the matured mRNA codon triplet, else it is marked as non-synonymous. Mutants denoted as xyz (x = base/residue type in W, y = base/residue location, z = base/residue type in M); x, z is a base in the set (A, C, G, U) or any one of 20 common amino acid residues.
- **Cycle Start (CS)**: Cycle start values for a CAM (CA machine) cell location modelling a base; ith base modelled by ith cell location.
- **Cycle Length (CL)**: Cycle length of a CA cell of CAM.
- **Cycle Length Difference (CLD)**: Cycle Length (CL) difference between W and M.
- **Cycle Start Difference (CSD)**: Cycle Start (CS) difference between a pair of adjacent cell locations ($i - 1$) and i of ComRCAM/NBRCAM.
- **CSSG ComRCAM-W (or M)**: CS signal graph derived out of ComRCAM designed for W (or M).
- **CSSG NBRCAM-W (or M)**: CS signal graph derived out of NBRCAM designed for W (or M).
- **CSD Values in ComRCAM**: Mostly negative values with rare cases of positive values; negative CSD values in ComRCAM mostly varies between 0 and -4 with high negative values for some of the cell locations.
- **CSD Values in NBRCAM**: Displays both positive and negative values.

The tables in the case studies reported below also report a parameter referred to as 'critical pair' which is explained next. The cell locations for which CSD values are outside the normal range are viewed as relevant information that are processed in the algorithm.

- **Critical Pair (CrP)**: Critical pair $(i - 1)$ and i marked as CrP(i) (denoted as CrP), where CSD value lies outside the normal range of +1 to −4; CSD table displays CrP locations and value for ComRCAM/NBRCAM
- **Positive/Negative CrP**: It denotes positive/negative value of CrP. All positive/ negative CrPs are marked in the CSD table
- **Difference Frame (DF)**: It is derived by subtracting CSD/CL values of M from that of W for each cell location. The DF is defined as frame/region covering the locations (say) j to k, $(k > j)$, where the difference of CSD and CLD is zero from location 1 to $(j - 1)$ and from location $(k + 1)$ to last base of W/M. CSD in certain locations from j to k may be also zero followed by non-zero values.
- **Sum of Difference (SoD)**: Sum of difference of CrP values between W and M at CrP locations in a difference frame where either W or M display a CrP at a location (say p). Subtract CSD value of M at a CrP location p from the corresponding CSD value of pth location of W. SoD is marked as Positive (P)/ Negative (N) depending on non-zero Positive/Negative CrP of W. If CrP of W is zero, then it is marked as (P)/(N) as per the positive/negative CrP value of M.
- Computation of SoD-P/N, as explained below, demands due consideration with respect to the presence of CrPs on W and M at the same base location. Simultaneous presence of same type of CrPs on same base location indicates that M does not deviate from W for this location and so we consider the difference value only if that is above some threshold as noted below.
- **SoD-P**: SoD for positive CrP values in W and M (ignore if the difference is less than 4 for positive CrPs that exist in same location of W and M.
- **SoD-N**: SoD for negative CrP values in W and M (ignore if the difference is less than 2 for negative CrPs that exist in same location of W and M.

Algorithm: Predict mutational effect of SNP in a gene

Input: mRNA transcript Wild (W) version (coding or non-coding region) derived out of an input gene and its Mutant (M) on insertion of SNP in W.
Output: Prediction of mutational effect in a qualitative scale as high (Hi)/medium (Me)/low (Lo) difference of CA model parameters of M compared to that of W. Signal graph analytics methodology enables mapping of the High/Medium/Low (Hi/Med/Lo) difference of CA parameter values to Hi/Med/Low deviation of physical domain feature(s) of mutated biomolecule compared to wild.
Steps:

1. Evaluate the CSD/CLD output of NBRCAM for W and M to define Difference Frame (DF). For non-zero DF length, go to next step; else evaluate CSD/CLD output of ComRCAM for W and M. If DF length is zero for both NBRCAM and ComRCAM, the given mutation does not affect the gene.

2. Note the CrPs—P/N (positive and negative) for M and W in the DF region.
3. Compute SoD-P/N in DF and derive 'Net Value' on subtracting the value of SoD-N from SoD-P. In case CrP exists for both W and M for same base location, then compute the difference between W and M, and add the difference value to SoD only if difference is (i) more than 3 for positive CrPs for W and M, and (ii) difference is more than 1 for two negative CrPs on W and M.
4. Derive a CA parameter by dividing Net Value by DF length. For evaluation of CA parameter out of ComRCAM, the parameter value is divided by 2.
5. CLD of M and W in the DF adds another CA parameter.
6. The deviation of M from W is marked as high/medium/low based on the difference of CA parameter values between wild and the mutant.
7. Map Hi/Med/Lo deviation of CA parameter values to physical domain features of the biomolecule of M compared to that of W retrieved from wet lab/clinical study results.
8. Predict the results for any arbitrary SNP mutants based on the experimental setup formulated in earlier steps.

The algorithmic steps are illustrated in the next two case studies.

Case Study 1: Mutational Study of Precursor ID pre-miR-30c (111 bases)
In the paper [18], the authors reported the secondary structure of the precursor for the wild and mutant version G27A (base G at base location 27 replaced with base A) of the precursor having 111 bases. This mutation has been observed for breast and gastric cancer patients due to an increased level of mature miRNA. The authors focused on this wet lab study in vitro and cultured cells to assess the effect of mutation on the secondary structure of mutant compared to that of wild. The large difference in secondary structure of wild and mutant, the authors concluded, leads to increased level of miRNA. In the case study reported in Sect. 3.6.3, we reported the secondary structure of the precursor wild and mutant based on the CA model. The secondary structure for wild is derived from ComRCAM. However, for the G27A mutant version, the ComRCAM output for wild and mutant has been found to be identical as reported in Fig. 4.13. Hence, the secondary structure of the mutant has been derived out of NBRCAM. The structures are expressed as a sequence of structural motifs—stem, bulge and Hairpin Loop (HPL). The sequence of secondary structural motifs for wild and mutant is reported in Tables 3.18 and 3.19 of Sect. 3.6.3 of Chap. 3 on execution of the 'Algorithm: Predict miRNA precursor secondary structure' reported in that section. Deviation of G27A mutant structure (Table 3.19) from that of the wild (Table 3.18) established the validation of the in silico results derived out of CA model.

In the current case study, we report the mutational effect of SNP for pre-miR-30c precursor molecule for three different mutations in the same base location. We show that the mutational effect for the mutant G27A is significantly different from that of G27C and G27T. The relevant points we establish through this case study are summarized as follows.

The wet lab results are reported in published literature only for G27A since this mutation has been reported for breast and gastric cancer patients from clinical study followed by cell line wet lab experiment reported by Fernandez et al. [18]. No published results are reported for the other two mutants G27C and G27T. So we assume that these two mutations do not show up for any cancer patient. Assuming that base location 27 is a hot spot for mutation, we undertake a study of all three mutants, G27A, G27C and G27T, with our in silico CA-based model. We establish that the CA parameter for G27A differs from the other two mutants. While for the mutants G27C and G27T, there are low or no deviation of CA model parameters from that of wild, the deviation for G27A is significantly higher compared to the other two mutants. We conclude from this case study that higher deviation for CA parameters of G27A mutant represents the physical domain features that are different from that of the wild. The deviation of this physical domain feature for the mutant, we assume, corresponds to the higher level of expression for matured miRNA resulting from the precursor molecule pri-mir30c biogenesis.

Table 4.5 reports the CS/CL table derived out of NBRCAM wild and three mutants (G27A, G27C, G27T) along with the marking of Cycle Start Difference (CSD) and CrP (outside the CSD range of 0 to −3). The signal graphs for wild and mutant G27A derived out of NBRCAM and ComRCAM are shown earlier in Figs. 4.13 and 4.14. The Cycle Start (CS) and Cycle Length (CL) variations of NBRCAM for wild and its mutants in Table 4.5 are shown on the graphs in Figs. 4.15 and 4.16. The full table of Table 4.5 is provided in Annexure 4.3.

The signal graphs for wild and mutant G27A derived out of NBRCAM and ComRCAM are shown earlier in Figs. 4.13 and 4.14. The Cycle Start (CS) and Cycle Length (CL) variations of NBRCAM across wild and its mutants (G27A, G27C and G27T) in Table 4.5 are shown on the graphs in Fig. 4.15a–c and Fig. 4.16a–c, respectively. It is to be noted that there are significant variations between wild and G27A for both CS and CL, while wild and G27C differs only in CS and signal graphs of wild and G27T are identical.

Wild (W) pre-miR-30c is the miRNA precursor molecule with 111 bases. Table 4.5 reports the partial NBRCAM CS/CL values for the wild and its three mutants. The full table is reported in Annexure 4.3. The base at location 27 is G with the Hairpin Loop (HPL) stem–loop–stem at base location (23 to 26)—(27 to 32)—(33 to 36)—these refer to the base serial numbers noted in the paper by Fernandez [18] that reports study of the mutant G27A. For our study, we have assigned base location numbers from 1 to 111. Base location number 27 in the paper refers to our base location number 52, while HPL location numbers are 48–61.

Mutational study is initiated with the Mutant (M) noted as xyz where y refers to the mutation site (residue/base location); x refers to the residue/base type in the wild

Table 4.5 NBRCAM table of pre-miR-30c WILD, G27A, G27C, G27T

Position	Wild					G27A					G27C					G27T				
	NB	CS	CL	CSD		NB	CS	CL	CSD		NB	CS	CL	CSD		NB	CS	CL	CSD	
1	T	11	1			T	11	1			T	11	1			T	11	1		
2	G	10	1	−1		G	10	1	−1		G	10	1	−1		G	10	1	−1	
3	G	10	1	0		G	10	1	0		G	10	1	0		G	10	1	0	
4	G	9	1	−1		G	9	1	−1		G	9	1	−1		G	9	1	−1	
5	C	9	1	0		C	9	1	0		C	9	1	0		C	9	1	0	
6	T	7	1	−2		T	7	1	−2		T	7	1	−2		T	7	1	−2	
7	A	7	1	0		A	7	1	0		A	7	1	0		A	7	1	0	
8	T	7	1	0		T	7	1	0		T	7	1	0		T	7	1	0	
9	A	7	1	0		A	7	1	0		A	7	1	0		A	7	1	0	
10	A	7	1	0		A	7	1	0		A	7	1	0		A	7	1	0	
11	C	7	1	0		C	7	1	0		C	7	1	0		C	7	1	0	
12	C	12	1	5	CrP	C	12	1	5	CrP	C	12	1	5	CrP	C	12	1	5	CrP
13	A	13	1	1	CrP	A	13	1	1	CrP	A	13	1	1	CrP	A	13	1	1	CrP
14	T	13	1	0		T	13	1	0		T	13	1	0		T	13	1	0	
15	G	14	1	1	CrP	G	14	1	1	CrP	G	14	1	1	CrP	G	14	1	1	CrP
16	C	15	1	1	CrP	C	15	1	1	CrP	C	15	1	1	CrP	C	15	1	1	CrP
17	T	15	1	0		T	15	1	0		T	15	1	0		T	15	1	0	
18	G	16	1	1	CrP	G	16	1	1	CrP	G	16	1	1	CrP	G	16	1	1	CrP
19	T	17	1	1	CrP	T	17	1	1	CrP	T	17	1	1	CrP	T	17	1	1	CrP
20	A	17	1	0		A	17	1	0		A	17	1	0		A	17	1	0	
	•	•	•	•		•	•	•	•		•	•	•	•		•	•	•	•	
	•	•	•	•		•	•	•	•		•	•	•	•		•	•	•	•	
	•	•	•	•		•	•	•	•		•	•	•	•		•	•	•	•	

(continued)

Table 4.5 (continued)

Position	Wild					G27A					G27C					G27T				
	NB	CS	CL	CSD		NB	CS	CL	CSD		NB	CS	CL	CSD		NB	CS	CL	CSD	
99	A	6	1	0		A	6	1	0		A	6	1	0		A	6	1	0	
100	A	1	1	-5	CrP	A	1	1	-5	CrP	A	1	1	-5	CrP	A	1	1	-5	CrP
101	A	2	1	1	CrP	A	2	1	1	CrP	A	2	1	1	CrP	A	2	1	1	CrP
102	C	2	1	0		C	2	1	0		C	2	1	0		C	2	1	0	
103	A	2	1	0		A	2	1	0		A	2	1	0		A	2	1	0	
104	T	2	1	0		T	2	1	0		T	2	1	0		T	2	1	0	
105	C	2	1	0		C	2	1	0		C	2	1	0		C	2	1	0	
106	A	8	1	6	CrP	A	8	1	6	CrP	A	8	1	6	CrP	A	8	1	6	CrP
107	G	8	6	0		G	8	6	0		G	8	6	0		G	8	6	0	
108	C	8	6	0		C	8	6	0		C	8	6	0		C	8	6	0	
109	T	8	6	0		T	8	6	0		T	8	6	0		T	8	6	0	
110	G	8	6	0		G	8	6	0		G	8	6	0		G	8	6	0	
111	A	8	6	0		A	8	6	0		A	8	6	0		A	8	6	0	

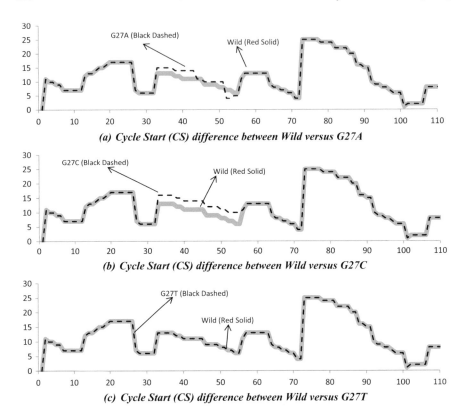

(a) *Cycle Start (CS) difference between Wild versus G27A*

(b) *Cycle Start (CS) difference between Wild versus G27C*

(c) *Cycle Start (CS) difference between Wild versus G27T*

Fig. 4.15 Cycle Start (CS) difference of NBRCAM evolution for wild versus mutants (*Y*-axis: CS value; *X*-axis: nucleotide base location)

with mutant residue/base noted as z. As per our serial numbering, it is G52A in place of G27A reported in [18]. Base triplet at location 52–54 is GAG (residue Glu). Assuming that base location 52 is a hot spot for possible SNP, in this case study we first evaluate the mutational effect for three non-synonymous mutants G52A (AAG—residue Lys), G52C (CAG—residue Gln) and G52T (TAG—stop codon) to investigate the difference in CA model parameter values for G52A compared to the other two mutants. Next, we investigate the effect of synonymous mutations in the Hairpin Loop (HPL) region 48–61. The HPL region is noted in the case study reported in Sect. 3.7.3 of Chap. 3. We have undertaken mutational effect in the mRNA precursor secondary structural motif HPL since it exerts considerable influence in the biogenesis of precursor.

The result table for these mutational studies based on the CA model is reported in Table 4.6.

Difference Frame (DF) is the region where Cycle Start Difference (CSD) and/or Cycle Length Difference (CLD) values for W and M differ. Figures 4.15 and 4.16 are derived out of full CS/CL table reported in Annexure 4.3. Figure 4.15 displays

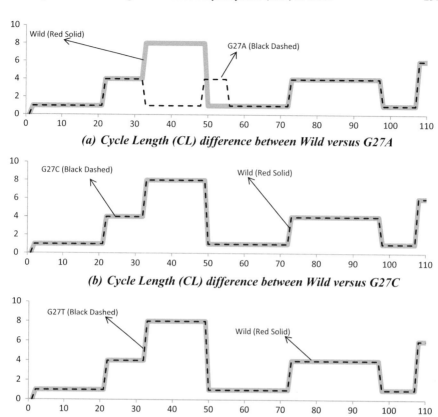

(a) Cycle Length (CL) difference between Wild versus G27A

(b) Cycle Length (CL) difference between Wild versus G27C

(c) Cycle Length (CL) difference between Wild versus G27T

Fig. 4.16 Cycle Length (CL) difference of NBRCAM evolution for wild versus mutants (Y-axis: CS value; X-axis: nucleotide base location)

the difference in CSD values for three mutants. Figure 4.16 shows the difference of CL values for the mutants compared to wild. G27A displays CLD values which differ significantly from that of wild (Fig. 4.16a). On the other hand, the CL values of other two mutants do not differ from that of wild (Fig. 4.16b, c).

The CA parameter values for each mutant are noted in columns 6 and 7 of the result Table 4.6. It is computed as per the algorithmic steps noted in the 'Algorithm: Predict mutational effect of SNP in a gene'. The algorithm computes CA model parameter values based on signal graph analytics of CS and CL signal graphs of W and M derived out of NBRCAM. The inputs for this analytics study are CSD and CLD value table and graph. The signal graphs for wild and three mutants G27A, G27C, G27T are reported in Figs. 4.15 and 4.16.

Results derived out of CA model on execution of the algorithm confirm difference of CA parameter value for CSD and also CLD value for G52A (G27A) compared to the other two mutants. The combined effect of these deviations is

Table 4.6 Results derived out of NBRCAM on execution of the algorithm for different mutants noted in row 1

Mutant (1)	SoD-P (2)		SoD-N (3)		NVD(4)	DF (5)			PV (6) (NVD/DF length)	CLD (7)	Inference (8)
	Location	Value	Location	Value		Start	End	Length			
G52A	–	–	51	5	5	32	55	24	0.21	Large difference with wild	High
G52C	55	4	–	–	4	32	55	24	0.17	Similar to wild	Low
G52T	55	4	–	–	4	32	55	24	0.17	Similar to wild	Low
T51A	32	12	51	4	2	32	63	32	0.06	Similar to wild	Low
	55	5	58	5							
	56	3	63	9							
T51C	51	15	11	18	101	2	69	68	0.74	Differ from wild	High
			13	23							
			22	3							
			32	22							
			35	31							
			52	13							
			69	6							
T51G	–	–	–	–	0	32	55	24	0.0	Similar to wild	Low
G54A	–	–	–	–	0	32	55	24	0.0	Similar to wild	Low
G54C	–	–	–	–	0	32	55	24	0.0	Similar to wild	Low
G54T	–	–	–	–	0	32	55	24	0.0	Similar to wild	Low
C57G	32	4	44	11	1	32	70	39	0.03	Similar to wild	Low
	36	3	52	6							
	38	2									
	55	5									
	56	2									
	60	2									

(continued)

Table 4.6 (continued)

| Mutant (1) | SoD-P (2) | | SoD-N (3) | | NVD(4) | DF (5) | | | PV (6) (NVD/DF length) | CLD (7) | Inference (8) |
	Location	Value	Location	Value		Start	End	Length			
C57A	32	4	49	8	2	32	70	39	0.05	Similar to wild	Low
	35	3	60	8							
	38	5									
	55	4									
	56	2									
G57T	35	3	61	10	5	32	64	33	0.15	Similar to wild	Low
	40	4									
	55	6									
	56	2									
G60A	56	2	–	–	2	56	63	8	0.25	Similar to wild	Low
G60C	38	6	49	7	1	33	70	38	0.03	Similar to wild	Low
G60T	38	6	49	7	1	33	70	38	0.03	Similar to wild	Low

likely to be co-related to the physical domain features that lead to higher level of matured miRNA expression for G52A (G27A). These three mutants are non-synonymous with the changes in residue in the mRNA codon. The mutant location noted within () refers to the locations noted in the paper.

The results of a few synonymous and non-synonymous mutations other than G27A/C/T are reported in Table 4.6 from fourth row onwards. Deviation for all the mutants are zero or low compared to wild, excepting the mutant T51C(T26C)—the associated codon for wild at location 49–51 is TGT (Cys), its mutant TGC is also Cys, and hence, it is a synonymous mutation for which high deviation of CA model parameter has been noticed compared to wild. Results of a few synonymous and non-synonymous mutations are reported as per our base numbering. Mutations are inserted on location 51 and 54. The mutations on these two locations are synonymous. Mutations on 57 and 60 cover both types—synonymous and non-synonymous. Predicted deviations for all mutants are 'Low' excepting two instances G52A and T51C (G52, T51 serial number refers to G27 and T26 in the referenced paper).

Case Study 2—KRAS Gene Mutations (570 bases)

In view of the phenomenal growth of cancer patients around the globe, researchers have focused on the detailed study of KRAS—an oncogene homologue of (Ki-ras2 Kirsten rat sarcoma viral oncogene).

The KRAS protein performs the essential function of normal tissue signalling including cell division and transfer of external signal to cell nucleus. KRAS mutation may impair function in respect of switching its state between active and inactive states. A large number of recent publications have been reported on the KRAS protein [19, 20]. The review paper [20] summarizes the mechanism reported in these publications with respect to gene regulation. Somatic KRAS mutations have been reported to be associated with different classes of cancers including leukaemia and colorectal/pancreatic/lung cancers.

Brief about KRAS

In the above context, we report the CA model for mRNA transcript derived out of KRAS human gene AAB59445 (KRAS Protein Uniprot ID P01116) having 570 bases. Wet lab results based on the cell line study have been reported for ten instances of mutations in the paper by Waters et al. [20]. The authors reported major deviation of biological features for mutants—protein expression, minimum doubling time and maximum cell line count. Based on the CA model for SNP we report the results on executing the 'Algorithm: Predict Mutational Effect of SNP in a Gene' reported in Sect. 4.5.2. We report the deviation of CA model parameters for each of the mutants compared to that of wild version of the mRNA transcript. The descending order of CA parameter values, in general, agrees with the protein expression results reported in the paper (reproduced in Table 4.10). What we tried to establish through this case study on KRAS is: in silico study of CA Machine

(CAM) evolution of the proposed CA model can be explored to predict wide varieties of mutations on KRAS gene. In the process, the high cost and time required in wet lab study of KRAS (or any other protein) can be reduced significantly.

The acronyms used to derive CA parameter values are recapitulated for easy readability of the results.

- **W**: The locations of nucleotide base triplets for a residue location at z are $(3z - 2)$, $(3z - 1)$ and $3z$. Wild KRAS protein with 570 bases has residue Gly at locations 12, 13 and 60; the corresponding base locations are (i) 34–36 with base triplets GGT (ii) 37–39 with triplets GGC and (iii) 178–180 with triplet GGT.
- **M**: Mutational study with mutation noted as xyz: y refers to the mutation site (residue/base location); x refers to the residue/base type in the wild with mutant residue/base noted as z at the location y.

  - Non-synonymous mutation—G12V (Gly codon GGT in W got replaced with codon GGT at base location 34–36 that represents residue Valine).
  - Synonymous Mutations—For third base of residue Gly at location 12, 13, 60: (i) codon GGT of W at base location 34–36 replaced with GGG, GGA, GGC, that is T36G, T36A, T36C; (ii) codon GGC at base location 37–39 replaced with GGT, GGG, GGA, that is, C39T, C39G, C39A; and (iii) codon GGT of W at base location 178–180 replaced with GGG, GGA, GGC, that is, T180G, T180A, T180C.

- **DF**: Difference frame where Cycle Start Difference (CSD) and/or Cycle Length Difference (CLD) values for W and M differ. In this case, there is no difference in CLD values between W and M and hence CLDs are excluded from the result table.
- **CrP**: Critical pair where CSD between a pair of adjacent CAM cells lies outside the range 0 to −3; CrP values of M are subtracted from CrP/non-CrP values of W; Positive (P)/Negative (N) CrP refers to instances where CrP value is positive/negative; CrP may exist either—(i) on different residue locations of W or M, or (ii) in the same residue location of W and M.
- **CrP–P/N**: A CrP is marked as positive/negative as per positive/negative for W; for non-CrP location of W, if CrP exists for M, it is marked as positive/negative as per M's positive/negative value.
- **SoD**: Sum of difference (CSD values at CrP locations in the difference frame) between W and M.
- **SoD-P**: Sum of Difference (SoD) for positive CrP values in W and M (ignore if the difference is less than 4 for positive CrPs existing in same location of W and M).

- **SoD-N**: SoD for negative CrP values in W and M (ignore if the difference is less than 2 for negative CrPs existing in same location of W and M).

The inputs for this analytics study are NBRCAM CS Difference (CSD) value table and graph. The CS/CL tables and graphs for wild and mutants G12V, T180G, C39T are noted in Table 4.7 and Fig. 4.17. While Table 4.7 is a partial table of the 200 nucleotide locations, Annexure 4.4 provides the detailed table of all bases.

Table 4.8 reports the results derived on implementing the algorithm on the wild and mutant versions of the KRAS gene. The CA parameter values for each mutant are noted in the last column. It is computed as per the algorithmic steps noted in the 'Algorithm: Predict mutational effect of SNP in a gene'. The algorithm computes the model parameter values based on signal graph analytics of CS signal graphs of wild and mutant.

To refresh the reader on the computations in the algorithm used in Table 4.8

- Column 2 shows Sum of Difference-Positive (SoD-P), column 3 is Sum of Difference-Negative (SoD-N) and column 4 is the net value difference (NVD) computed subtracting SoD-N from SoD-P.
- Column 5 is the Difference Frame (DF) from—to (length expressed as number of bases covered in the DF), and column 6 displays the CA model parameter values derived out of signal graph analytics on CS signal graph for W and M. CA parameter value above 1.0 indicates high (Hi) deviation, medium (Med) deviation for parameter value in the range 0.1–1.0, and Low (Lo) if the value is less than 0.1.
- Predicted deviation of CA parameter values for M compared to W marked as high/medium/low corresponds to the deviation in physical domain features of M compared to W. The features noted as a three-element vector in columns 2, 3, 4 reproduced in Table 4.10 are 'minimum doubling time', 'KRAS protein expression' and 'maximum cells ($\times$ 10^6)'. The main determinant factor is 'KRAS Protein Expression' noted in column 3.

The CA parameter values are listed within bracket in descending order for different mutants: G12V (1.48), C39G (1.42), T36C (1.38), C39T (1.33), T180C (1.17), T36G (1.10), C39A (0.88), T180A (0.59), T36A (0.38) and T180G (0.08). The highest and least parameter values 1.48 and 0.08 correspond to the highest and least protein expression values 16.6 and 2.4 noted in the paper [20]. For other mutants, there is a general agreement of the descending order of CA parameter values with the features reported out of wet lab experimental results reproduced in Table 4.10.

For two mutants T180C and T36C, the Cycle Start (CS) and Cycle Length (CL) values derived out of NBRCAM evolution for W and M are identical with zero difference of the CS signal graph. Hence, the results for these two mutants are derived out of ComRCAM output. ComRCAM CS/CL partial table for W and M (T36C and T180C) are shown in Table 4.9 for 200 bases. The signal graphs showing the difference between wild and mutants are shown in Fig. 4.18 and Table 4.8b.

The CS parameters for NBRCAM of KRAS protein are reported in Table 4.8. It is interesting to note that in the NBRCAM, the base location 38 displays high

Table 4.7 NBRCAM for KRAS WILD, G12 V, T180G and C39T for 1–200 base locations

Position	Wild					G12V					T180G					C39T				
		CS	CL	CsD	CrP		CS	CL	CsD	CrP		CS	CL	CsD	CrP		CS	CL	CsD	CrP
1	A	9	11	0		A	9	11	0		A	9	11	0		A	9	11	0	
2	T	9	11	0		T	9	11	0		T	9	11	0		T	9	11	0	
3	G	9	11	0		G	9	11	0		G	9	11	0		G	9	11	0	
4	A	9	11	0		A	9	11	0		A	9	11	0		A	9	11	0	
5	C	8	11	−1		C	8	11	−1		C	8	11	−1		C	8	11	−1	
6	T	8	11	0		T	8	11	0		T	8	11	0		T	8	11	0	
7	G	6	11	−2		G	6	11	−2		G	6	11	−2		G	6	11	−2	
8	A	6	11	0		A	6	11	0		A	6	11	0		A	6	11	0	
9	A	6	11	0		A	6	11	0		A	6	11	0		A	6	11	0	
10	T	6	11	0		T	6	11	0		T	6	11	0		T	6	11	0	
11	A	1	2	−5		A	1	2	−5		A	1	2	−5		A	1	2	−5	
12	T	0	2	−1		T	0	2	−1		T	0	2	−1		T	0	2	−1	
13	A	0	2	0		A	0	2	0		A	0	2	0		A	0	2	0	
14	A	1	2	1		A	1	2	1		A	1	2	1		A	1	2	1	
15	A	2	2	1		A	2	2	1		A	2	2	1		A	2	2	1	
16	C	2	1	0		C	2	1	0		C	2	1	0		C	2	1	0	
17	T	2	1	0		T	2	1	0		T	2	1	0		T	2	1	0	
18	T	2	1	0		T	2	1	0		T	2	1	0		T	2	1	0	
19	G	4	1	2	CrP	G	4	1	2	CrP	G	4	1	2	CrP	G	4	1	2	CrP
20	T	4	1	0		T	4	1	0		T	4	1	0		T	4	1	0	
		• • •					• • •					• • •					• • •			

(continued)

Table 4.7 (continued)

Position	Wild					G12V					T180G					C39T				
		CS	CL	CsD	CrP		CS	CL	CsD	CrP		CS	CL	CsD	CrP		CS	CL	CsD	CrP
180	T	4	1	−42	CrP	T	4	1	−42	CrP	G	7	1	−7	CrP	T	4	1	−42	CrP
181	C	4	1	0		C	4	1	0		C	3	1	−4		C	4	1	0	
182	A	2	1	−2		A	2	1	−2		A	3	1	0		A	2	1	−2	
183	A	2	1	0		A	2	1	0		A	3	1	0		A	2	1	0	
184	G	4	1	2	CrP	G	4	1	2	CrP	G	4	1	1	CrP	G	4	1	2	CrP
185	A	5	1	1		A	5	1	1		A	5	1	1		A	5	1	1	
186	G	5	1	0		G	5	1	0		G	5	1	0		G	5	1	0	
187	G	14	9	9	CrP	G	14	9	9	CrP	G	14	9	9	CrP	G	14	9	9	CrP
188	A	14	9	0		A	14	9	0		A	14	9	0		A	14	9	0	
189	G	14	9	0		G	14	9	0		G	14	9	0		G	14	9	0	
190	T	14	9	0		T	14	9	0		T	14	9	0		T	14	9	0	
191	A	14	9	0		A	14	9	0		A	14	9	0		A	14	9	0	
192	C	14	9	0		C	14	9	0		C	14	9	0		C	14	9	0	
193	A	14	9	0		A	14	9	0		A	14	9	0		A	14	9	0	
194	G	14	9	0		G	14	9	0		G	14	9	0		G	14	9	0	
195	T	13	9	−1		T	13	9	−1		T	13	9	−1		T	13	9	−1	
196	G	13	9	0		G	13	9	0		G	13	9	0		G	13	9	0	
197	C	12	9	−1		C	12	9	−1		C	12	9	−1		C	12	9	−1	
198	A	12	9	0		A	12	9	0		A	12	9	0		A	12	9	0	
199	A	13	9	1		A	13	9	1		A	13	9	1		A	13	9	1	
200	T	13	9	0		T	13	9	0		T	13	9	0		T	13	9	0	

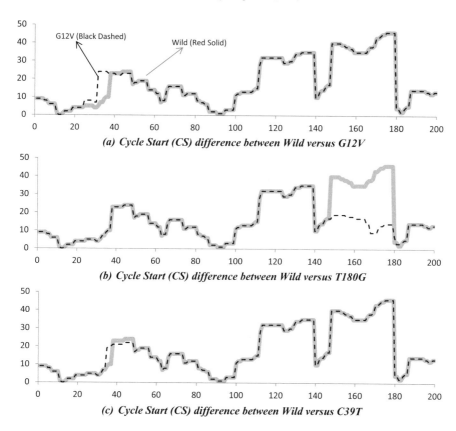

Fig. 4.17 Cycle Start (CS) difference of NBRCAM evolution for wild versus mutants (Y-axis: CS value; X-axis: nucleotide base location)

positive CrP values +13, while base location 180 displays large negative CrP value (−42). That is, the mutational results reported in the paper in the region covering base location 34–39 and 180 have some co-relation with the high Cycle Start Difference (CSD) values reported for the CS/CL table output for the wild transcript of KRAS gene. Based on this observation, we report the mutational effect on base locations 148 and 235 displaying CrP value of +23 and +26, respectively. Table 4.11 shows the effect of synonymous mutations on two locations 144 (Gly–GGA at location 142–144) and 225 (Gly–GGG at location 223–225) in the vicinity of 148 and 235 displaying high positive CrP values derived out of NBRCAM. High and medium deviations of CA parameter values have been observed for mutants A144G and A144C, respectively.

The major focus of this case study is to establish that the mutational effect of a gene SNP can be investigated in silico from CA models. Based on the program code developed for the algorithm reported in this section, we have started experimenting with the SNPs listed in different databases including the one noted in [21].

Table 4.8 Result table for ten mutants noted in column 1

Mutant (1)	SoD-P (2)		SoD-N (3)		NVD (4)	DF (5)			PV (6) (NVD/DF length)	Inference (7)
	Location	Value	Location	Value		Start	End	Length		
(a) KRAS gene mutation and mutational effects employing NBRCAM										
G12 V	25	3	–	–	36	25	49	25	1.44	High
	32	16								
	34	2								
	35	2								
	38	13								
T180G	148	23	168	5	3	148	184	37	0.08	Low
	149	2	180	35						
	166	5								
	170	4								
	72	2								
	184	1								
C39T	35	10	49	2	20	35	49	15	1.3	High
	38	12								
C39G	25	3	–	–	20	25	38	14	1.42	High
	32	5								
	38	12								
T36A	33	4	49	2	9	26	49	24	0.38	Medium
	35	3								
	38	4								

(continued)

Table 4.8 (continued)

Mutant (1)	SoD-P (2)		SoD-N (3)		NVD (4)	DF (5)			PV (6) (NVD/DF length)	Inference (7)
	Location	Value	Location	Value		Start	End	Length		
T180A	148	23	168	5	32	148	201	54	0.59	Medium
	149	2	180	42						
	166	5								
	170	4								
	172	5								
	179	3								
	184	12								
	187	6								
	201	19								
C39A	25	3	49	2	21	25	49	25	0.84	Medium
	32	1								
	35	7								
	38	12								
T36G	28	9	49	6	26	26	49	24	1.10	High
	31	5								
	33	2								
	35	2								
	38	14								

(continued)

Table 4.8 (continued)

Mutant (1)	SoD-P (2)		SoD-N (3)		NVD (4)	DF (5)			PV (6) (NVD/DF length)	Inference (7)
	Location	Value	Location	Value		Start	End	Length		
(b) KRAS gene mutations and mutational effects employing ComRCAM										
T180C	–	–	Many	Many	221	2	188	187	1.17	High
T36C	–	–	6	8	54	2	40	39	1.38	High
			15	12						
			20	6						
			30	51						
			31	9						
			38	22						

Note For next two mutants, NBRCAM CS/CL outputs for W and M are identical with zero difference of CS signal graph. Hence, the results for these two mutants are derived out of ComRCAM output. ComRCAM CS/CL table for Wand mutants T36C and T180C is shown in Table 4.9. Figure 4.18 displays signal graphs for T36C and T180C

Net Value Difference (NVD) = (Sum of SoD-*N* values)/2 for these two cases

4.6 cLife Model for sgRNA Binding on Target Gene Editing Site for CRISPR/Cas9

The acronym CRISPR stands for clustered regularly interspaced short palindromic repeats, while Cas9 refers to a DNA nuclease enzyme. In last two decades, these two terminologies generated considerable interest as these provide the basis for the promising gene editing technology referred to as CRISPR/Cas9. This technology, as noted in Fig. 4.19, allows editing of a double-stranded DNA at the level of a nucleotide base. Within the cell body, the Cas9 enzyme is guided by a synthetic short guide RNA (sgRNA) to the editing site. Once the sequence in the edit site is located, Cas9 unwinds the DNA helix and creates a double-stranded break (Fig. 4.19) in cell's DNA. Another enzyme next repair the DNA break by integrating the new experimental DNA.

The historical background of the development of this technology is reported in (https://en.wikipedia.org/wiki/CRISPR) along with the contribution of major researchers for its development. Efficient and ease of editing (knock down or knock in) of any gene with this versatile platform is preferred compared to the earlier technologies like zinc finger, transcription activator like effectors nucleases (TALEN), etc. In view of the high promise of this technology for treatment of genetic disorder including disease-causing mutations, a large number of publications are reported in recent years [22]. The technology is getting matured in the last few years even though there are few major challenges, specifically in respect of off-target binding of the CRISPR/Cas9 nuclease that may lead to disastrous adverse effect. The principle based on which this technology has been developed is briefly introduced next.

Virus is an infective agent consisting of a nucleic acid molecule in a protein coat. It is able to multiply only within the living cells of a host. The hepatitis B virus is a typical example. Viruses are a common threat to all forms of cellular life— prokaryotic (including bacteria, archaea) and eukaryotic cells. Consequently, a variety of adaptive immune system has evolved to protect against virus infection, as elaborated next. The cluster of regulatory palindromic DNA sub-sequences of CRISPR has been observed in bacterial genome. It constitutes a crucial component of the bacterial immune system against viral infection. On getting invaded by a new virus not encountered earlier, CRISPR spacer sequence of bacteria gets extended with a new spacer sequence derived from the virus gene. During the next attack by a virus, the CRISPR sequence of bacterial genome is transcribed to a CRISPR RNA molecule. This RNA transcript guides its bacterial molecular machinery to the target sequence of the invading viral gene to cut it off and destroy. Bacterial CRISPR has been found to contain a protospacer adjacent sequence (PAM) of 2–6 base length; a typical PAM is of the type 5′ end—NGG—3′ end with minimum of twin G bases, where N can be any base. Gene cleavage is performed by the Cas9 on recognizing the PAM. Variation in PAM sequences has been observed for different Cas9 enzymes.

Table 4.9 ComRCAM for KRAS WILD and mutant T36C and T180C 1–200 base locations

Wild					T36C					T180C				
Position		CS	CL	CsD		CS	CL	CsD			CS	CL	CsD	
1	A	993	272			A	1000	272			A	1205	272		
2	T	992	272	−1		T	1000	272	0		T	1205	272	0
3	G	992	272	0		G	1000	272	0		G	1204	272	−1
4	A	988	272	−4		A	999	272	−1		A	1204	272	0
5	C	988	272	0		C	999	272	0		C	1200	272	−4
6	T	980	272	−8		T	999	272	0		T	1200	272	0
7	G	980	272	0		G	998	272	−1		G	1188	272	−12	CrP
8	A	980	272	0		A	998	272	0		A	1188	272	0
9	A	980	272	0		A	997	272	−1		A	1188	272	0
10	T	980	272	0		T	997	272	0		T	1188	272	0
11	A	980	272	0		A	996	272	−1		A	1187	272	−1
12	T	980	272	0		T	996	272	0		T	1186	272	−1
13	A	980	272	0		A	992	272	−4		A	1186	272	0
14	A	980	272	0		A	992	272	0		A	1186	272	0
15	A	980	272	0		A	980	272	−12	CrP	A	1184	272	−2
16	C	980	272	0		C	980	272	0		C	1184	272	0
17	T	980	272	0		T	980	272	0		T	1184	272	0
18	T	980	272	0		T	979	272	−1		T	1184	272	0
19	G	979	272	−1		G	979	272	0		G	1182	272	−2
20	T	978	272	−1		T	972	272	−7		T	1182	272	0
		•	•	•		•	•	•			•	•	•
		•	•	•		•	•	•			•	•	•
		•	•	•		•	•	•			•	•	•
186	G	759	272	−1		G	759	272	−1		G	743	272	0
187	G	759	272	0		G	759	272	0		G	743	272	0
188	A	743	272	−16	CrP	A	743	272	−16	CrP	A	743	272	0
189	G	741	272	−2		G	741	272	−2		G	741	272	−2
190	T	734	272	−7		T	734	272	−7		T	734	272	−7
191	A	734	272	0		A	734	272	0		A	734	272	0
192	C	734	272	0		C	734	272	0		C	734	272	0
193	A	734	272	0		A	734	272	0		A	734	272	0
194	G	734	272	0		G	734	272	0		G	734	272	0
195	T	734	272	0		T	734	272	0		T	734	272	0
196	G	734	272	0		G	734	272	0		G	734	272	0
197	C	733	272	−1		C	733	272	−1		C	733	272	−1
198	A	733	272	0		A	733	272	0		A	733	272	0
199	A	732	272	−1		A	732	272	−1		A	732	272	−1
200	T	732	272	0		T	732	272	0		T	732	272	0

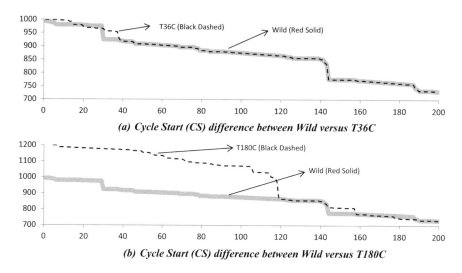

(a) Cycle Start (CS) difference between Wild versus T36C

(b) Cycle Start (CS) difference between Wild versus T180C

Fig. 4.18 Cycle Start (CS) difference of ComRCAM evolution for wild versus mutants (Y-axis: CS value; X-axis: nucleotide base location)

Table 4.10 Three features of wet lab experimental results on KRAS protein (reproduced from Table 2 of the paper [20])

	Minimum doubling time (h)	KRAS protein expression[a]	Maximum cells ($\times 10^5$)
WT	44.0	1.0	7.3
G12 V	20.5	16.6	51.9
36 T → A	28.7	5.4	15.9
36 T → C	22.5	14.3	18.7
36 T → G	34.2	3.2	18.8
39 C → A	30.0	3.4	16.3
39 C → G	28.5	5.8	16.1
39 C → T	27.1	6.1	16.8
180 T → A	25.1	4.0	18.1
180 T → C	24.6	13.0	19.2
180 T → G	28.4	2.4	21.4

Normalized to WT = 1.0

The CRISPR/Cas9 technology has been developed from the study of the virus–host interaction. The technology demands artificial design of a short guide RNA (sgRNA) inserted in cell's nucleus for gene editing. The sgRNA sequence of length (say) X is copied from the target gene, where X is around 20 nucleotide bases of the target. The sgRNA guides the nuclease enzyme Cas9 to the gene editing site, where the enzyme edits the target gene. A concise description of interaction of sgRNA

Table 4.11 Predicted mutational effect for synonymous mutations for base locations 144 and 225

Mutant	SoD-P		SoD-N		NVD	DF			PV (NVD/ DF length)	Inference
	Location	Value	Location	Value		Start	End	Length		
A144G	127	1	140	25	23	126	148	23	1.0	High
	129	2								
	130	2								
	133	2								
	142	3								
	143	26								
	145	2								
	148	10								
A144C	142	3	140	6	24	140	180	41	0.59	Medium
	145	8	180	25						
	148	23								
	150	2								
	152	13								
	166	2								
	170	4								
A144T	126	4	138	6	2	126	148	23	0.09	Low
	127	1	140	25						
	129	2								
	130	2								
	142	3								
	143	15								
	145	2								
G225A	–	–	–	–	0	–	–	–	0.00	Low
G225C	–	–	–	–	0	222	232	11	0.00	Low
G225T	201	6	217	6	6	201	232	32	0.19	Low
	208	2	222	8						

Fig. 4.19 CRISPR/Cas9 technology for gene editing

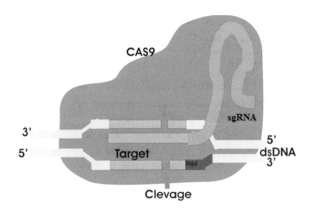

with Cas9 and target gene is available in many publications including the one noted in Tufts University website [23].

From the above scenario, we can conclude that Mother Nature has provided us with a voluminous encyclopaedia from which we can learn new lessons. In the above background, we propose a CA model to address the problem of on-off target binding of CRISPR/Cas9 machinery for editing a target gene. The editing site of the target gene has a specific nucleotide base sequence. Earlier brief introduction points to two important issues.

- Design of short guide RNA (sgRNA) base sequence to guide the Cas9 nuclease enzyme to the correct location of the gene having identical base sequence for editing target gene—this is usually referred to as on-target binding.
- In order to avoid the possible deleterious effect of gene editing on a wrong site, the off-target binding of Cas9 enzyme has to be reduced to nil. One of the major avenues to achieve this goal is to exploit the use of protospacer adjacent sequence (PAM) since Cas9 activation demands presence of PAM.

Several recent publications on sgRNA design cover both the above issues. Several commercial organizations like Synthego (synthego.com), Clontech (Clonetech.com) and Synbio-tech (synbio-tech.com) design sgRNA for different classes of target genes. The current focus of research is directed for reduction of off-target binding of the sgRNA on a target gene.

We encountered the problem similar to the earlier one in Sect. 3.6.5 of Chap. 3 for prediction of binding of another small molecule referred to as siRNA (small interfering RNA) on the mRNA transcript of a target gene. The siRNA is so designed that it can guide the RISC (RNA Induced Silencing Complex) on the target gene mRNA transcript to control the biological process of transcription. In the present scenario instead of siRNA, the synthetically designed sgRNA guides the nuclease enzyme Cas9 to the correct site of DNA for its editing. So we use ComRCAM for sgRNA sequence to derive its signal graph. On the other hand, we use ComDCAM for the target gene sequence to derive its signal graph. We design our algorithm based on analytics we perform on sgRNA and target gene signal graphs. The experimental setup for design and testing of the algorithm is next reported.

The experimental setup is designed by utilizing genome CRISPR, a database for high throughput CRISPR/Cas9 screens available [24].

This database contains data on more than 550,000 single guide RNAs (sgRNA) derived from 84 different experiments performed in 48 different human cell lines, comprising all screens in human cells using CRISPR/Cas9 published data. We have assumed that a sgRNA ID and its target gene ID reported in the database [25] as true positive instances. For our experimental setup, we also need true negative cases. In order to collect true negative instances, we analyse the output of the service available at database [26] by selecting those cases identified by this service with high score [22] but not covered by genome CRISPR database.

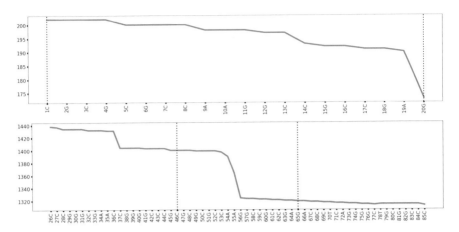

Fig. 4.20 ComRCAM signal graph—synthetic sgRNA versus target gene

As reported earlier, for each sgRNA of length 20 bases covered in the database, there is a target gene having bases sequence identical to that of sgRNA. A pair of signal graphs is noted in Fig. 4.20 with the x-axis displaying the nucleotide bases and their serial location both for sgRNA (the upper graph) and target gene. The base sequence of the gene locations 46–65 (marked as X to Y) matches with that of the sgRNA sequence. The sequence match region is shown within a pair of vertical lines on the lower graph for target gene. In addition to match region, we have included two regions on its left and right, each covering 20 bases. So signal graph for target gene covers 60 bases. In subsequent discussions, we have noted the target gene ID and sgRNA of the database as RRM2_ X_Y and sgRNA_X_Y, respectively. The variables X and Y illustrated in Fig. 4.20 refer to the base location marked with vertical lines.

A signal graph is derived out of Cycle Start (CS) table generated from evolution of CAM. Each ith location of signal graph refers to CS Difference (CSD) of $(i - 1)$th and ith CS values. Usually, CSD values range from 0 to −2, while for some positions signal value may have significantly high positive or negative. We set a threshold limit $T = -3$. The CSD value different from the range 0 to −3 are processed in the algorithmic steps. The highlight of the algorithm follows.

In Fig. 4.20, the maximum signal value change between a pair of adjacent bases is referred to as Primary Signal (PS). Secondary Signal (SS) has signal change value lesser than PS. The PS of sgRNA signal graph is aligned with that of target gene signal graph. Subsequent to alignment, the number of SS on 5′ and 3′ end of each signal graph are compared and if these are equal, the sgRNA is predicted to bind on target gene—an instance of 'true positive' case, else it is marked as 'true negative' case (the sgRNA is predicted not to bind on target gene).

In Fig. 4.20, the top signal graph is derived out of ComRCAM output of a synthetic sgRNA, and the bottom one is from the ComDCAM output of a target gene. Note that the base sequence at location $X = 46$ to $Y = 65$ of target gene match with that of the sgRNA base sequence, the IDs for gene and sgRNA are noted as

RRM2_46_65 and sgRNA_46_65. In general, the ID of the target gene is noted as RRM2_X_Y.

The variables and acronyms used in the algorithmic steps are defined below.

- The sgRNA is denoted as M′, while M refers to the target gene region having the same base sequence. This identical base sequence for sgRNA, and the target region of gene is a prerequisite for binding of synthetic sgRNA on the target gene.
- We next define two regions adjacent to M referred to as LoM (Left of M) and RoM (Right of M) each of length 20 bases.
- PS refers to primary signal that has the maximum value (PSVal) at the primary signal location (PSLoc).
- For M′, the corresponding parameters are marked as PSVal′, PSLoc′, while PSVal and PSLoc are for target gene M for transcript region (LoM-M-RoM).
- Secondary Signals (SSs) in the signal graph are noted as two vectors on 5′/3′ end of PSLoc: 5′-SSV and 3′-SSV within a length of 20 bases around primary signal location (PSLoc) identified on alignment of PSLoc′ with PSLoc.
- If a negative secondary signal (of any value) exists at ith location, while a PS exists at $(i - 1)$ or $(i + 1)$ location, then the SS value is added to the PS value with PSLoc marked as $i - 1$ to i or i to $i + 1$.
- PSLoc′ is aligned with PSLoc to select 20 bases out of (LoM-M-RoM) on either end of PSLoc (marked as 5′ and 3′ end).
- The Secondary Signals (SSs) outside the range 0 to −3 and less than PS value are marked in M′ and (LoM-M-ROM).
- The parameters for M′ are noted as PSVal′, PSLoc′, 5′-SSV′, 3′-SSV′, while for the target gene region covering (LoM-M-RoM) these variables are noted as PSVal, PSLoc, 5′-SSV and 3′-SSV.
- Cardinality of SSVs for M′ is noted as 5′-CSSV′ and 3′-CSSV′, while for gene region, these are noted as 5′-CSSV and 3′-CSSV.

The algorithm aligns M′ PSLoc′ with gene (LoM-M-RoM) primary signal location (PSLoc) and computes the cardinality of 5′ and 3′ secondary signal vectors both for M′ and transcript (LoM-M-RoM). If the cardinality of 5′/3′ secondary signal vectors for M′ and gene (LoM-M-RoM) signal graphs are identical, we predict it as a case of 'true positive'. Exception handling, as noted in Step 5 of the algorithm (noted below), handles the cases of zero PS/SS for the signal graphs. Degree of Match (DoM) of M′ signal graph with that of gene region is defined with a score of 0–2, 0 signifying best match and 2 as worst. If Degree of Match (DOM) is more than 2, it is marked as a case of 'true negative' instance.

Algorithm: Predict true positive and true negative instances of binding a candidate sgRNA on target gene

Input: Signal graph for the candidate sgRNA M' of 20 bases and the signal graph of mRNA transcript for the region (LoM-M-RoM) of the target gene, where sgRNA base sequence matches with the base sequence of M region.
Output: True positive/true negative instances with respect of binding candidate sgRNA on target gene.
Steps

1. Analyse sgRNA signal graph M' to mark PSVal', PSLoc, and compute 5'-CSSV', 3'-CSSV' (cardinality of secondary signals on 5' and 3' end of PSLoc').
2. Analyse mRNA transcript graph for (LoM-M-RoM) to mark PSVal and PSLoc and the corresponding 5'-CSSV and 3'-CSSV.
3. Align primary signals of sgRNA and transcript (LoM-M-RoM) to select 20 bases and identify number of secondary signals on 5' and 3' ends of either graph.
4. For the cases of match of cardinality of secondary signals on 5'/3' end for sgRNA and target gene, predict it as an instance of 'true positive' case, while for no match predict it as an instance of 'true negative' case.
5. Exception Handling: In the event of no PS/SS (primary signal/secondary signal) for signal graph, these are treated as true positive cases if any one of the following two conditions is valid—signal graph of sgRNA M' has no PS/SS, while mRNA Transcript (LoM-M-RoM) region has (i) a very high valued PS or (ii) (LoM-M-RoM) has a region with no PS/SS.

Case Studies

We tested our algorithm to predict the results for 10 cases of true positive and true negative instances picked up from the experimental setup reported earlier out of genome CRISPR database [26]. Signal graph pairs for each case and the result tables are noted below. The entries are left blank if there is no PS/SS. The exception case with no PS/SS instances for sgRNA M' are marked with '* nil' in the result tables. There are three such cases for true positive table results.

The algorithm correctly identifies each of the ten true positive cases with Degree of Match (DOM) value as 0. As noted before if an SS (of any value) exists at a location adjacent to a PSLoc, the SS value is added to the SS value to derive the PS. We have coded the algorithm and ran exhaustive experimentation running blind tests on a large number of cases taken from the database and refined the algorithm to achieve the targeted goal. The experimental results are reported in Annexure 4.5.

Reported below are all true positive cases in Table 4.12. For each of the following cases, subsequent of alignment of Primary Signal (PS) of the signal graphs, the count of Secondary Signals (SSs) on 5' and 3' end of sgRNA signal graph and

Table 4.12 Prediction result for binding of sgRNA on its target gene—true positive

sgRNA/Gene ID	M' signal Location	M' signal Value	LoM signal Location	LoM signal Value	M signal Location	M signal Value	RoM signal Location	RoM signal Value	sgRNA PSLoc'	sgRNA PSVal'	Gene PSLoc	Gene PSVal	Gene match region for sgRNA sequence		LoM'-M'-RoM' Cardinality	LoM-M-RoM Cardinality	True positive
RRM2_46_65	14	-4	37	-27	55	-26	78	1	20	-17	55-56	-66	LoM-M	37-56	1 and 0	1 and 0	Yes
	20	-17			56	-40											
RRM2_61_80	7	-4	46	-9	61	-17			17	-15	61	-17	LoM-M	45-64	2 and 0	2 and 0	Yes
	15	-4	53	-3													
	17	-15															
RRM2_70_89					76	-4	98	-17					LoM	50-69	0 and 0	0 and 0	Yes
					82	-4											
RRM2_154_173	6	-16	144	-38	164	-12			6	-16	144	-38	LoM-M	138-157	0 and 1	0 and 1	Yes
	17	-6	151	-10													
RRM2_217_236			213	-17	225	-41	240	-12			225-226	-59	Very high value	225-226	0 and 0	M has -59	Yes
					226	-18	241	-8									
					228	-16											
RRM2_270_289	7	-4			270	-4			9	-22	272	-38	LoM-M	264-283	1 and 1	1 and 1	Yes
	9	-22			272	-38											
	14	1			273	4											
					286	-4											
RRM2_292_311	20	-17			301	-46	320	-29	20	-17	301	-46	LoM-M	282-301	0 and 0	0 and 0	Yes
							330	-25									
RRM2_303_322	6	-21			319	-42	332	-13	6-7	-22	319	-42	M-RoM	312-331	0 and 0	0 and 0	Yes
	7	-1															
RRM2_331_350			315	-11	375	-4	353	-4					Very high value	326	0 and 0	LoM has -54	Yes
			316	-9	381	-4	355	-11									
			323	-25			356	-43									
			353	-54			463	-17									
RRM2_367_386	6	-13	353	-15	375	-4			14	-22	384-385	-42	LoM-M	372-391	2 and 0	2 and 0	Yes
	10	-20	359	-40	381	-4											
	14	-22	363	-13	384	-26											
					385	-16											

target gene signal graph are identical. A brief explanation follows after each of the ten 'true positive' instances..

True Positive Cases

The sgRNA Primary Signal (PS) between base locations 19 and 20 is aligned with the PS of target gene between base locations 55 and 56 with bases A and G, respectively. Subsequent to PS alignment, the number of secondary signals on 5′ and 3′ end of both signal graphs are 1 and 0. Hence, it is a true positive case (Fig. 4.21).

The sgRNA Primary Signal (PS) between base locations 16 and 17 is aligned with the PS of target gene between base locations 60 and 61 with bases G and C, respectively. Subsequent to PS alignment, the number of secondary signals on 5′

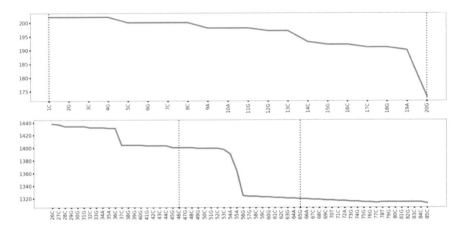

Fig. 4.21 sgRNA/Gene ID RRM2_46_65

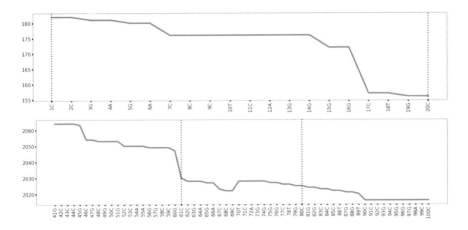

Fig. 4.22 sgRNA/Gene ID RRM2_61_80

and 3′ end of both signal graphs are 2 and 0. Hence, it is a true positive case (Fig. 4.22).

The signal graph for sgRNA and LOM (Left of M at location 70–89) do not have PS/SS signal. This is an instance of exception handling noted in Step 5 of the algorithm. So the number of SS signals for both the signal graphs is 0. Hence, it is a true positive case (Fig. 4.23).

The sgRNA Primary Signal (PS) between base locations 5 and 6 is aligned with the PS of target gene between base locations 143 and 144 with bases C and G, respectively. Subsequent to PS alignment, the number of secondary signals on 5′ and 3′ end of both signal graphs are 0 and 1. Hence, it is a true positive case (Fig. 4.24).

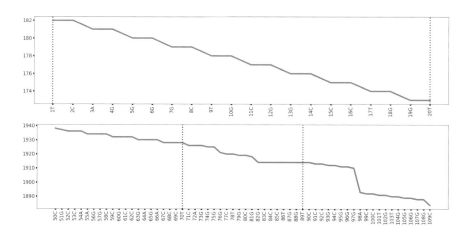

Fig. 4.23 sgRNA/Gene ID RRM2_70_89

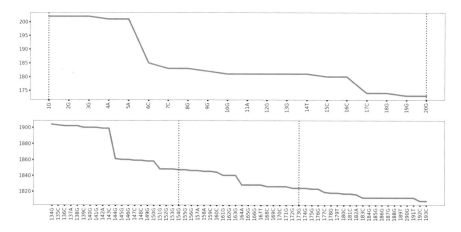

Fig. 4.24 sgRNA/Gene ID RRM2_154_173

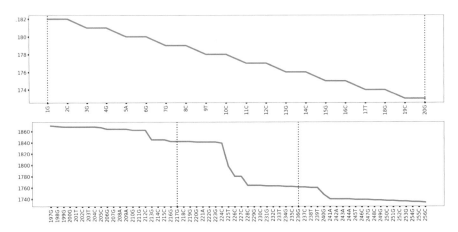

Fig. 4.25 sgRNA/Gene ID RRM2_217_236

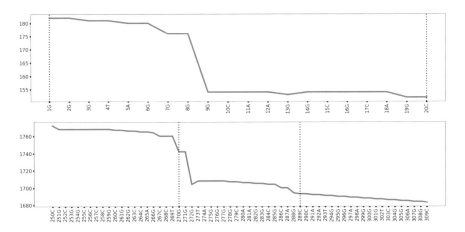

Fig. 4.26 sgRNA/Gene ID RRM2_270_289

The signal graph for sgRNA does not have PS/SS signal. This is an instance of exception handling noted in Step 5 of the algorithm. The number of SS signals for sgRNA is 0 with very high signal value of (−59) in location 224–226 of target gene signal graph. Hence, it is a true positive case (Fig. 4.25).

The sgRNA Primary Signal (PS) between base locations 8 and 9 is aligned with the PS of target gene between base locations 271 and 272 with bases G and G, respectively. Subsequent to PS alignment, the number of secondary signals on 5′ and 3′ end of both signal graphs are 1 and 1. Hence, it is a true positive case (Fig. 4.26).

The sgRNA Primary Signal (PS) between base location 19 and 20 is aligned with the PS of target gene between base location 300 and 301 with bases G and G, respectively. Subsequent to PS alignment, the number of secondary signals on 5′

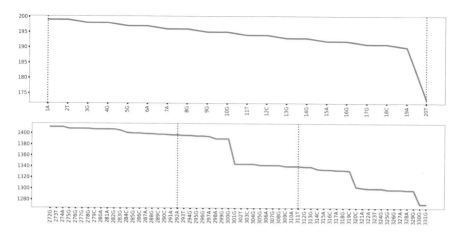

Fig. 4.27 sgRNA/Gene ID RRM2_292_311

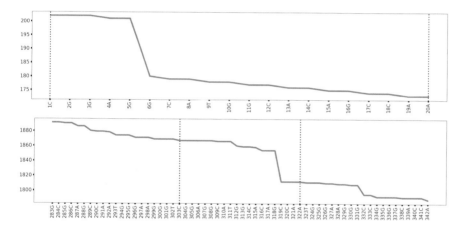

Fig. 4.28 sgRNA/Gene ID RRM2_303_322

and 3′ end of both signal graphs are 0 and 0. Hence, it is a true positive case (Fig. 4.27).

The sgRNA Primary Signal (PS) between base locations 5 and 6 is aligned with the PS of target gene between base locations 318 and 319 with bases G and C, respectively. Subsequent to PS alignment, the number of secondary signals on 5′ and 3′ end of both signal graphs are 0 and 0. Hence, it is a true positive case (Fig. 4.28).

The signal graph for sgRNA and M region of target gene do not have PS/SS signal. This is an instance of exception handling noted in Step 5 of the algorithm. The number of SS signals for both is 0. Hence, it is a true positive case (Fig. 4.29).

The sgRNA Primary Signal (PS) between base locations 13 and 14 is aligned with the PS of target gene between base locations 383 and 385 with bases T and A, respectively. Subsequent to PS alignment, the number of Secondary Signals on 5′

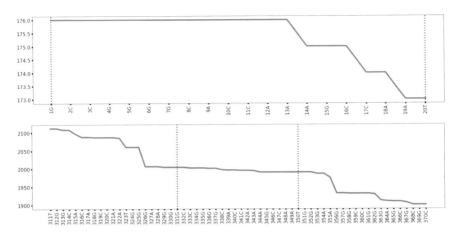

Fig. 4.29 sgRNA/Gene ID RRM2_331_350

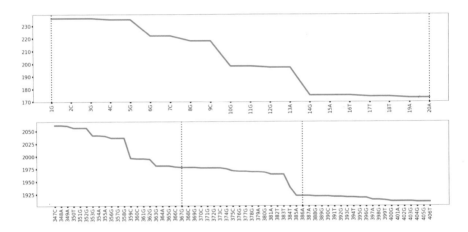

Fig. 4.30 sgRNA/Gene ID RRM2_367_386

and 3′ end of both signal graphs are 2 and 0. Hence, it is a true positive case. (Fig. 4.30).

True Negative Cases

Illustrations of true negative cases follow (Table 4.13). For each of the following cases, subsequent to alignment of Primary Signal (PS) of the signal graphs, the count of Secondary Signals (SSs) on 5′ and 3′ end of sgRNA signal graph and target gene signal graph differ. The result table along with the signal graph pairs for sgRNA and its target gene are reported.

The sgRNA Primary Signal (PS) between base locations 7 and (8, 9) is aligned with the PS of target gene between base locations 37 and 38 with bases C and G, respectively. Subsequent to PS alignment, the number of Secondary Signals on 5′

Table 4.13 Prediction result for binding of sgRNA on its target gene—true negative

sgRNA/Gene ID	M' signal		LoM signal		M signal		RoM signal		sgRNA		Gene		Gene match region for sgRNA sequence		LoM'-M'-RoM'	LoM-M-RoM	True negative
	Location	Value	Location	Value	Location	Value	Location	Value	PSLoc'	PSVal'	PSLoc	PSVal		sgRNA sequence	Cardinality	Cardinality	
RRM2_1_20	5, 8, 9	-10, -4, -8							8-9	-12					1 and 0	0 and 0	Yes
RRM2_7_26	7, 9, 13, 18	-4, -16, -4, 2			14	-5	32, 43	-12, -18	9	-16	43	-18	LoM	27–46	1 and 2	1 and 0	Yes
RRM2_8_27	6, 8, 15, 17	-4, -26, -4, -15	2	-16	14	-34			8	-26	14	-34	LoM-M	7–26	1 and 2	0 and 0	Yes
RRM2_9_28	7, 9	-4, -15			10, 11, 15	-7, -49, -4			9	-15	10-11	-56	LoM-M	1–20	1 and 0	0 and 1	Yes
RRM2_17_36	6	-21	14	-18	18, 19, 25, 34	-26, -121, -10, -19	48, 49	-7, -11	6	-21	18-19	-147	LoM-M	13–32	0 and 0	1 and 1	Yes
RRM2_18_37	4, 18, 20	-12, -57, -9	14	-9	34, 36	-4, -6	41, 43, 54, 55	-4, -12, -4, -6	18	-57	43	-12	M-RoM	26–45	1 and 1	3 and 0	Yes
RRM2_19_38	4, 2–	-4, -17	1, 2, 17	-12, -36, -13	19, 21, 34, 37	-4, -32, -4, -20	4, 43, 50, 53	-16, -30, -4, -30	20	-17	1-2	-48	LoM-M	1–20	1 and 0	0 and 1	Yes
RRM2_29_48	6	-16	20	-4	34	-35	50	-18	6	-16	34	-35	LoM-M		0 and 1	1 and 0	Yes

(continued)

Table 4.13 (continued)

sGRNA/Gene ID	M' signal		LoM signal		M signal		RoM signal		sgRNA		Gene		Gene match region for sgRNA sequence	LoM'-M'-RoM' Cardinality	LoM-M-RoM Cardinality	True negative
	Location	Value	Location	Value	Location	Value	Location	Value	PSLoc'	PSVal'	PSLoc	PSVal				
	8	−7	22	−12			52	−4					28–47			
			27	−4			54	−23								
							60	−4								
							63	−34								
RRM2_30_49	20	−17	20	−7	35	−13	52	−16	20	−17	46–47	−19	LoM-M	0 and 0	1 and 1	Yes
					46	−7	62	−4					29–48			
					47	−12										
RRM2_35_54	20	−17	20	−7	35	2	63	−8	20	−17	20–21	−27	LoM	0 and 0	0 and 2	Yes
			21	−20	51	−4							15–34			
			32	−12	53	−20										
			33	−4												

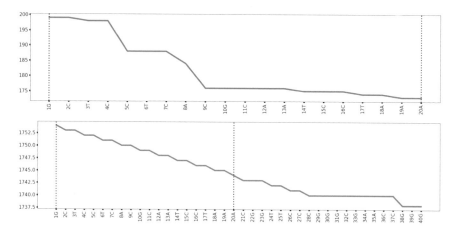

Fig. 4.31 sgRNA/Gene ID RRM2_1_20

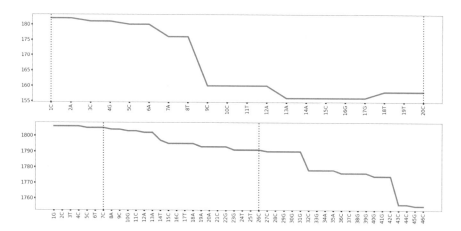

Fig. 4.32 sgRNA/Gene ID RRM2_7_26

and 3' end of sgRNA is (1 and 0), while gene signal graph displays (0 and 0). Due to the mismatch of Secondary Signal (SS) count, it is a true negative case (Fig. 4.31).

The sgRNA Primary Signal (PS) between base locations 8 and 9 is aligned with the PS of target gene between base locations 42 and 43 with bases C and C, respectively. Subsequent to PS alignment, the number of secondary signals on 5' and 3' end of sgRNA is (1 and 2), while gene signal graph displays (1 and 0). Due to the mismatch of Secondary Signal (SS) count, it is a true negative case (Fig. 4.32).

The sgRNA Primary Signal (PS) between base locations 7 and 8 is aligned with the PS of target gene between base locations 13 and 14 with bases A and T, respectively. Subsequent to PS alignment, the number of secondary signals on 5' and 3' ends of sgRNA is (1 and 2), while gene signal graph displays (0 and 0). Due

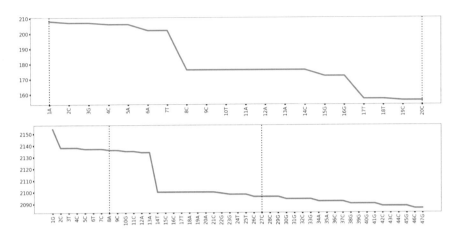

Fig. 4.33 sgRNA/Gene ID RRM2_8_27

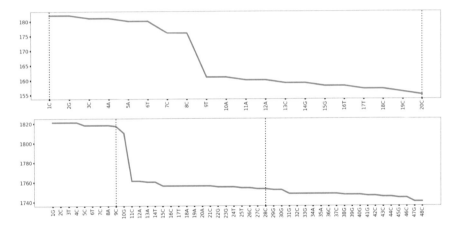

Fig. 4.34 sgRNA/Gene ID RRM2_9_28

to the mismatch of Secondary Signal (SS) count, it is a true negative case (Fig. 4.33).

The sgRNA Primary Signal (PS) between base locations 8 and 9 is aligned with the PS of target gene Primary Signal (PS) between base locations 9 and (10, 11) with bases C and (G, C), respectively. Subsequent to PS alignment, the number of secondary signals on 5′ and 3′ end of sgRNA is (1 and 0), while gene signal graph displays (0 and 1). Due to the mismatch of Secondary Signal (SS) count, it is a true negative case (Fig. 4.34).

The sgRNA Primary Signal (PS) between base locations 5 and 6 is aligned with the PS of target gene between base location 17 and (18, 19) with bases T and (A, A), respectively. Subsequent to PS alignment, the number of secondary signals on 5′ and 3′ end of sgRNA is (0 and 0), while gene signal graph displays (1 and 1).

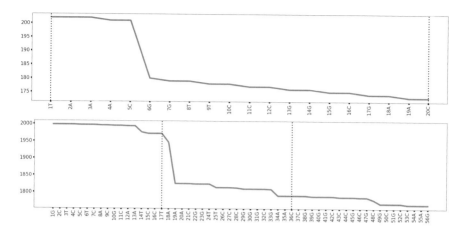

Fig. 4.35 sgRNA/Gene ID RRM2_17_36

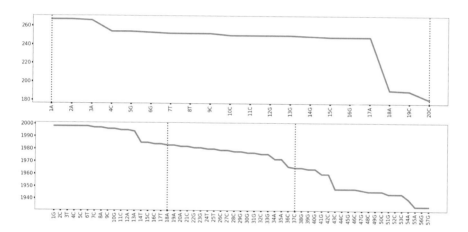

Fig. 4.36 sgRNA/Gene ID RRM2_18_37

Due to the mismatch of Secondary Signal (SS) count, it is a true negative case (Fig. 4.35).

The sgRNA Primary Signal (PS) between base locations 17 and 18 is aligned with the PS of target gene between base location 53 and (54, 55) with bases C and (A, A), respectively. Subsequent to PS alignment, the number of secondary signals on 5′ and 3′ end of sgRNA is (1 and 1), while gene signal graph displays (3 and 0). Due to the mismatch of Secondary Signal (SS) count, it is a true negative case (Fig. 4.36).

The sgRNA Primary Signal (PS) between base locations 18 and 19 is aligned with the PS of target gene between base locations 1 and (2, 3) with bases G and (C, T), respectively. Subsequent to PS alignment, the number of secondary signals on 5′ and 3′ end of sgRNA is (1 and 0), while gene signal graph displays (0 and 1). Due to the mismatch of Secondary Signal (SS) count, it is a true negative case (Fig. 4.37).

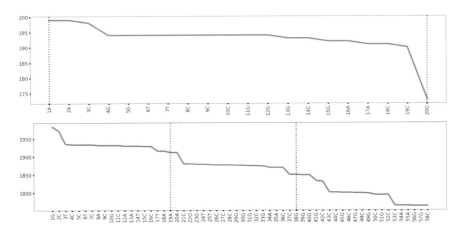

Fig. 4.37 sgRNA/Gene ID RRM2_19_38

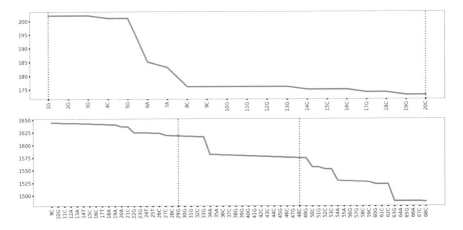

Fig. 4.38 sgRNA/Gene ID RRM2_29_48

The sgRNA Primary Signal (PS) between base locations 5 and 6 is aligned with the PS of target gene between base locations 62 and 63 with bases C and G respectively. Subsequent to PS alignment, the number of secondary signals on 5′ and 3′ end of sgRNA is (0 and 1), while gene signal graph displays (1 and 0). Due to the mismatch of Secondary Signal (SS) count, it is a true negative case (Fig. 4.38).

The sgRNA Primary Signal (PS) between base locations 19 and 20 is aligned with the PS of target gene between base base location 45 and 47 with bases (G, G), respectively. Subsequent to PS alignment, the number of secondary signals on 5′ and 3′ end of sgRNA is (0 and 0), while gene signal graph displays (1 and 1). Due to the mismatch of Secondary Signal (SS) count, it is a true negative case (Fig. 4.39).

The sgRNA Primary Signal (PS) between base location 19 and 20 is aligned with the PS of target gene between base location 18 and (19, 20) with bases A and (A, A),

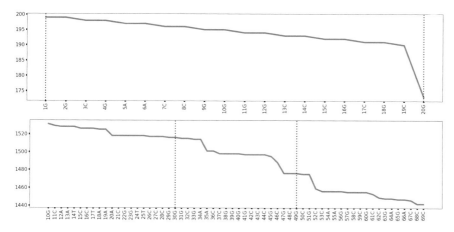

Fig. 4.39 sgRNA/Gene ID RRM2_30_49

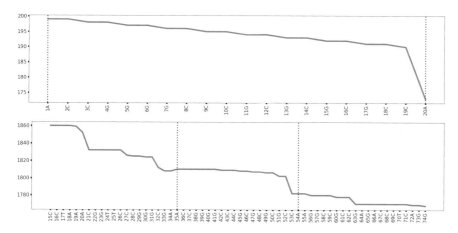

Fig. 4.40 sgRNA/Gene ID RRM2_35_54

respectively. Subsequent to PS alignment, the number of secondary signals on 5′ and 3′ end of sgRNA is (0 and 0), while gene signal graph displays (0 and 2). Due to the mismatch of Secondary Signal (SS) count, it is a true negative case (Fig. 4.40).

4.7 Summary

We initiated this chapter on DNA with the observation of James Gleick, a reputed science writer, "*DNA is the quintessential information molecule, the most advanced message processor at the cellular level—an alphabet and a code, 6 billion bits to form a human being*". A number of scientists have also strongly supported the viewpoint that the significant advances of information processing paradigm should be explored for analysis of the underlying biological process of current life form.

DNA fundamentals introduced in Chap. 1 have been extended in this chapter with the CA model for DNA strands—sense and antisense. The output of CA Machine (CAM) designed with CA rules presented in Chap. 2 is reported for a sample Gene ID J00272 (Human metallothionein-II pseudogene (mt-IIps)) with 607 nucleotide bases. The foundation of CA model for DNA (information storage molecule) has been laid down with the design of three CAMs—backbone DNA CA machine modelling sugar–phosphate backbone (BBDCAM), nucleotide base DNA CA machine modelling base sequence of DNA (NBDCAM) and composite DNA CA machine designed through co-evolution of BBDCAM and NBDCAM (ComDCAM). The Cycle Start (CS) and Cycle Length (CL) table and graph for each CAM (sense and antisense) are presented. The signal graphs derived out CS/CL table output are utilized in subsequent sections to model different characteristics of DNA strand.

The BBDCAM CS/CL output for both sense and antisense are identical since the same five-neighbourhood CA (5NCA) rule is used to model sugar–phosphate molecule of two strands. Irrespective of the length of DNA strand, it can be observed that the CS value of BBDCAM cells uniformly decreases as follows after every four bases. If (say) from base location i to $(i + 3)$ the CS value is x, the CS value for the next 4 cells at location $(i + 4)$ to $(i + 7)$ is $(x - 2)$, next block of 4 CA cells displays CS value is $(x - 4)$ and so on. Section 4.2 of the chapter investigates how this regularity of spatio-temporal characteristics of CA model for DNA sense and antisense strands is co-related with the regularity of DNA helical structure noted in Fig. 4.2. The DNA structure shows four bases in each helical loop of sense and antisense strands crossing each other at a regular interval of four bases with spatial distance covering two bases. An alternative representation of this structural feature can be observed in the BBDCAM output.

Section 4.3 presents the functionally important regions of a DNA strand. The regions are marked on the mRNA transcript derived out of a DNA antisense strand. Transcription is a highly complex biological process in all living organisms. Proteins/enzymes interacting with different functionally important regions of DNA are getting identified through biochemical analysis, genetics and recent advances in imaging of living cells. We have presented relevant information on the functionally important regions in Sect. 4.3 prior to focusing on the specific region associated with exon–intron boundary. Introns are removed out of transcript to generate mature mRNA string. Rather than building CA model to predict intron–exon boundary, our major emphasis in Sect. 4.4 is to study the process of alternative splicing of exons. Such differential splicing refers to the biological process of merging different combinations of exons in a transcript to make different mature mRNAs out of the same gene; depending on the prevailing physical domain conditions, each combination of exons generates different protein product. Splicing is a highly regulated biological process controlled by a large molecular machine complex called 'splicesome', an assembly of small RNAs and proteins. Any deviation of this differential splicing is frequently associated with different diseases. In

Sect. 4.4, we have reported CA model to define valid TG/AG boundary of an intron in an mRNA transcript in addition to predicting start and stop codon location in a gene. The challenging task is to pick up the correct TG/AG pair for the prevailing physical domain condition. We have presented the algorithm to predict exon–intron–exon boundary along with the provision for skipping an exon. Large-scale experimentation on running program code for the algorithm is in progress. We expect that the model and the associated parameters derived out of CAM evolution would help for in silico analysis of physical domain condition relevant for alternative splicing.

Sections 4.5 reports the CA model of Single Nucleotide Polymorphism (SNP) in cLife. The focus is to determine the effect of the mutant due to SNP on the expression level of mature mRNA or protein product. The algorithm is reported in this section to derive CA parameter values out of the CAMs for mutant and its wild version. The difference of the parameter values between wild and mutant are categorized as high/medium/low difference. Signal graph analytics methodology enables mapping of high/medium/low difference of CA model parameters to the physical domain features of high/medium/low difference of physical domain features in mutant compared to that for its wild version. Two case studies are reported for disease-related SNPs: one for miRNA precursor molecule and other for the mRNA transcript of a KRAS gene. The wet lab results for both of these case studies are recently reported. We have also presented predicted results for some mutants derived out of the wild version of these two cases studies with results similar to that reported in published literature. Large-scale wet lab experiments for mutational study of a biomolecule involves high cost and time. In this context, we hope the experimentalists can narrow down to a limited cases to be analysed through wet lab experiment on extracting results of our in silico platform based on CA model.

Finally, the CA model in cLife for recent gene editing technology of CRISPR/Cas9 is reported in Sect. 4.6. In view of high interest generated for this gene editing methodology, a number of databases report sgRNAs designed for its binding on target gene. We have reported CA model parameters to predict binding of a single guide RNA (sgRNA) on its target gene. The parameters are derived out of ComDCAM designed for DNA strand base sequence and ComRCAM for RNA transcript sgRNA sequence. Algorithm for prediction of true positive and true negative instances of binding of sgRNA on target gene are reported followed by validation of the results for the sgRNA and target gene pair available in databases. Design of sgRNA for a target gene minimizing false positive instances is the major challenge for application of this technology. We expect that CA model parameter reported in this section will add value for in silico analysis of false positive instances.

Questions

1. Note the characteristic features of the double helix structure of DNA molecule. 'DNA molecules are helical with two periodicities along their long axis, a primary one of 3.4A and a secondary one of 34A'—True or False? If 'True', justify your answer.

2. What is a gene? What problem do you anticipate if instead of DNA, an RNA molecule is entrusted with the task of storing hereditary information?

3. What are the functionally important regions in a gene? Note the underlying principle of 'Transcription' process. Explain 5' and 3' UTR and their role in the biological process of protein synthesis.

4. An mRNA molecule is transcribed from the DNA sequence and is next translated into a protein. Elaborate the sequential steps of biological synthesis of protein out of a gene.

5. How is matured mRNA derived out of mRNA transcript? What is the role of 'spliceosome' protein complex?

6. Explain how 'alternative splicing' enables synthesis of multiple protein products out of the same gene?

7. 'SNP can be used as a biological marker' True or False? If 'True', describe the functional role of such markers on disease-associated genes.

8. Synonymous SNP can have deleterious effect—True or False? If 'True', describe the reasons for such adverse effect from the study of recent published literature in this field.

9. Define the acronym CRISPR. What does Cas9 represent? Note the underlying principle of CRISPR/Cas9 gene editing technology along with the historical background of its development.

10. From the study of recently published materials, note the methodology employed for design of synthetic sgRNA for editing a target gene, while reducing off-target effect.

11. Select a partial sequence of one gene of NCBI database. Design the two CAMs (BBRCAM, NBRCAM) and derive the CS/CL table for the mRNA transcript for the gene. Draw the CS Difference (CSD) graph for the CAMs.

12. For the example worked for question 11, derive ComRCAM CS/CL graph and CSD graph.

13. Note the sequential steps implemented in the CA model to predict binding of a sgRNA and its target gene.

14. Select one sgRNA from the database and note the signal graph output derived out of its ComRCAM.

Annexure

Annexure 4.1: Signal Graph of a Full-Length DNA with 607 Nucleotide Bases

In Sect. 4.2, we have included the partial table and signal graphs for Gene ID J00272.

The signal graphs derived out of different CA Machines (CAMs) are reported in this annexure. Results derived with signal graph analytics methodologies are covered in the main text. Readers can get a clear picture in respect of the evolution of different CAMs for sense and antisense strands of DNA with the similarity of their signal graphs.

The Cycle Start (CS) signal graph for full-length DNA with 607 nucleotide bases (sense and antisense strands) is reported for three CAMs—BBDCAM, NBDCAM and ComDCAM below. The *Y*-axis shows the CS values and *X*-axis gives the nucleotide base locations (Figs. 4.41, 4.42, 4.43, 4.44, 4.45 and 4.46).

Annexure 4.2: Full ATG list derived out of gene IDS from database

Figure 4.3 shows a partial set of Splicing Rule (SR) set to identify valid W (W ∈ {ATG, xGT, AGy, stop codon}). Different rules of the SR set are illustrated with reference to valid W in NCBI Genes (https://www.ncbi.nlm.nih.gov/gene) noted in column 1. Table 4.3b reports a partial list of ATGs derived out of signal graph for transcripts of the genes. The full table is shown below after Fig. 4.46.

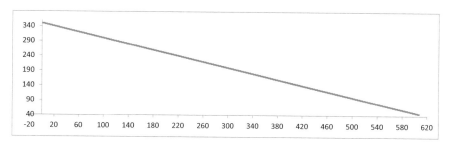

Fig. 4.41 CS graph for BBDCAM Gene ID J00272 sense strand

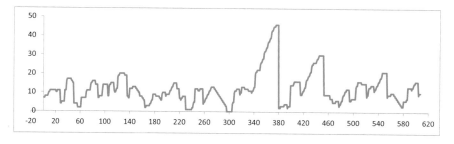

Fig. 4.42 CS graph for NBDCAM Gene ID J00272 sense strand

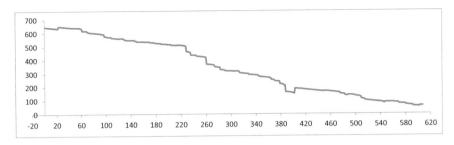

Fig. 4.43 CS graph for ComGene ID J00272 sense strand

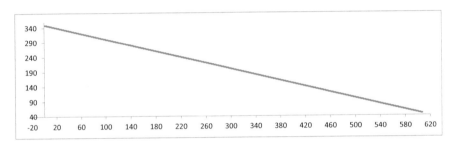

Fig. 4.44 CS graph for BBDCAM Gene ID J00272 antisense strand

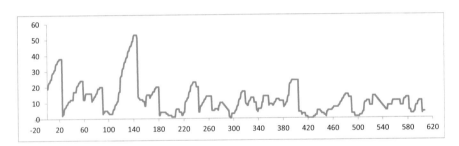

Fig. 4.45 CS graph for NBDCAM Gene ID J00272 antisense strand

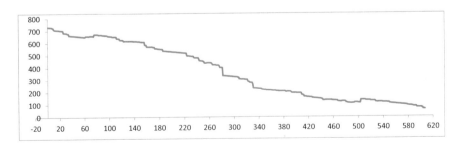

Fig. 4.46 CS graph for ComDCAM Gene ID J00272 antisense strand

ID	ATG	Location of ATG	CSDF start	CSDF end	5' end val	3' end val	Pattern 1	Pattern 2	Pattern 3	Pattern 4	CL value	Loc of CL	ATG pos in CSDF	DF CSD sign
J00271	−1:0:−1	45	28	49	8	−4	0	73	18	9	272	CLF0	3'	+ive +ive
J00271	−1:0:−1	166	155	175	−6	−14	0	86	0	14	272	CLF0	5'/3'	+ive +ive
J00271	0:−4:0	193	189	196	−8	−8	0	0	50	50	272	CLF1	3'	+ive +ive
J00271	0:−1:−3	317	307	323	−12	−8	18	24	41	18	272	CLF1	3'	+ive +ive
J00277	−1:0:−1	154	151	166	−4	−35	0	88	0	13	272	CLF0	5'	+ive +ive
J00277	−8:0:0	353	353	381	−8	1	0	34	66	0	272	CLF0	5'	+ive +ive
J00277	0:0:0	367	353	381	−8	1	0	34	66	0	272	CLF0	5'/3'	+ive +ive
J00277	0:−1:0	488	482	512	−23	−4	0	90	0	10	272	CLF0	5'	+ive +ive
J00277	−1:0:−1	1082	1073	1093	−6	−9	0	86	0	14	272	CLF0	5'/3'	+ive +ive
J00277	−2:0:0	1536	1527	1545	−4	−15	47	11	21	21	272	CLF0	3'	+ive +ive
X00806	0:−4:0	49	34	52	−4	−29	0	42	42	16	272	CLF0	3'	+ive +ive
X00806	−7:−8:0	132	132	138	−7	−4	0	0	57	43	272	CLF0	5'	+ive +ive
X00806	0:0:0	187	185	201	−4	−4	0	47	47	6	272	CLF0	5'	+ive +ive
X00806	−1:0:−4	199	185	201	−4	−4	0	47	47	6	272	CLF0	3'	+ive +ive
X00806	−4:0:0	228	228	243	−4	−11	0	13	63	25	272	CLF0	5'	+ive +ive
X00806	0:0:0	269	262	285	−10	−11	25	25	50	0	272	CLF1	5'	+ive +ive
X00806	0:−1:0	355	349	378	−6	−4	0	93	0	7	272	CLF1	5'	+ive +ive
X00806	−1:0:−1	441	428	449	9	−4	0	73	18	9	272	CLF1	3'	+ive +ive
X00806	1:0:0	488	488	493	1	1	0	0	67	33	272	CLF2	5'	+ive +ive
X00806	0:−1:0	619	593	627	−4	1	0	86	11	3	272	CLF2	3'	+ive +ive
X00806	−1:0:0	643	635	654	−10	−11	30	20	45	5	272	CLF2	5'/3'	+ive +ive
X00806	−1:0:−1	709	706	713	−4	−4	0	75	0	25	272	CLF3	5'	+ive +ive
X00806	−1:0:−1	747	729	761	−4	−4	0	67	30	3	272	CLF3	5'/3'	+ive +ive
X00806	0:−4:0	784	772	801	−42	−4	10	33	47	10	272	CLF3	5'/3'	+ive +ive
X00806	0:0:0	869	867	888	1	−46	0	36	55	9	272	CLF3	5'	+ive +ive

(continued)

(continued)

1	2	3	4		5		6				7	8	9	10
ID	ATG	Location of ATG	CSDF start	CSDF end	5' end val	3' end val	Pattern 1	Pattern 2	Pattern 3	Pattern 4	CL value	Loc of CL	ATG pos in CSDF	DF CSD sign
X00806	0:0:0	914	913	922	-4	-4	0	0	80	20	272	CLF3	5'	+ive +ive
X03821	0:-4:0	13	5	23	-4	-6	16	11	42	32	272	CLF0	5'/3'	+ive +ive
X03821	0:0:0	42	40	70	-4	-4	0	52	42	6	272	CLF0	5/	+ive +ive
X03821	0:-4:0	69	40	72	-4	-6	0	48	39	12	272	CLF0	3'	+ive +ive
X03821	-1:0:-1	89	77	93	-4	-4	0	47	47	6	272	CLF0	3'	+ive +ive
X03821	0:-6:-1	110	109	115	-4	-4	0	29	0	71	272	CLF0	5'	+ive +ive
X03821	0:-1:-4	130	127	132	-6	-4	0	33	0	67	272	CLF0	3'	+ive +ive
X03821	0:-1:0	158	151	173	-8	-4	0	70	26	4	272	CLF1	5'	+ive +ive
X03821	-1:0:-1	169	151	173	-8	-4	0	70	26	4	272	CLF1	3'	+ive +ive
X15227	-1:-1:0	21	6	79	-4	-4	0	81	16	3	272	CLF0	5'	+ive +ive
X15227	0:0:-1	122	116	126	-22	-4	55	18	0	27	272	CLF0	5'/3'	+ive +ive
X15227	-1:1:0	184	179	189	-5	2	0	18	45	36	272	CLF0	5'/3'	+ive +ive
X15227	-2:0:0	332	328	340	-11	-27	46	0	23	31	272	CLF1	5'	+ive +ive
X15227	-1:0:-1	375	369	388	-16	-4	0	60	25	15	272	CLF1	5'	+ive +ive
X15227	-1:0:-1	399	388	404	-4	-4	0	35	47	18	272	CLF2	3'	+ive +ive
X54518	-1:0:-1	62	60	68	-15	-4	0	67	0	33	272	CLF0	5'	+ive +ive
X54518	0:-1:0	65	60	68	-15	-4	0	67	0	33	272	CLF0	3'	+ive +ive
X54518	0:0:0	205	200	209	-4	-4	0	0	80	20	272	CLF0	5'/3'	+ive +ive
X54518	0:0:0	225	224	240	-4	-26	0	35	47	18	272	CLF0	5'	+ive +ive
X54518	-1:0:-1	311	299	315	-8	-4	0	47	47	6	272	CLF0	3'	+ive +ive
X54518	0:-1:0	325	317	333	-12	1	0	47	47	6	272	CLF0	5'/3'	+ive +ive
X54518	0:0:0	423	414	437	-12	-57	50	0	29	21	272	CLF0	5'	+ive +ive
X54518	0:0:0	519	517	531	-4	-4	0	0	87	13	272	CLF0	5'/3'	+ive +ive
X54518	0:-1:0	550	542	556	2	1	0	53	27	20	272	CLF1	5'3'	+ive +ive

(continued)

(continued)

1	2	3	4		5		6				7	8	9	10
ID	ATG	Location of ATG	CSDF start	CSDF end	5' end val	3' end val	Pattern 1	Pattern 2	Pattern 3	Pattern 4	CL value	Loc of CL	ATG pos in CSDF	DF CSD sign
X54518	0:-2:0	562	556	574	1	-4	16	11	63	11	272	CLF1	5'	+ive +ive
X54518	0:0:0	566	556	574	1	-4	16	11	63	11	272	CLF1	5'	+ive +ive
X54518	0:0:0	576	574	620	-4	11	0	68	30	2	272	CLF1	5'	+ive +ive
X54518	0:0:0	744	740	748	-54	-9	0	22	56	22	272	CLF1	5'/3'	+ive +ive
X54518	-4:0:0	817	817	820	-4	-6	0	0	50	50	272	CLF1	5'	+ive +ive
X54518	0:-1:-22	894	888	896	-34	-22	0	67	0	33	272	CLF1	3'	+ive +ive
X54518	0:0:0	934	932	954	1	-4	0	43	57	0	272	CLF1	5'	+ive +ive
X54518	0:-1:0	1034	1010	1040	1	-19	0	45	45	10	272	CLF1	3'	+ive +ive
X54518	-2:0:0	1203	1109	1249	-11	-31	70	1	4	25	272	CLF2	3'	+ive +ive
X54518	-1:0:-1	1268	1260	1275	-15	-4	0	75	0	25	272	CLF2	5'/3'	+ive +ive
X54518	0:0:0	1278	1275	1284	-4	-4	0	0	80	20	272	CLF2	5'	+ive +ive
X54518	-1:-4:0	1329	1304	1333	-4	-12	0	67	20	13	272	CLF2	3'	+ive +ive
X54518	-4:0:-18	1363	1363	1365	-4	-18	0	0	0	100	272	CLF2	5'	+ive +ive
X54518	0:0:0	1520	1509	1530	-4	-4	0	27	59	14	272	CLF2	5'/3'	+ive +ive
X54518	0:-1:0	1641	1636	1652	-15	-4	0	82	0	18	272	CLF3	5'	+ive +ive
X65606	0:0:-1	13	0	24	0	-4	12	32	52	4	272	CLF0	5'/3'	+ive +ive
X65606	-4:0:-6	66	66	68	-4	-6	0	0	0	100	272	CLF0	5'	+ive +ive
X65606	0:0:0	108	106	112	-4	1	0	29	57	14	272	CLF0	5'	+ive +ive
X65606	-1:-1:0	125	112	135	1	-4	0	8	83	8	272	CLF0	5'/3'	+ive +ive
X65606	0:0:0	136	135	143	-4	-13	0	0	78	22	272	CLF0	5'	+ive +ive
X65606	0:-1:0	156	155	164	-9	-4	0	40	50	10	272	CLF0	5'	+ive +ive
X65606	-1:1:0	166	164	176	-4	1	0	15	54	31	272	CLF0	5'	+ive +ive

(continued)

(continued)

1	2	3	4		5		6				7	8	9	10
ID	ATG	Location of ATG	CSDF start	CSDF end	5' end val	3' end val	Pattern 1	Pattern 2	Pattern 3	Pattern 4	CL value	Loc of CL	ATG pos in CSDF	DF CSD sign
X65606	0:−1:−5	271	263	273	1	−5	0	36	36	27	272	CLF0	3'	+ive +ive
X65606	0:1:0	291	284	300	−6	−26	0	35	59	6	272	CLF0	5'/3'	+ive +ive
X65606	−1:1:0	340	339	355	−6	−8	18	0	59	24	272	CLF0	5'	+ive +ive
X65606	0:−1:0	365	355	368	−8	−4	0	57	36	7	272	CLF0	3'	+ive +ive
X65606	0:0:2	408	406	410	1	2	0	0	60	40	272	CLF0	5'/3'	+ive +ive
X65606	0:0:0	95	79	108	−49	−4	0	33	63	3	272	CLF3	3'	+ive +ive
X69215	−1:0:−1	127	124	134	−4	1	0	55	36	9	136	CLF0	5'	+ive +ive
X69215	−60:0:−4	210	210	212	−60	−4	0	0	0	100	136	CLF0	5'	+ive +ive
X69215	0:−26:−21	306	303	308	−15	−21	0	33	0	67	136	CLF0	3'	+ive +ive
X69215	0:−1:0	353	352	356	−20	−4	60	40	0	0	136	CLF0	5'	+ive +ive
Z12021	0:−2:0	44	19	58	−62	−11	68	5	0	28	544	CLF0	3'	+ive +ive
Z12021	0:0:−15	97	96	99	−4	−15	0	0	50	50	544	CLF0	5'	+ive +ive
Z12021	0:0:−1	118	115	147	2	−6	9	30	58	3	544	CLF0	5'	+ive +ive
Z12021	0:0:0	190	187	211	4	−9	24	16	48	12	544	CLF0	5'	+ive +ive
Z12021	0:0:0	804	802	810	−4	−4	0	0	78	22	544	CLF1	5'	+ive +ive
Z12021	0:0:0	871	866	875	−4	−9	0	20	50	30	544	CLF1	5'/3'	+ive +ive
Z12021	0:0:0	878	875	890	−9	−7	56	13	0	31	544	CLF1	5'	+ive +ive
Z12021	0:−1:0	944	940	948	1	−6	0	22	67	11	544	CLF1	5'/3'	+ive +ive
Z12021	−1:0:−1	969	948	988	−6	1	0	83	15	2	544	CLF1	5'/3'	+ive +ive
Z12021	0:0:−1	1040	1035	1048	−8	−10	0	57	36	7	544	CLF1	5'	+ive +ive
Z12021	−1:0:0	1110	1099	1115	−7	−9	35	12	29	24	544	CLF2	3'	+ive +ive
Z12021	0:0:0	1134	1115	1140	−9	−6	58	0	15	27	544	CLF2	3'	+ive +ive

(continued)

(continued)

ID	ATG	Location of ATG	CSDF start	CSDF end	5' end val	3' end val	Pattern 1	Pattern 2	Pattern 3	Pattern 4	CL value	Loc of CL	ATG pos in CSDF	DF CSD sign
ZI2021	−1:0:−1	1285	1271	1296	−30	−4	12	54	35	0	544	CLF2	5'/3'	+ive +ive
ZI2021	0:−1:0	1331	1317	1338	12	−4	0	73	18	9	544	CLF2	3'	+ive +ive
ZI2021	−8:−4:−1	1584	1584	1591	−8	−9	0	50	0	50	544	CLF2	5'	+ive +ive
ZI2021	0:−1:0	1646	1644	1660	−40	1	0	71	24	6	544	CLF3	5'	+ive +ive
ZI2021	0:0:0	1656	1644	1660	−40	1	0	71	24	6	544	CLF3	3'	+ive +ive
AF092923	0:−3:0	22	21	28	−27	−4	38	0	0	63	136	CLF0	5'	+ive +ive
AF092923	−1:0:−1	303	289	312	2	−4	0	58	29	13	68	CLF0	5'/3'	+ive +ive
AF092923	0:−1:0	306	289	312	2	−4	0	58	29	13	68	CLF0	3'	+ive +ive
BC093995	0:−1:0	2	0	77	0	−4	0	95	3	3	68	CLF0	5'	+ive +ive
DQ168992	−1:0:−1	22	17	29	−8	−31	0	46	31	23	136	CLF0	5'/3'	+ive +ive
DQ168992	0:−15:0	48	46	63	−4	1	0	67	11	22	136	CLF0	5'	+ive +ive
DQ168992	0:−1:0	181	162	230	−15	−12	0	96	0	4	136	CLF0	5'/3'	+ive +ive
J00272	−1:0:−1	56	0	144	0	−4	0	95	3	2	136	CLF0	5'/3'	+ive +ive

Annexure 4.3: NBRCAM Table of Pre-miR-30c Wild, G27A, G27C, G27T

This annexure reports the full table for NBRAM for wild and its mutants (G27A, G27C and G27T). A partial version of the table is presented in Table 4.5. Critical pairs are shown in grey shaded rows.

Position	Wild				G27A				G27C				G27T			
	NB	CS	CL	CSD	NB	CS	CL	CSD	NB	CS	CL	CSD	NB	CS	CL	CSD
1	T	11	1		T	11	1		T	11	1		T	11	1	
2	G	10	1	-1	G	10	1	-1	G	10	1	-1	G	10	1	-1
3	G	10	1	0	G	10	1	0	G	10	1	0	G	10	1	0
4	G	9	1	-1	G	9	1	-1	G	9	1	-1	G	9	1	-1
5	C	9	1	0	C	9	1	0	C	9	1	0	C	9	1	0
6	T	7	1	-2	T	7	1	-2	T	7	1	-2	T	7	1	-2
7	A	7	1	0	A	7	1	0	A	7	1	0	A	7	1	0
8	T	7	1	0	T	7	1	0	T	7	1	0	T	7	1	0
9	A	7	1	0	A	7	1	0	A	7	1	0	A	7	1	0
10	A	7	1	0	A	7	1	0	A	7	1	0	A	7	1	0
11	C	7	1	0	C	7	1	0	C	7	1	0	C	7	1	0
12	C	12	1	5	C	12	1	5	C	12	1	5	C	12	1	5
13	A	13	1	1	A	13	1	1	A	13	1	1	A	13	1	1
14	T	13	1	0	T	13	1	0	T	13	1	0	T	13	1	0
15	G	14	1	1	G	14	1	1	G	14	1	1	G	14	1	1
16	C	15	1	1	C	15	1	1	C	15	1	1	C	15	1	1
17	T	15	1	0	T	15	1	0	T	15	1	0	T	15	1	0
18	G	16	1	1	G	16	1	1	G	16	1	1	G	16	1	1
19	T	17	1	1	T	17	1	1	T	17	1	1	T	17	1	1
20	A	17	1	0	A	17	1	0	A	17	1	0	A	17	1	0
21	G	17	4	0	G	17	4	0	G	17	4	0	G	17	4	0
22	T	17	4	0	T	17	4	0	T	17	4	0	T	17	4	0
23	G	17	4	0	G	17	4	0	G	17	4	0	G	17	4	0
24	T	17	4	0	T	17	4	0	T	17	4	0	T	17	4	0
25	G	17	4	0	G	17	4	0	G	17	4	0	G	17	4	0
26	T	7	4	-10	T	7	4	-10	T	7	4	-10	T	7	4	-10
27	G	6	4	-1	G	6	4	-1	G	6	4	-1	G	6	4	-1
28	T	6	4	0	T	6	4	0	T	6	4	0	T	6	4	0
29	A	6	4	0	A	6	4	0	A	6	4	0	A	6	4	0
30	A	6	4	0	A	6	4	0	A	6	4	0	A	6	4	0
31	A	6	4	0	A	6	4	0	A	6	4	0	A	6	4	0
32	C	13	8	7	C	15	1	9	C	16	8	10	C	13	8	7
33	A	13	8	0	A	15	1	0	A	16	8	0	A	13	8	0
34	T	13	8	0	T	15	1	0	T	16	8	0	T	13	8	0
35	C	13	8	0	C	15	1	0	C	16	8	0	C	13	8	0
36	C	13	8	0	C	15	1	0	C	16	8	0	C	13	8	0
37	T	12	8	-1	T	14	1	-1	T	15	8	-1	T	12	8	-1
38	A	12	8	0	A	14	1	0	A	15	8	0	A	12	8	0
39	C	11	8	-1	C	14	1	0	C	14	8	-1	C	11	8	-1
40	A	11	8	0	A	14	1	0	A	14	8	0	A	11	8	0
41	C	11	8	0	C	14	1	0	C	14	8	0	C	11	8	0
42	T	11	8	0	T	14	1	0	T	14	8	0	T	11	8	0
43	C	11	8	0	C	11	1	-3	C	14	8	0	C	11	8	0
44	T	11	8	0	T	11	1	0	T	14	8	0	T	11	8	0
45	C	9	8	-2	C	10	1	-1	C	12	8	-2	C	9	8	-2
46	A	9	8	0	A	10	1	0	A	12	8	0	A	9	8	0
47	G	9	8	0	G	10	1	0	G	12	8	0	G	9	8	0
48	C	9	8	0	C	10	4	0	C	12	8	0	C	9	8	0
49	T	8	1	-1	T	10	4	0	T	11	1	-1	T	8	1	-1
50	G	8	1	0	G	10	4	0	G	11	1	0	G	8	1	0

Pos																
51	T	7	1	-1	T	4	4	-6	T	10	1	-1	T	7	1	-1
52	G	7	1	0	A	4	4	0	C	10	1	0	G	7	1	0
53	A	6	1	-1	A	5	4	1	A	10	1	0	A	6	1	-1
54	G	6	1	0	G	5	4	0	G	10	1	0	G	6	1	0
55	C	11	1	5	C	11	1	6	C	11	1	1	C	11	1	5
56	T	13	1	2	T	13	1	2	T	13	1	2	T	13	1	2
57	C	13	1	0	C	13	1	0	C	13	1	0	C	13	1	0
58	A	13	1	0	A	13	1	0	A	13	1	0	A	13	1	0
59	A	13	1	0	A	13	1	0	A	13	1	0	A	13	1	0
60	G	13	1	0	G	13	1	0	G	13	1	0	G	13	1	0
61	G	13	1	0	G	13	1	0	G	13	1	0	G	13	1	0
62	T	13	1	0	T	13	1	0	T	13	1	0	T	13	1	0
63	G	9	1	-4	G	9	1	-4	G	9	1	-4	G	9	1	-4
64	G	8	1	-1	G	8	1	-1	G	8	1	-1	G	8	1	-1
65	C	8	1	0	C	8	1	0	C	8	1	0	C	8	1	0
66	T	7	1	-1	T	7	1	-1	T	7	1	-1	T	7	1	-1
67	G	7	1	0	G	7	1	0	G	7	1	0	G	7	1	0
68	G	6	1	-1	G	6	1	-1	G	6	1	-1	G	6	1	-1
69	G	6	1	0	G	6	1	0	G	6	1	0	G	6	1	0
70	A	4	1	-2	A	4	1	-2	A	4	1	-2	A	4	1	-2
71	G	4	1	0	G	4	1	0	G	4	1	0	G	4	1	0
72	A	25	4	21	A	25	4	21	A	25	4	21	A	25	4	21
73	G	25	4	0	G	25	4	0	G	25	4	0	G	25	4	0
74	G	25	4	0	G	25	4	0	G	25	4	0	G	25	4	0
75	G	25	4	0	G	25	4	0	G	25	4	0	G	25	4	0
76	T	25	4	0	T	25	4	0	T	25	4	0	T	25	4	0
77	T	24	4	-1	T	24	4	-1	T	24	4	-1	T	24	4	-1
78	G	24	4	0	G	24	4	0	G	24	4	0	G	24	4	0
79	T	24	4	0	T	24	4	0	T	24	4	0	T	24	4	0
80	T	24	4	0	T	24	4	0	T	24	4	0	T	24	4	0
81	T	22	4	-2	T	22	4	-2	T	22	4	-2	T	22	4	-2
82	A	22	4	0	A	22	4	0	A	22	4	0	A	22	4	0
83	C	22	4	0	C	22	4	0	C	22	4	0	C	22	4	0
84	T	22	4	0	T	22	4	0	T	22	4	0	T	22	4	0
85	C	20	4	-2	C	20	4	-2	C	20	4	-2	C	20	4	-2
86	C	20	4	0	C	20	4	0	C	20	4	0	C	20	4	0
87	T	16	4	-4	T	16	4	-4	T	16	4	-4	T	16	4	-4
88	T	16	4	0	T	16	4	0	T	16	4	0	T	16	4	0
89	C	15	4	-1	C	15	4	-1	C	15	4	-1	C	15	4	-1
90	T	15	4	0	T	15	4	0	T	15	4	0	T	15	4	0
91	G	10	4	-5	G	10	4	-5	G	10	4	-5	G	10	4	-5
92	C	9	4	-1	C	9	4	-1	C	9	4	-1	C	9	4	-1
93	C	9	4	0	C	9	4	0	C	9	4	0	C	9	4	0
94	A	8	4	-1	A	8	4	-1	A	8	4	-1	A	8	4	-1
95	T	8	4	0	T	8	4	0	T	8	4	0	T	8	4	0
96	G	7	4	-1	G	7	4	-1	G	7	4	-1	G	7	4	-1
97	G	6	1	-1	G	6	1	-1	G	6	1	-1	G	6	1	-1
98	A	6	1	0	A	6	1	0	A	6	1	0	A	6	1	0
99	A	6	1	0	A	6	1	0	A	6	1	0	A	6	1	0
100	A	1	1	-5	A	1	1	-5	A	1	1	-5	A	1	1	-5
101	A	2	1	1	A	2	1	1	A	2	1	1	A	2	1	1
102	C	2	1	0	C	2	1	0	C	2	1	0	C	2	1	0
103	A	2	1	0	A	2	1	0	A	2	1	0	A	2	1	0
104	T	2	1	0	T	2	1	0	T	2	1	0	T	2	1	0
105	C	2	1	0	C	2	1	0	C	2	1	0	C	2	1	0
106	A	8	1	6	A	8	1	6	A	8	1	6	A	8	1	6
107	G	8	6	0	G	8	6	0	G	8	6	0	G	8	6	0
108	C	8	6	0	C	8	6	0	C	8	6	0	C	8	6	0
109	T	8	6	0	T	8	6	0	T	8	6	0	T	8	6	0
110	G	8	6	0	G	8	6	0	G	8	6	0	G	8	6	0
111	A	8	6	0	A	8	6	0	A	8	6	0	A	8	6	0

Annexure 4.4: NBRCAM for KRAS Wild, G12V, T180G and C39T

Reported below is the full table for NBRCAM for wild and mutants (G12V, T180G and C39T). Critical pairs are shown in grey shaded rows. A partial table is shown in Table 4.7.

Position		WILD CS	CL	CsD		G12V CS	CL	CsD		T180G CS	CL	CsD		C39T CS	CL	CsD
1	A	9	11		A	9	11		A	9	11		A	9	11	
2	T	9	11	0	T	9	11	0	T	9	11	0	T	9	11	0
3	G	9	11	0	G	9	11	0	G	9	11	0	G	9	11	0
4	A	9	11	0	A	9	11	0	A	9	11	0	A	9	11	0
5	C	8	11	-1	C	8	11	-1	C	8	11	-1	C	8	11	-1
6	T	8	11	0	T	8	11	0	T	8	11	0	T	8	11	0
7	G	6	11	-2	G	6	11	-2	G	6	11	-2	G	6	11	-2
8	A	6	11	0	A	6	11	0	A	6	11	0	A	6	11	0
9	A	6	11	0	A	6	11	0	A	6	11	0	A	6	11	0
10	T	6	11	0	T	6	11	0	T	6	11	0	T	6	11	0
11	A	1	2	-5	A	1	2	-5	A	1	2	-5	A	1	2	-5
12	T	0	2	-1	T	0	2	-1	T	0	2	-1	T	0	2	-1
13	A	0	2	0	A	0	2	0	A	0	2	0	A	0	2	0
14	A	1	2	1	A	1	2	1	A	1	2	1	A	1	2	1
15	A	2	2	1	A	2	2	1	A	2	2	1	A	2	2	1
16	C	2	1	0	C	2	1	0	C	2	1	0	C	2	1	0
17	T	2	1	0	T	2	1	0	T	2	1	0	T	2	1	0
18	T	2	1	0	T	2	1	0	T	2	1	0	T	2	1	0
19	G	4	1	2	G	4	1	2	G	4	1	2	G	4	1	2
20	T	4	1	0	T	4	1	0	T	4	1	0	T	4	1	0
21	G	4	1	0	G	4	1	0	G	4	1	0	G	4	1	0
22	G	4	1	0	G	4	1	0	G	4	1	0	G	4	1	0
23	T	4	1	0	T	4	1	0	T	4	1	0	T	4	1	0
24	A	4	1	0	A	4	1	0	A	4	1	0	A	4	1	0
25	G	5	1	1	G	8	1	4	G	5	1	1	G	5	1	1
26	T	5	1	0	T	8	1	0	T	5	1	0	T	5	1	0
27	T	5	1	0	T	8	1	0	T	5	1	0	T	5	1	0
28	G	5	1	0	G	8	1	0	G	5	1	0	G	5	1	0
29	G	5	1	0	G	8	1	0	G	5	1	0	G	5	1	0
30	A	4	1	-1	A	7	1	-1	A	4	1	-1	A	4	1	-1
31	G	4	1	0	G	7	1	0	G	4	1	0	G	4	1	0
32	C	5	1	1	C	24	1	17	C	5	1	1	C	5	1	1
33	T	7	1	2	T	24	1	0	T	7	1	2	T	7	1	2
34	G	7	1	0	G	24	1	0	G	7	1	0	G	7	1	0
35	G	9	1	2	T	24	1	0	G	9	1	2	G	19	1	12
36	T	10	1	1	T	24	1	0	T	10	1	1	T	20	1	1
37	G	10	1	0	G	23	1	-1	G	10	1	0	G	20	1	0
38	G	23	1	13	G	23	1	0	G	23	1	13	G	21	1	1
39	C	23	1	0	C	23	1	0	C	23	1	0	T	21	1	0
40	G	23	1	0	G	23	1	0	G	23	1	0	G	21	1	0

#	B				B				B				B			
41	T	23	1	0	T	22	1	-1	T	23	1	0	T	21	1	0
42	A	23	1	0	A	22	1	0	A	23	1	0	A	21	1	0
43	G	23	1	0	G	22	1	0	G	23	1	0	G	21	1	0
44	G	24	1	1	G	23	1	1	G	24	1	1	G	22	1	1
45	C	24	1	0	C	23	1	0	C	24	1	0	C	22	1	0
46	A	24	1	0	A	23	1	0	A	24	1	0	A	22	1	0
47	A	24	1	0	A	23	1	0	A	24	1	0	A	22	1	0
48	G	24	1	0	G	23	1	0	G	24	1	0	G	22	1	0
49	A	17	1	-7	A	17	1	-6	A	17	1	-7	A	17	1	-5
50	G	18	1	1	G	18	1	1	G	18	1	1	G	18	1	1
51	T	18	1	0	T	18	1	0	T	18	1	0	T	18	1	0
52	G	19	3	1	G	19	3	1	G	19	3	1	G	19	3	1
53	C	19	3	0	C	19	3	0	C	19	3	0	C	19	3	0
54	C	19	3	0	C	19	3	0	C	19	3	0	C	19	3	0
55	T	19	3	0	T	19	3	0	T	19	3	0	T	19	3	0
56	T	19	3	0	T	19	3	0	T	19	3	0	T	19	3	0
57	G	14	3	-5	G	14	3	-5	G	14	3	-5	G	14	3	-5
58	A	14	3	0	A	14	3	0	A	14	3	0	A	14	3	0
59	C	14	3	0	C	14	3	0	C	14	3	0	C	14	3	0
60	G	14	3	0	G	14	3	0	G	14	3	0	G	14	3	0
61	A	12	1	-2	A	12	1	-2	A	12	1	-2	A	12	1	-2
62	T	12	1	0	T	12	1	0	T	12	1	0	T	12	1	0
63	A	7	1	-5	A	7	1	-5	A	7	1	-5	A	7	1	-5
64	C	8	1	1	C	8	1	1	C	8	1	1	C	8	1	1
65	A	8	1	0	A	8	1	0	A	8	1	0	A	8	1	0
66	G	14	1	6	G	14	1	6	G	14	1	6	G	14	1	6
67	C	16	1	2	C	16	1	2	C	16	1	2	C	16	1	2
68	T	16	1	0	T	16	1	0	T	16	1	0	T	16	1	0
69	A	16	1	0	A	16	1	0	A	16	1	0	A	16	1	0
70	A	16	1	0	A	16	1	0	A	16	1	0	A	16	1	0
71	T	16	1	0	T	16	1	0	T	16	1	0	T	16	1	0
72	T	16	1	0	T	16	1	0	T	16	1	0	T	16	1	0
73	C	16	1	0	C	16	1	0	C	16	1	0	C	16	1	0
74	A	11	1	-5	A	11	1	-5	A	11	1	-5	A	11	1	-5
75	G	11	1	0	G	11	1	0	G	11	1	0	G	11	1	0
76	A	12	7	1	A	12	7	1	A	12	7	1	A	12	7	1
77	A	12	7	0	A	12	7	0	A	12	7	0	A	12	7	0
78	T	12	7	0	T	12	7	0	T	12	7	0	T	12	7	0
79	C	12	7	0	C	12	7	0	C	12	7	0	C	12	7	0
80	A	12	7	0	A	12	7	0	A	12	7	0	A	12	7	0
81	T	8	7	-4	T	8	7	-4	T	8	7	-4	T	8	7	-4
82	T	8	7	0	T	8	7	0	T	8	7	0	T	8	7	0
83	T	7	7	-1	T	7	7	-1	T	7	7	-1	T	7	7	-1

84	T	7	7	0	T	7	7	0	T	7	7	0	T	7	7	0
85	G	6	7	-1	G	6	7	-1	G	6	7	-1	G	6	7	-1
86	T	6	7	0	T	6	7	0	T	6	7	0	T	6	7	0
87	G	2	7	-4	G	2	7	-4	G	2	7	-4	G	2	7	-4
88	G	2	7	0	G	2	7	0	G	2	7	0	G	2	7	0
89	A	2	7	0	A	2	7	0	A	2	7	0	A	2	7	0
90	C	2	7	0	C	2	7	0	C	2	7	0	C	2	7	0
91	G	1	7	-1	G	1	7	-1	G	1	7	-1	G	1	7	-1
92	A	1	7	0	A	1	7	0	A	1	7	0	A	1	7	0
93	A	1	7	0	A	1	7	0	A	1	7	0	A	1	7	0
94	T	1	7	0	T	1	7	0	T	1	7	0	T	1	7	0
95	A	3	1	2	A	3	1	2	A	3	1	2	A	3	1	2
96	T	3	1	0	T	3	1	0	T	3	1	0	T	3	1	0
97	G	3	1	0	G	3	1	0	G	3	1	0	G	3	1	0
98	A	3	1	0	A	3	1	0	A	3	1	0	A	3	1	0
99	T	3	1	0	T	3	1	0	T	3	1	0	T	3	1	0
100	C	11	1	8	C	11	1	8	C	11	1	8	C	11	1	8
101	C	12	1	1	C	12	1	1	C	12	1	1	C	12	1	1
102	A	12	1	0	A	12	1	0	A	12	1	0	A	12	1	0
103	A	13	1	1	A	13	1	1	A	13	1	1	A	13	1	1
104	C	13	1	0	C	13	1	0	C	13	1	0	C	13	1	0
105	A	13	1	0	A	13	1	0	A	13	1	0	A	13	1	0
106	A	13	1	0	A	13	1	0	A	13	1	0	A	13	1	0
107	T	13	1	0	T	13	1	0	T	13	1	0	T	13	1	0
108	A	13	1	0	A	13	1	0	A	13	1	0	A	13	1	0
109	G	14	1	1	G	14	1	1	G	14	1	1	G	14	1	1
110	A	15	1	1	A	15	1	1	A	15	1	1	A	15	1	1
111	G	15	1	0	G	15	1	0	G	15	1	0	G	15	1	0
112	G	28	1	13	G	28	1	13	G	28	1	13	G	28	1	13
113	A	32	1	4	A	32	1	4	A	32	1	4	A	32	1	4
114	T	32	1	0	T	32	1	0	T	32	1	0	T	32	1	0
115	T	32	1	0	T	32	1	0	T	32	1	0	T	32	1	0
116	C	32	1	0	C	32	1	0	C	32	1	0	C	32	1	0
117	C	32	1	0	C	32	1	0	C	32	1	0	C	32	1	0
118	T	32	1	0	T	32	1	0	T	32	1	0	T	32	1	0
119	A	32	1	0	A	32	1	0	A	32	1	0	A	32	1	0
120	C	32	1	0	C	32	1	0	C	32	1	0	C	32	1	0
121	A	32	1	0	A	32	1	0	A	32	1	0	A	32	1	0
122	G	32	1	0	G	32	1	0	G	32	1	0	G	32	1	0
123	G	32	1	0	G	32	1	0	G	32	1	0	G	32	1	0
124	A	29	1	-3	A	29	1	-3	A	29	1	-3	A	29	1	-3
125	A	29	1	0	A	29	1	0	A	29	1	0	A	29	1	0
126	G	29	1	0	G	29	1	0	G	29	1	0	G	29	1	0

#												
127	C	30	1	1	C	30	1	1	C	30	1	1	C	30	1	1
128	A	30	1	0	A	30	1	0	A	30	1	0	A	30	1	0
129	A	32	1	2	A	32	1	2	A	32	1	2	A	32	1	2
130	G	34	1	2	G	34	1	2	G	34	1	2	G	34	1	2
131	T	34	1	0	T	34	1	0	T	34	1	0	T	34	1	0
132	A	34	1	0	A	34	1	0	A	34	1	0	A	34	1	0
133	G	35	1	1	G	35	1	1	G	35	1	1	G	35	1	1
134	T	35	1	0	T	35	1	0	T	35	1	0	T	35	1	0
135	A	35	1	0	A	35	1	0	A	35	1	0	A	35	1	0
136	A	35	1	0	A	35	1	0	A	35	1	0	A	35	1	0
137	T	35	1	0	T	35	1	0	T	35	1	0	T	35	1	0
138	T	35	1	0	T	35	1	0	T	35	1	0	T	35	1	0
139	G	35	1	0	G	35	1	0	G	35	1	0	G	35	1	0
140	A	10	1	-25	A	10	1	-25	A	10	1	-25	A	10	1	-25
141	T	10	1	0	T	10	1	0	T	10	1	0	T	10	1	0
142	G	13	1	3	G	13	1	3	G	13	1	3	G	13	1	3
143	G	14	1	1	G	14	1	1	G	14	1	1	G	14	1	1
144	A	14	1	0	A	14	1	0	A	14	1	0	A	14	1	0
145	G	16	1	2	G	16	1	2	G	16	1	2	G	16	1	2
146	A	17	1	1	A	17	1	1	A	17	1	1	A	17	1	1
147	A	17	1	0	A	17	1	0	A	17	1	0	A	17	1	0
148	A	40	1	23	A	40	1	23	A	17	1	0	A	40	1	23
149	C	40	1	0	C	40	1	0	C	19	1	2	C	40	1	0
150	C	40	1	0	C	40	1	0	C	19	1	0	C	40	1	0
151	T	40	1	0	T	40	1	0	T	19	1	0	T	40	1	0
152	G	40	1	0	G	40	1	0	G	19	1	0	G	40	1	0
153	T	39	1	-1	T	39	1	-1	T	19	1	0	T	39	1	-1
154	C	39	1	0	C	39	1	0	C	19	1	0	C	39	1	0
155	T	38	1	-1	T	38	1	-1	T	19	1	0	T	38	1	-1
156	C	38	1	0	C	38	1	0	C	18	1	-1	C	38	1	0
157	T	37	1	-1	T	37	1	-1	T	18	1	0	T	37	1	-1
158	T	37	1	0	T	37	1	0	T	18	1	0	T	37	1	0
159	G	35	1	-2	G	35	1	-2	G	18	1	0	G	35	1	-2
160	G	35	1	0	G	35	1	0	G	17	1	-1	G	35	1	0
161	A	35	1	0	A	35	1	0	A	17	1	0	A	35	1	0
162	T	35	1	0	T	35	1	0	T	17	1	0	T	35	1	0
163	A	35	1	0	A	35	1	0	A	17	1	0	A	35	1	0
164	T	35	1	0	T	35	1	0	T	17	1	0	T	35	1	0
165	T	35	1	0	T	35	1	0	T	17	1	0	T	35	1	0
166	C	37	1	2	C	37	1	2	C	14	1	-3	C	37	1	2
167	T	38	1	1	T	38	1	1	T	14	1	0	T	38	1	1
168	C	38	1	0	C	38	1	0	C	9	1	-5	C	38	1	0
169	G	39	1	1	G	39	1	1	G	9	1	0	G	39	1	1

170	A	43	1	4	A	43	1	4	A	9	1	0	A	43	1	4
171	C	43	1	0	C	43	1	0	C	9	1	0	C	43	1	0
172	A	44	1	1	A	44	1	1	A	12	1	3	A	44	1	1
173	C	45	1	1	C	45	1	1	C	13	1	1	C	45	1	1
174	A	45	1	0	A	45	1	0	A	13	1	0	A	45	1	0
175	G	46	1	1	G	46	1	1	G	14	1	1	G	46	1	1
176	C	46	1	0	C	46	1	0	C	14	1	0	C	46	1	0
177	A	46	1	0	A	46	1	0	A	14	1	0	A	46	1	0
178	G	46	1	0	G	46	1	0	G	14	1	0	G	46	1	0
179	G	46	1	0	G	46	1	0	G	14	1	0	G	46	1	0
180	T	4	1	-42	T	4	1	-42	G	7	1	-7	T	4	1	-42
181	C	4	1	0	C	4	1	0	C	3	1	-4	C	4	1	0
182	A	2	1	-2	A	2	1	-2	A	3	1	0	A	2	1	-2
183	A	2	1	0	A	2	1	0	A	3	1	0	A	2	1	0
184	G	4	1	2	G	4	1	2	G	4	1	1	G	4	1	2
185	A	5	1	1	A	5	1	1	A	5	1	1	A	5	1	1
186	G	5	1	0	G	5	1	0	G	5	1	0	G	5	1	0
187	G	14	9	9	G	14	9	9	G	14	9	9	G	14	9	9
188	A	14	9	0	A	14	9	0	A	14	9	0	A	14	9	0
189	G	14	9	0	G	14	9	0	G	14	9	0	G	14	9	0
190	T	14	9	0	T	14	9	0	T	14	9	0	T	14	9	0
191	A	14	9	0	A	14	9	0	A	14	9	0	A	14	9	0
192	C	14	9	0	C	14	9	0	C	14	9	0	C	14	9	0
193	A	14	9	0	A	14	9	0	A	14	9	0	A	14	9	0
194	G	14	9	0	G	14	9	0	G	14	9	0	G	14	9	0
195	T	13	9	-1	T	13	9	-1	T	13	9	-1	T	13	9	-1
196	G	13	9	0	G	13	9	0	G	13	9	0	G	13	9	0
197	C	12	9	-1	C	12	9	-1	C	12	9	-1	C	12	9	-1
198	A	12	9	0	A	12	9	0	A	12	9	0	A	12	9	0
199	A	13	9	1	A	13	9	1	A	13	9	1	A	13	9	1
200	T	13	9	0	T	13	9	0	T	13	9	0	T	13	9	0

Annexure 4.5: Predicted Results for Binding of Synthetic sgRNA on Target Gene Reported in Database

The experimental setup has been designed with the sgRNAs retrieved from the breakingcas database [26] for a given input gene name RRM2. We took those sgRNAs and tested it with the target gene RRM2 with our algorithm whether the sgRNAs binds or not with the corresponding gene. Predicted results are reported below.

Gene and sgRNA ID refers to gene region between two locations X and Y (marked as X_Y)—its base sequence match with the sgRNA base sequence. The results have been derived on executing the code implementing the 'Algorithm: Predict true positive and true negative instances of binding a candidate sgRNAs on target gene' reported in Sect. 4.6. We work with two parameters referred to as 5′ end 3′ end secondary signal count. Degree of Match (DOM) is set as '0' if two parameters derived out of signal graphs for sgRNA and its target gene are exactly identical—this refers to full binding of sgRNA on target gene. DOM is set as 1 if only one of the two parameters of sgRNA and its target gene is identical while other parameters differ—this refers to partial binding. In the main text of Sect. 4.6 we have marked partial binding as 'no binding'. DOM has a value of 2 if both parameters differ and these cases imply no binding according to the algorithm.

sgRNA ID/Gene ID	M′	LOM	M	ROM	LOM′/ROM′	LOM/ROM	DOM
sgRNA_482_501/ RRM2_482_501	20	447–465	466		1/0	1/0	0
sgRNA_1402_1421/ RRM2_1402_1421	10	975–983	984	985–994	1/1	1/0	1
sgRNA_678_697/ RRM2_678_697	20	657–675	676		0/0	3/0	1
sgRNA_61_80/ RRM2_61_80	17	45–60	61	62–64	0/0	1/0	1
sgRNA_46_65/ RRM2_46_65	20	37–55	56		0/0	1/0	0
sgRNA_502_521/ RRM2_502_521	9	518–525	526	527–537	0/0	0/1	1
sgRNA_1435_1454/ RRM2_1435_1454	20	967–985	986		0/0	0/0	0
sgRNA_703_722/ RRM2_703_722	–	718–723	724	725–737	0/0	0/0	0
sgRNA_679_698/ RRM2_679_698	20	654–672	673		0/0	0/0	0
sgRNA_1117_1136/ RRM2_1117_1136	–	969–985	986	987–988	0/0	1/0	1
sgRNA_1390_1409/ RRM2_1390_1409	12	1020–1030	1031	1032–1039	0/1	1/0	0
sgRNA_684_703/ RRM2_684_703	20	702–720	721		0/0	0/0	0
sgRNA_270_289/ RRM2_270_289	9	264–271	272	273–283	0/0	1/0	1
sgRNA_667_686/ RRM2_667_686	–	657–660	661	662–676	0/0	0/1	1
sgRNA_1179_1198/ RRM2_1179_1198	4	1005–1007	1008	1009–1024	0/0	0/1	1
sgRNA_4808_4827/ RRM2_4808_4827	–	990–991	992	993–1009	0/0	0/0	0
sgRNA_681_700/ RRM2_681_700	17	657–672	673	674–676	0/0	1/0	1
sgRNA_839_858/ RRM2_839_858	–	838–843	844	845–857	0/0	1/1	2

(continued)

(continued)

sgRNA ID/Gene ID	M′	LOM	M	ROM	LOM′/ ROM′	LOM/ ROM	DOM
sgRNA_1158_1177/ RRM2_1158_1177	–	996–1000	1001	1002–1015	0/0	1/1	2
sgRNA_217_236/ RRM2_217_236	–	223–224	225	226–242	0/0	0/2	1
sgRNA_1416_1435/ RRM2_1416_1435	6	1008–1012	1013	1014–1027	0/1	0/0	1
sgRNA_694_713/ RRM2_694_713	20	659–677	678		0/0	1/0	1
sgRNA_292_311/ RRM2_292_311	20	282–300	301		0/0	0/0	0
sgRNA_303_322/ RRM2_303_322	6	314–318	319	320–333	0/0	0/1	1
sgRNA_70_89/ RRM2_70_89	–	96–97	98	99–115	0/0	0/1	1
sgRNA_486_505/ RRM2_486_505	6	473–477	478	479–492	0/1	0/1	0
sgRNA_690_709/ RRM2_690_709	20	676–694	695		0/0	0/0	0
sgRNA_503_522/ RRM2_503_522	–	–	–	–	0/0	0/0	0
sgRNA_435_454/ RRM2_435_454	17	448–463	464	465–467	0/0	1/0	1
sgRNA_699_718/ RRM2_699_718	20	670–688	689		0/0	2/0	0
sgRNA_549_568/ RRM2_549_568	20	515–533	534		0/0	1/0	1
sgRNA_367_386/ RRM2_367_386	14	371–383	384	385–390	2/0	0/0	1
sgRNA_722_741/ RRM2_722_741	20	739–757	758		0/0	1/0	1
sgRNA_1152_1171/ RRM2_1152_1171	–	–	–	–	0/0	0/0	0
sgRNA_1484_1503/ RRM2_1484_1503	13	1021–1032	1033	1034–1040	0/0	0/0	0
sgRNA_331_350/ RRM2_331_350	–	313–325	326	327–332	0/0	2/0	1
sgRNA_1196_1215/ RRM2_1196_1215	20	979–997	998		0/0	1/0	1
sgRNA_154_173/ RRM2_154_173	6	139–143	144	145–158	0/1	0/1	0
sgRNA_666_685/ RRM2_666_685	8	677–683	684	685–696	0/0	0/2	1
sgRNA_1116_1135/ RRM2_1116_1135	–	992–993	994	995–1011	0/0	0/1	1

References

1. Watson, J.D., Crick, F.H.C.: Molecular structure of nucleic acids. Nature **171**(4356), 737–738 (1953)
2. Clancy, S.: DNA transcription. Nat. Educ. **1**(1), 41 (2008)
3. Bateson, W., Gregor, M.: Mendel's Principles of Heredity. Courier Corporation (2013)
4. https://www.ncbi.nlm.nih.gov/genome/guide/gnomon.shtml Gnomon - the NCBI eukaryotic gene prediction tool - NIH https://www.ncbi.nih.gov/genome/guide/genome
5. Shafee, T., Lowe, R.: Eukaryotic and prokaryotic gene structure. Wiki. J. Med. **4**(1), 002 (2017)
6. https://www.promega.in/resources/pubhub/enotes/what-is-a-kozak-consensus-sequence/
7. Godbey, W.T.: An Introduction to Biotechnology: The Science, Technology and Medical Applications. Elsevier (2014)
8. Will, C.L., Lührmann, R.: Spliceosome structure and function. Cold Spring Harb. Perspect. Biol. **3**(7), a003707 (2011)
9. Monlong, J., et al.: Identification of genetic variants associated with alternative splicing using sQTLseekeR. Nat. Commun. **5**, 4698 (2014)
10. Bolognini, R., et al.: Characterization of two novel intronic OPA1 mutations resulting in aberrant pre-mRNA splicing. BMC Med. Genet. **18.1**, 22 (2017)
11. Li, Y., Xu, Y., Ma, Z.: Comparative analysis of the exon-intron structure in eukaryotic genomes. Yangtze Med. **1.01**, 50 (2017)
12. Brown, T.: Introduction to Genetics: A Molecular Approach. Garland Science (2011)
13. Sakharkar, M., et al.: ExInt: an exon intron database. Nucleic Acids Res. **30**(1), 191–194 (2002)
14. Patricia, A.R., et al.: Before it gets started: regulating translation at the 5′ UTR. Comp. Funct. Genomics **2012** (2012)
15. Smith, C.W.J., Juan, V.: Alternative pre-mRNA splicing: the logic of combinatorial control. Trends Biochem. Sci. **25.8**, 381–388 (2000)
16. Yates, B., et al. Genenames.org: the HGNC and VGNC resources in 2017. Nucleic Acids Res. gkw1033 (2016)
17. Wang, Y., et al.: Mechanism of alternative splicing and its regulation. Biomed. Rep. **3.2**, 152–158.4.18 (2015)
18. Fernandez, N., et al.: Genetic variation and RNA structure regulate microRNA biogenesis. Nat. Commun. **8**, 15114 (2017)
19. Jančík, S., et al.: Clinical Relevance of KRAS in Human Cancers. BioMed Research International 2010 (2010)
20. Waters, A.M., et al.: Single synonymous mutations in KRAS cause transformed phenotypes in NIH3T3 cells. PloS one **11.9**, e0163272 (2016)
21. Forbes, S.A., et al.: The catalogue of somatic mutations in cancer (COSMIC). Curr. Protocols Hum. Genet. **57.1**, 10–11 (2008)
22. Staniland, M.: The Top CRISPR Papers of 2017. Nature News, Nature Publishing Group, 16 Jan 2018. blogs.nature.com/ofschemesandmemes/2018/01/16/our-10-most-popular-crispr-papers-of-2017
23. CRISPR Mechanism. CRISPRCas9, Tufts University (2017). sites.tufts.edu/crispr/crispr-mechanism/
24. Rauscher, B., et al.: GenomeCRISPR-a database for high-throughput CRISPR/Cas9 screens. Nucleic Acids Res. gkw997 (2016)
25. Park, J., Kim, J.-S., Bae, S.: Cas-database: web-based genome-wide guide RNA library design for gene knockout screens using CRISPR-Cas9. Bioinformatics **32**(13), 2017–2023 (2016)
26. Oliveros, J.C., et al.: Breaking-Cas—interactive design of guide RNAs for CRISPR-Cas experiments for ENSEMBL genomes. Nucleic Acids Res. **44.W1**, W267-W271 (2016)

Chapter 5
Cellular Automata (CA) Model for Protein

"Genes are effectively one-dimensional. If you write down the sequence of A, C, G and T, that's kind of what you need to know about that gene. But proteins are three-dimensional. They have to be because we are three-dimensional, and we're made of those proteins. Otherwise we'd all sort of be linear, unimaginably weird creatures".

—Francis Collins

5.1 Introduction

Protein is the workhorse of living cells of any organism. The discipline of Proteomics involves large-scale study of structure, function and interactions of protein molecules with other biomolecules. This chapter reports the Cellular Automata (CA) model for a protein molecule. Different sections of the chapter are organized as follows. The CA model is presented in Sect. 5.4 subsequent to the introduction of the background and survey of CA models proposed in the fields of Genomics and Proteomics.

- Bioinformatics Tools for Proteomics
- Cellular Automata Applications in Genomic and Proteomics—A Comprehensive Survey
- Design of Protein Modelling CA Machine (PCAM)
- Protein Modelling CA Machine (PCAM) Evolution
- Feature Extraction
- Modelling Protein–Protein Interaction
- Study of Monoclonal Antibodies (MAbs)
- Study of Mutational Effects on Binding
- The Scope of Future Study

.

© Springer Nature Singapore Pte Ltd. 2018
P. P. Chaudhuri et al., *A New Kind of Computational Biology*,
https://doi.org/10.1007/978-981-13-1639-5_5

5.2 Bioinformatics Tools for Proteomics

In the post-genomic era, the discipline of Proteomics got enriched in the past decade with the addition of a large number of protein sequences displaying wide varieties. However, such an explosion in the number of uncharacterized protein sequences demands high-throughput tools for rapidly and reliably identifying various attributes based on their sequence information alone. The knowledge thus obtained can help us to utilize these newly found protein sequences for both basic research and drug discovery.

Many bioinformatics tools have been developed employing machine learning methods [1–5]. The recent surge of interest in Artificial Intelligence (AI) is primarily driven by a phenomenal increase in computing power and a decrease of storage cost. In parallel, the associated support for AI has come from the disciplines of computer science, mathematics, psychology and cognitive science with respect to reasoning, learning and problem-solving. All these developments have provided the renewed emphasis for development of AI-based tools in different fields. Structure–function relationship of protein molecules are being reinvestigated [2–6] with AI-based methodologies. However, while developing AI-based tools for genomics and proteomics, it is worthwhile to take note of the following observation: 'The majority of data in biology are still atypical for Machine Learning: they are too sparse and incomplete, too biased and too noisy.'—Irina S. Moreira et al. [7] Consequently, extraction of noise-free unbiased data is a challenging task for building AI applications for Genomics and Proteomics.

On completion of the Human Genome project in 2003, the next major challenge was to understand when and where the encoded proteins are expressed, and to generate a map of the protein complexes, interconnected pathways, networks that control the working of all cells, tissues, organs of all living organisms. The discipline of Proteomics is fundamental for such studies. Consequently, it is essential to design a new kind of high speed computational modelling tools in addition to the conventional Molecular Dynamics (MD) simulation that involves high computational resources and time. The new kind of simulation we have proposed in this book employs 'Information' content of biomolecules and their building blocks rather than physics-based modelling of molecular interactions.

In the earlier two chapters (Chaps. 3 and 4), we have addressed the problems of Genomics by developing CA models for RNA and DNA. This chapter continues with the similar approach of designing CA rule for amino acid and PCAM (Protein modelling CA Machine) followed by top-down validation.

In Chap. 2, we reported the basic differences in the CA model for nucleic acids and protein. We also outlined how our CA modelling tool for nucleic acids differs from that of the protein molecule. Although both the classes of molecules have a backbone, the sugar–phosphate backbone of the nucleic acid chain has much higher stability compared to the peptide backbone of the protein chain. Further, the nucleic acid has limited number of four bases connected to its backbone and the four bases are derived out of two basic molecules; purine and pyrimidine. On the other hand, a protein chain displays a wide variety in the molecular structure for residue side chain bonded to C-alpha atom of its backbone. For example, amino acid glycine (Gly) has a single hydrogen atom in its side chain, while tryptophan (Trp) has 18 hydrogen and non-hydrogen atoms (carbon, nitrogen, oxygen) in its side chain. Prior to going to the details of our CA model for protein chain, the next section reports a brief survey on the application of CA models proposed earlier in the field of Genomics and Proteomics.

5.3 Cellular Automata Applications in Genomics and Proteomics—A Comprehensive Survey

A Cellular Automata (CA) provides a discrete dynamic structure that evolves over time displaying wide varieties of interesting patterns. CA evolution holds enormous potentials for modelling complex systems. Researchers, scientists and practitioners from different fields have utilized CA for visualizing protein sequences, investigating their evolution patterns, and predicting their various attributes. Extensive study of CA rule evolutions to analyse the behaviour of complex systems is a well-researched area [8–15].

The idea of employing CA model to the central dogma of molecular biology is not new. Many attempts have been reported to model the central dogma in terms of structure, function, and evolution of biomolecules. CA framework is used for modelling DNA sequences [8], evolution [9], mutation prediction [10] and gene networks [11]. For proteomic applications, Amino Acid Coding (AAC) language is proposed [12] to design the initial configuration of the elementary CA. This model is used to predict protein sub-cellular location [13], the G-protein-coupled receptor functional classes [14], and protein structural classes [15, 16]. Pseudo Amino Acid code (PseAA) proposed in [17] has been used for protein modelling. This modelling tool differs from the schemes employing AAC in that it takes into consideration the effect of amino acid sequence of a protein chain as a set of discrete numbers. Surveys on the methods utilizing pseudo amino acid composition are reported in [18] and [19]. Such CA applications are implemented both for two-dimensional (2D) [20] and three-dimensional (3D) [21] CA. The work reported in [22] combines CA with genetic algorithms to predict the protein secondary structure, where the genetic algorithm is used to optimize the parameters (CA Rules). The authors arrived at the conclusion that prediction quality of CA model depends on three factors—neighbourhood, weights assigned to the neighbourhood and the number of generations of CA evolution.

A new phase of development can be observed in the models surveyed next. Cristea [23] proposed a representation of genetic code, which converts the DNA sequences into digital signals and used a base four representation of the nucleotides. It leads to the conversion of the codons into numbers in the range 0–63 and the amino acids in the range 0–20. In this model, single-letter representation of amino acids are coded as: F = 0, L = 1, S = 2, Y = 3, end = 4, C = 5, W = 6, P = 7, H = 8, Q = 9, R = 10, I = 11, M = 12, T = 13, N = 14, K = 15, V = 16, A = 17, D = 18, E = 19, G = 20. Cristea [23] model reflects better amino acid structure and degeneracy, and the genetic signals built from genes on this model show low auto-correlation. Pan et al. [24] also proposed another amino acid coding scheme: A = 10, C = 20, D = 30, E = 40, F = 50, G = 60, H = 70, I = 80, K = 90, L = 100, M = 110, N = 120, P = 130, Q = 140, R = 150, S = 160, T = 170, V = 180, W = 190, Y = 200. Though both the procedures can encode a protein sequence to a sequence of digital signals, the physicochemical properties of the amino acids were ignored. Xiao et al. [12] proposed a model of digital coding for amino acids based on rule similarity, complimentarity of rule, molecular recognition theory and information theory. The model reflects better amino acid physicochemical properties and degeneracy. Xiao amino acid codes are 5-bit binary pattern: P = 00001, Q = 00100, R = 00110, Y = 01100, W = 01110, T = 10000, M = 10011,   N = 10101,   V = 11010,   E = 11101,   L = 00011,   H = 00101, S = 01001,   F = 01011,   C = 01111,   I = 10010,   K = 10100,   A = 11001, D = 11100, G = 11110, end = 11111. It can transform the symbolic DNA sequences into digital genetic signals of amino acids which can be employed to build CA model. The concept of pseudo amino acid composition was also proposed by Chou et al. [17, 18]. In pseudo amino acid composition model, the formulation is based on kth sequence order where ($k = 1, 2, 3…$). Thus, the behaviour of protein depends on the characteristics and interaction of neighbouring amino acids; $k = 1$ means the closest neighbour, $k = 2$ means neighbour with a gap of 1, and so on. Pseudo amino acid composition is used in various studies related to proteomics. Xiao et al. [14–16] first introduced cellular automata as the interaction modelling tool in pseudo amino acid composition model.

   While introducing CA in Chap. 1 and subsequent CA rule design in Chap. 2, we proposed our viewpoint that CA model should be designed with simple rule structure in one dimension to represent a linear chain of biomolecules. Use of complex CA rule structures for genomics and proteomics usually lead to an upheaval task for the designer to match CA evolution parameters with the features extracted from the system being modelled. This problem, in our view, becomes more complex than CA rule design itself, specifically with the variation of seeds for CA cells. In this background Chap. 2 reports our CA model for the building blocks of macromolecules DNA, RNA and protein. In Sect. 2.7 of Chap. 2, we have reported design of 3-neighborhood CA (3NCA) rules for amino acids based on the information retrieved from their molecular structure. The model does not require searching capabilities of evolutionary algorithms to design the CA rule structure. Next section of the current chapter elaborates the design of three-neighbourhood null-boundary two-state per cell CA for Protein Modelling CA Machine (PCAM) we introduced in Sect. 2.7 of Chap. 2. This model of PCAM is employed to address the problems in the field of proteomics reported in subsequent sections of this chapter.

5.4 Design of Protein Modelling CA Machine (PCAM)

Cellular Automata, as noted in Chap. 1, is a discrete model consisting of a set of cells, which occupy some or all sites of a regular lattice. These cells have one or more internal state variables and a set of rules specifying the evolution of their state. The change of a cell state depends on the current state of the cell and those of neighbouring cells. The simplest type of cellular automata is one dimensional, three/five neighbourhood and two-state per cell. We have used such a CA for modelling biomolecules. The model for RNA and DNA strands are reported in Chaps. 3 and 4.

Section 2.7 of Chap. 2 reports the CA model for amino acid—the building block of a protein chain. A brief recapitulation of CA model of amino acid and protein chain follows in the subsequent discussions.

Each amino acid has a common backbone interconnected with its neighbours through peptide bonds. A peptide bond can be viewed as a chemical bond between carboxyl group of one amino acid and amino group of its neighbour releasing a water molecule H_2O. The model for the peptide chain of amino acid residues is reported in Sect. 2.7 of Chap. 2 while taking into consideration the effect of the diverse side chain of amino acids. The basic structure of the amino acid backbone and the formation of protein are highlighted in Fig. 5.1. The 3D structure and primary structure of Human Insulin are shown in Fig. 5.1a, b, while the molecular structure of one of the amino acids—valine is shown in Fig. 5.1c.

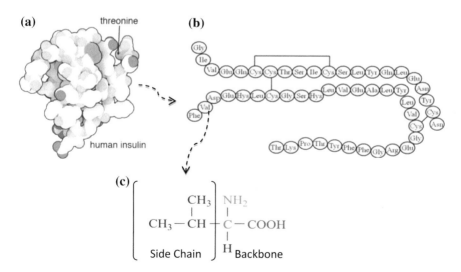

Fig. 5.1 Protein structure **a** 3D structure of human insulin, **b** Primary structure of Human Insulin, **c** Molecular structure of valine—the side chain atoms are shown in the bracket. There are three non-hydrogen atoms in the side chain

Proteins intrinsically interact with different micro and macromolecules in widely diverse ways since its peptide backbone, unlike sugar–phosphate of DNA, can undergo different structural modifications. The complexity and compartmentalization of its structural modification affect the protein function. Such a wide diversity in protein structure and function, as per our view, cannot be modelled by a single PCAM to model all types of interactions. Consequently, multiple PCAMs are designed to model the wide spectrum of protein interactions with different biomolecules. However, this bottom-up design of CA rule structure adds the higher complexity of the problem of top-down validation of the model. Fortunately, as we have discussed earlier, availability of large number of experimental results of protein interactions are currently available. Consequently, top-down validation can be undertaken by using known protein interaction data available in a large number of databases in different websites and publications.

In Sect. 2.7.2 of Chap. 2, we reported the implementation of the following two steps for PCAM design:

(i) An amino acid residue of a protein chain is represented by 8 CA cells which are initialized with the 8-bit seed derived out of its side chain.

(ii) The backbone is designed with a uniform 3NCA out of 64 CA rules. Sixty-four 3NCA rules model 64 different variations of an amino acid backbone. The motivation of providing 64 variants for backbone model has been elaborated in earlier discussions while we emphasized the difference between backbone of the nucleic acid strand and that of the protein chain.

For an input protein chain, we have 64 different PCAMs. Based on the similarity of evolutionary patterns of different PCAMs, the 64 PCAMs for a protein are grouped into (say) k number of classes. Each of these k classes represents different features of a protein for interaction with different micro- and macromolecules under normal temperature and pH value. Any variation of these two environmental parameters (temperature and pH) adds further complexity for analysis of protein structure and function.

A protein chain with p amino acid residues is represented by an n ($n = 8p$) cell uniform 3NCA. Each 8 cell block is initialized with the seed derived out 8-bit pattern of the residue side chain. Next two subsections elaborate the methodologies of CA rule design for amino acids.

5.4.1 CA Model for Amino Acid Backbone

The input for the biological process of protein synthesis is the codon string of mRNA introduced in Chap. 3. Section 3.2 reports design of 64 3NCA rules for 64 codons, triplet of nucleotide bases. Next, Sect. 3.3 of Chap. 3 reports the CA model for ribosomal translation of a codon to an amino acid through matching of codon with anticodon of tRNA molecule carrying the amino acid for the codon.

In the above background, we have employed 64 3NCA rules to model backbone of an amino acid. Each of these rules, as explained in Sect. 3.2 of Chap. 3, is a balanced rule with two 1's in 1-major and 0-major representation of a rule. We designed CA rules for the building blocks of DNA and RNA molecules in Chap. 2 based on the information retrieved from the molecular structure of micromolecules. We continue with the similar approach for rule design for amino acid backbone and encoding of side chain based on the analysis of their molecular structure.

As reported in Sect. 2.7.1, for design of CA rules for amino acid residues, we consider only heavy atoms (C, N, O) and exclude Hydrogen atom (H) of molecular structure of 20 different amino acids shown in Fig. 2.22 of Chap. 2. On exclusion of the H atom, the molecular structure of the common backbone of all amino acid residues has 4 atoms—O, N and two C atoms as shown in Fig. 5.1c. We design 8-bit pattern of 3NCA rules for backbone as follows. Two 1's are assigned to represent O and N atoms in the next state values of 1-major Rule Min Term (RMT) s, while two 1's are assigned to represent two carbon atoms in the next state value of 0-major RMTs. In the process following 64 3NCA rules are designed to model different variants of amino acid backbone.

[141, 197, 177, 163, 75, 89, 210, 154, 45, 101, 180, 166, 58, 114, 92, 78, 139, 209, 153, 195, 57, 99, 156, 198, 46, 116, 60, 102, 77, 178, 90, 165, 135, 149, 225, 169, 27, 83, 216, 202, 39, 53, 228, 172, 30, 86, 120, 106, 147, 201, 150, 105, 15, 85, 240, 170, 29, 71, 184, 226, 51, 204, 54, 108]

5.4.2   Encoding Amino Acid Side Chain

Digital encoding of amino acid side chain proceeds on considering non-H atoms in the molecular structure of side chains. We consider an 8-bit pattern of a 3NCA rule as 1-major and 0-major format. Representation of non-H atoms of a side chain proceeds on assigning 1's as the next of RMTs elaborated next. The acronyms PRMT, CoP, etc., used in subsequent discussions are explained in Chap. 2.

The side chain encoding is based on the assignment of atoms to the next state of PRMT (palindromic RMT: 7 (111), 5 (101), 2 (010), 0 (000) which remain unaltered on reversal of its sequence) and other non-palindromic RMT pairs called CoP (Conjugate Pair): (6, 3) and (4, 1), where one is derived out of the other on reversing the 3-bit string (Table 5.1). Atoms are assigned to the next state of different RMTs based on the following assumptions with respect to the covalent bond between a pair of atoms of the molecular structure of amino acid (Fig. 2.22 of Chap. 2):

Assumption 1 Only non-H atoms C, N, O, S of the side chain are considered in the design process.

Table 5.1 Amino acid side chain encoding. The decimal value and binary pattern of RMTs and their Class (PRMT or CoP) are shown. $y = CP(x)$ means RMT y is the Conjugate Pair (CoP) of RMT x. An example side chain encoding for amino acid valine is also shown with the assigned atoms and covalent bonds

RMT	7	6	5	4	3	2	1	0
Binary	111	110	101	100	011	010	001	000
Class	PRMT	CoP(3)	PMRT	CoP(1)	CoP(6)	PRMT	CoP(4)	PRMT
Valine	0	0	0	1	0	1	1	0
Atom				C_β		C_δ	$C\gamma$	
Bond								

Assumption 2 The 6-carbon ring (aromatic ring) is considered as a single entity and it is placed as the next state of the RMT 7 for the residues having side chain ring.

Assumption 3 The design ensures that the covalent bond between a pair of side chain atoms, as elaborated next, are mapped while assigning atoms to next states of RMTs $\ll 7, 6, 5, 3 > <4, 2, 1, 0 \gg$:

- A covalent bond is implicit between atoms placed on CP RMT pairs (6, 3) and (4, 1).
- An atom placed on a PRMT (7, 5, 2, 0), whenever necessary, is assumed to make a covalent bond with any atom placed as the next state of any other RMT.
- C atoms, in general, are placed as the next state of 0-major, while O, N, S atoms are mapped to the next state of 1-major RMTs. In case the number of C atoms is more than 2, it is placed as the next state of a 1-major RMT as illustrated in Table 5.1.

In Table 5.1, the encoding of amino acid valine (Fig. 5.1c) is shown in the fourth row along with the decimal values and binary patterns of RMTs in the first and second row, respectively. The third row shows the class of RMT—PRMT or CoP. The side chain of Valine has three non-hydrogen atoms—three Carbon (C) atoms. There are two C–C covalent bonds in the side chain. One C atom (C_β) is connected to other two C atoms—$C\gamma$ and C_δ by these bonds. The encoding of Valine in 4th row shows three 1's representing three C atoms—CoP RMT 4 and 1 are assigned to C_β and $C\gamma$, respectively, while PRMT 2 is assigned to C_δ. Thus, under Assumption 3, CoP RMT 1 and 4 are connected by covalent bond ($C\gamma$–C_β), and PRMT 2 is connected to CoP RMT 4 by covalent bond (C_δ–C_β). In Table 5.1, the atom assignment and covalent bonds are shown in fifth and sixth rows, respectively. This atom assignment is interchangeable between C_δ and $C\gamma$, similar to the configuration change among L/R isomers.

It is worthwhile to note the point that it is possible to assign atoms differently while still maintaining the assumptions. Thus, the amino acid encoding used here is one of such possible configurations. This is an open area of discussion and development to find the best configuration to be used. For the current study, we are using the encoding reported in Table 5.2 showing RMTs as <7 6 5 4 3 2 1 0> with marking of 1-major RMTs in black and 0-major RMTs as non-black (white).

Table 5.2 Amino acid encoding—Assigned next state of RMTs marked black for 1-major: and white for 0-major; Total non-H atoms and compositions are shown—R = six carbon aromatic ring, C = Carbon, O = Oxygen, N = Nitrogen and S = Sulphur

Amino Acids	Abbreviations	RMTs 76543210	Total non-H Atom	Exception
Glycine	G (gly)	00000000	0	
Alanine	A (ala)	00000100	1 (1C)	
Proline	P (pro)	00100110	3 (3C)	Exceptionto Assumption 3
Valine	V (val)	00010110	3 (3C)	
Methionine	M (met)	00110110	4 (3C + 1S)	
Tryptophan	W (trp)	10110110	R+4 (R + 3C +1N)	
Phenylalanine	F (phe)	10000100	R+1 (R + 1C)	
Isoleucine	I (ile)	00011110	4 (4C)	Exception to Assumption 3
Leucine	L (leu)	00010111	4 (4C)	
Serine	S (ser)	00100100	2 (1C + 1O)	
Cysteine	C (cys)	01000100	2 (1C + 1S)	
Threonine	T (thr)	00110100	3 (2C + 1O)	
Asparagine	N (asn)	00101110	4 (2C + 1N + 1O)	
Glutamine	Q (gln)	00101111	5 (3C + 1N + 1O)	
Tyrosine	Y (tyr)	10100100	R+2 (R + 1C + 1O)	
Histidine	H (his)	01111110	6 (4C + 2N)	Exception to Assumption 3
Lysine	K (lis)	00110111	5 (4C + 1N)	
Arginine	R (arg)	01111111	7 (4C + 3N)	
Aspartic Acid	D (asp)	01110100	4 (2C + 2O)	
Glutamic Acid	E (glu)	01110110	5 (3C + 2O)	

The first two columns of Table 5.2 show the name and abbreviation of 20 amino acids. The 8-bit binary patterns for each amino acid side chain are reported in third column. The 1-Major RMTs <7 6 5 3> are marked in black background. While next state value of each RMT from 0 to 6 represent an individual non-hydrogen (non-H) atom, the RMT 7 represents the 6-carbon aromatic ring (R)—all amino acids containing such ring structure will have 1 in RMT 7. The 4th column shows the count of non-hydrogen (non-H) atoms and their compositions.

For example, amino acid Asparagine (Asn) has total 4 non-H atoms—2 Carbon (C), 1 Nitrogen (N) and 1 Oxygen (O). It's encoding is <00101110 >. There are two 1's in 1-Major RMTs representing 1 N and 1 O, while two C atoms are placed in 0-Major RMTs, maintaining Assumption 3 noted above representing covalent bond structures. For 17 amino acids out of 20, all the assumptions are maintained. For three amino acids, Pro, Ile and His, Assumption 3 is not maintained to implement the covalent bonds—these are noted as exceptions.

Next section analyses the patterns derived out of evolution of PCAMs designed for protein chains.

5.5   Protein Modelling CA Machine (PCAM) Evolution

A PCAM modelling a protein has two components—(i) CA seed modelled by amino acid side chain, and (ii) CA rule modelled by amino acid backbone. The side chain encoding is discussed in the previous subsection. The amino acid backbone, which is similar for all of the 20 amino acids, is represented by 64 balanced 3NCA rules. As discussed earlier, the choice of backbone 3NCA rule will depend on the intrinsic correlation with the functionality to be modelled. This means that we need to choose the rule depending on what we want to model and predict. For example, the rule for analysing localization of a protein will be different from the rule for predicting its binding affinity with a ligand.

The method of PCAM evolution is reported in Chap. 2.5 with illustration of evolution of an example protein chain. In order to model peptide chain of a protein we employ eight CA cells for each residue. In this process, a protein chain having p number of amino acid residues is encoded as an n cell CA, where $n = 8 \times p$. The 8-bit pattern of an amino acid side chain is assigned as the seed (initial binary state) of 8 cells assigned for the backbone of the residue. The CA is evolved with each of 64 rules as null-boundary 3-neighborhood uniform CA for 999 time steps. The resulting binary evolution matrices also known as CA-image are stored for further texture analysis.

PCAM is designed to be a platform to experiment with any length, naturally occurring or synthetic protein. Protein sequence can be mutated, shortened or extended. To analyse the importance of a specific amino acid, we can introduce an ALA-mutagenesis at PCAM seed by replacing the amino acid side chain's 8-bit binary pattern with ALA's binary pattern and evolve the PCAM to generate CA-image of the modified chain, with each of the 64 rules. Next, we can compare the effect of the mutagenesis by comparing the derived CA images. From the known data, we can map single or multiple rules, for which the CA image can efficiently model the experimental findings in respect of difference observed between wild and mutant proteins. Those rules can be used for the prediction of the unknown or test cases.

The PCAM is implemented as an in-house Python module. To make it faster, the Python code is translated into C++ code by Shed Skin utility, an experimental (restricted-Python) to C++ compiler. The compiled module is used as a Python extension in larger complex code. Many open-source Python libraries, including MatPlotLib for graphics, Scikit-learn for data analysis and skimage for texture analysis are also used. The extension runs through the following steps.

Program: PCAM Module

Input: Amino acid sequence of n-length in FASTA format and 64 3NCA rules designed for backbone
Output: CA Evolution Matrices for 64 non-trivial balanced 3NCA rules
Steps

1. Parse the input amino acid sequence. Check for any symbol other than 20 common amino acids. If passed, proceed to step 2
2. Encode the amino acid string using amino acid side chain encoding table (Table 5.2). This 8n length encoded binary string will be used as the CA seed.
3. For each of the 64 balanced 3NCA rules,

   (i)   Create a binary matrix of (8n columns × 1000 rows)
   (ii) Store the CA seed as the first row in the matrix
   (iii) Evolve the next state as null-boundary 3-neighbourhood CA and repeat the step for 999 times. Store the binary-output of each step in a consecutive row in the matrix

   Thus for 64 rules, 64 CA evolution matrices are generated.

Figure 5.2 reports sample evolution patterns derived with the python code for a few PCAMs designed for human pre-pro-insulin (INS_HUMAN) having 110 residues. The amino acid sequence is shown in Fig. 5.2a, while PCAM output pattern of 12 rules is reported in Fig. 5.2b). Two rule pairs (Rule 77, Rule 78) and (Rule 102, 105) show similar patterns, which is a common characteristics of CA evolution. So, we can cluster these patterns and study only the unique patterns.

The method to analyse the CA-image texture is discussed next.

(a)

```
102030405060
MALWMRLLPL LALLALWGPD PAAAFVNQHL CGSHLVEALY LVCGERGFFYTPKTRREAED

708090100110
LQVGQVELGG GPGAGSLQPL ALEGSLQKRG IVEQCCTSICSLYQLENYCN
```

(b)

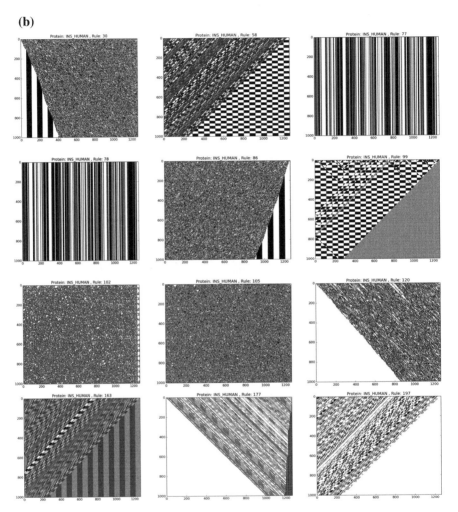

Fig. 5.2 The PCAM evolution of human insulin, **a** Amino acid sequence of Human pro-insulin sequence, **b** PCAM evolution image for 12 rules—30, 58, 77, 78, 86, 99, 102, 105, 120, 163, 177, and 197

5.6   Feature Extraction

The PCAM generates CA evolution image referred to as CA-image. It is *an* $n \times m$ binary matrix where n is the number of CA cells and m represents the number of generations the PCAM is operated. For the current study, we have assumed $m = 1000$ and we employ the texture analysis scheme detailed next. Image texture represents the information with respect to the spatial arrangement of binary data in the CA image.

A statistical method of examining texture that considers the spatial relationship of pixels is the Gray-Level Co-occurrence Matrix (GLCM), also known as the grey-level spatial dependence matrix. The GLCM functions characterize the texture of an image by calculating how often pairs of the pixel with specific values and in a specified spatial relationship occur in an image, creating a GLCM, and then extracting statistical measures from this matrix. Our CA texture analysis scheme is based on GLCM method [19].

The following steps are executed to analyse the texture of CA evolution.

Algorithm: Analyse texture of CA Evolution

Input: Binary Matrix of CA Evolution
Output: Texture Features
Steps

1. Generate Co-occurrence Matrices (GLCM):

   (a) Row-wise GLCM (rGLCM)—Count the occurrence of four transitions (00, 01, 10, 11) from left to right and right to left for each of the rows. Derive a row-wise GLCM matrix of $t \times 4$ size from $t \times n$ CA evolution matrix, where t represents the number CA evolution generations.

   (b) Column-wise GLCM (cGLCM)—Count the occurrence of four transitions (00, 01, 10, 11) from top to bottom and bottom to top for each of the columns. Derive a column-wise GLCM matrix of $4 \times n$ size from $1000 \times n$ CA evolution matrix, where n is the number of CA cells.

Calculate feature functions from rGLCM and cGLCM: Four feature functions Angular Second Moment (ASM), Contrast (CON), Correlation (COR) and Entropy (ENT) are calculated from the matrices. The method of calculating the parameters are discussed next. In the process, we get four numeric values (ASM, CON, COR, ENT) for each of the matrices, forming a set of eight features.

Row-wise GLCM (rGLCM) is sequence length independent, which means any length of protein sequence will form a $t \times 4$ rGLCM matrix with the time step of *t*. Thus, rGLCM will be particularly useful in comparing different length sequences for an experiment like classification of proteins. Further, rGLCM captures the spatial nature of CA evolution. For our PCAM texture analysis purpose, we have used rGLCM exclusively.

In the next subsection, the feature functions of GLCM are explained followed by an illustration of the entire process.

5.6.1 GLCM Feature Functions

Texture classification is usually accomplished by using certain characteristic features of co-occurrence matrices. This means that the values of a number of feature functions can be used to summarize the content of the matrices. Fourteen such functions were introduced by Haralick et al. [25], four of which appear to be the most widely used in practice. These functions are listed below since they are used for the classification experiments described in Sects. 5.4, 5.5 and 5.6.

$$\text{Angular second moment (ASM)} = \sum_i \sum_j \phi_{ij}^2$$

$$\text{Contrast (CON)} = \sum_i \sum_j (i-j)^2 \phi_{ij}$$

$$\text{Correlation (COR)} = \sum_i \sum_j \left(ij\phi_{ij} - \mu_x\mu_y\right)^2 (\sigma_x\sigma_y)$$

$$\text{Entropy (ENT)} = -\sum_i \sum_j \phi_{ij} \log \phi_{ij}$$

ϕ_{ij} denotes the (i, j)th element of a normalized co-occurrence matrix (GLCM) ϕ_N; μ_x, and σ_x are the mean and standard deviation of the marginal probability vector, obtained by summing over the rows of ϕ_N; and μ_y, and σ_y are the corresponding statistics of the column sums.

In our applications, ASM values from GLCM analysis found to be the most useful parameter. Angular Second Moment (ASM) is a measure of homogeneity of an image. A homogeneous scene will contain only a few grey levels, giving a GLCM with only a few but relatively high values of $P(i, j)$. Thus, the sum of squares will be high. To analyse the effect of a mutation on a protein sequence, the difference of overall homogeneity is reflected through ASM difference. ASM is calculated for four distances and angles of $\theta = 0$, $\pi/4$, $\pi/2$, and $3\pi/4$ radians under the assumption of angular symmetry. Thus, each GLCM generates four ASM values.

For analyzing the effect of the mutation on a specific residue position of a protein, we can substitute the amino acid at that position with Alanine. This method is widely used and known as in silico Ala-mutagenesis. Subsequently, we run PCAM on both wild and mutant (with Ala-mutagenesis) sequences to generate rGLCM of each. Next, the ASM values are derived from both the wild and mutant rGLCM. The 'Difference Score' for a mutant is the Mahalanobis Distance [26] between the mutant and wild ASM values. Higher difference score signifies a higher effect of mutation at that position.

As each CA Rule has its own characteristics, difference scores of different mutations should be compared from the PCAM output of a specific rule. For example, if we want to compare the effect of the mutation on residue position 120 (A_{120}) and 190 (A_{190}) of protein A, we will compare the PCAM output of both the mutants with the wild type under the same rule say Rule 201. But we cannot compare between A_{120} PCAM output with Rule 172 and A_{90} PCAM output with Rule 201. Further, multiple rules may be required to model different functions of protein due to the complex nature of protein functionalities.

In the next subsection, the process of feature extraction is explained with an illustration.

5.6.2 Illustration of Feature Extraction and Difference Score Calculation

In Table 5.3, a CA evolution matrix of a sample protein Wild type (W) is shown on the left column, while the Row-wise GLCM (rGLCM) matrix is shown on the right column. The evolution matrix of CA size 8 and evolution time step $t = 8$ are used for the illustration. The four binary transitions are 00, 01, 10, 11. The co-occurrence values of 00, 01, 10, 11 are calculated from both directions—left to right and right to left in each row of rGLCM. The same matrices of a mutant of the protein are

Table 5.3 CA evolution matrix, row-wise GLCM and column-wise GLCM matrices for a wild type protein

CA evolution matrix								Row-wise GLCM (rGLCM)			
								00	01	10	11
1	0	0	1	1	0	1	1	2	4	4	4
1	1	0	0	1	0	0	1	4	4	4	2
1	0	0	0	0	1	0	0	8	3	3	0
0	1	1	1	1	0	0	0	4	2	2	6
1	0	0	0	0	1	0	1	6	4	4	0
0	1	0	1	0	0	1	0	2	6	6	0
1	1	1	0	1	0	1	0	0	5	5	4
0	1	1	0	0	0	1	1	4	3	3	4

Table 5.4 CA evolution matrix, row-wise GLCM and column-wise GLCM matrices for a mutant of the sample protein reported in Table 5.1

CA evolution matrix								Row-wise GLCM (rGLCM)			
								00	01	10	11
1	1	1	1	1	0	0	1	2	2	2	6
1	1	0	0	1	0	0	1	4	4	4	2
1	0	1	0	0	1	0	0	4	5	4	0
0	1	1	1	1	0	0	0	4	2	2	6
1	0	1	1	0	1	0	1	0	6	6	2
0	1	0	1	0	0	1	0	2	5	6	0
1	1	0	0	1	0	1	0	2	5	5	2
0	1	1	0	0	0	1	1	4	3	2	4

Table 5.5 Angular Second Moment (ASM) of rGLCM of wild type (Table 5.3) and mutant (Table 5.4) with the difference score—Mahalanobis distance between ASM (wild versus mutant)

rGLCM of wild type (Table 5.3)					rGLCM of mutant (Table 5.4)			
ASM					ASM			
50	29	58	29		40	27	66	31
Difference score (Mahalanobis distance)				13.114				

shown in Table 5.4. In Table 5.3, the ASM parameters (for $\theta = 0$, $\pi/4$, $\pi/2$ and $3\pi/4$) for both wild type and mutant rGLCM matrix are shown. In the last row of Table 5.5, the difference score of the mutant compared to the wild type is also reported.

5.7   Modelling Protein–Protein Interaction

The focus of this section is to design generic methodology that can be tuned for different applications. The term 'Interactome' in the domain of molecular biology refers to the set of molecular interactions in a biological cell.

Extensive protein interaction maps are constructed for the human genome to know how the proteins organize into pathways and systems. There are more than 40,000 protein interactions (PPI) found in the human genome [27–29]. The structural knowledge of these PPIs is crucial for understanding the molecular function in human diseases at the atomic level [30, 31]. The accurate identification of key residues participating in PPIs is necessary to design new drug inhibitors and understand disease-associated mutations. The binding forces for protein coupling are not evenly distributed across their interaction surfaces. Instead, a small number of residues play the critical role for binding—these are referred to as the binding Hot-Spots (HS). From the experimental view, HS residues, upon alanine mutation, generate a binding free energy difference (ΔGbinding) ≥ 2.0 kcal/mol. On the other hand, Null-Spots (NS) correspond to residues with (ΔGbinding) <2.0 kcal/mol when mutated to alanine [30]. In the subsequent discussions we shall use these two acronyms HS and NS to refer to hot-spots and null-spots.

5.7.1 Hot-Spot Prediction—Machine Learning (ML) Method

Experimental molecular biology methods including Ala-mutagenesis produce accurate HS identification. However, expressing and purifying each individual protein is complex, time-consuming and expensive [32]. Hence, a computational method for HS prediction can be a viable alternative to experimental techniques. Furthermore, high-throughput computational methods can render valuable insight of PPIs. There are numerous statistical and Machine-Learning-based (ML) methods [33, 34] for HS prediction, which fall into two broad categories: (i) Sequence-based methods using sequence-derived features including physicochemical properties, predicted solvent accessibility, Position-Specific Scoring Matrices (PSSMs), conservation in evolution and interface propensities. The second category of ML-based HS identification is (ii) Structure-based methods, which use structure-based features like interface size, geometry, chemical composition, SASA (Solvent Accessible Surface Area) and atomic interactions [27–36]. Both methods can be combined for better accuracy [34]. A detailed review of current ML algorithms applied to HS detection can be found in Moreira's review [29].

The accuracy of any ML method mainly depends on the quality of the feature sets and the experimental data available to train the model. The knowledge of experimentally determined HS is limited, while a representative dataset can be constructed with a vast coverage of all relevant type of interactions. However, these datasets are incomplete, biased and noisy [37]. Thus, the ML-based methods for HS identification face major challenges to predict HS in proteins, where experimental data are lacking. In the next subsection, PCAM model is employed to overcome this challenge.

5.7.2 Hot-Spot Prediction—PCAM Model

Considering all the relevant issues, we curate relevant sample data from different databases and published literature. We next design the in silico experimental set up for the CA-based model through an iterative scheme till we are comfortable to design the algorithm which can be coded to generate large-scale in silico results for validation of the model. The dataset used for this experimental set up covers 36 residues from 4 protein–protein complexes with 12 hot-spots and 24 null-spots (12 HS and 24 NS). The algorithm design for this study is next reported.

Algorithm: Predict Hot-Spots (HS) from a mixture of HS and NS (Null-Spots)

Input: Protein Sequences and residue positions (mixture of HS and NS)
Output: Hot-spot residue positions
Steps

1. Run PCAM for each of the sequence for all residue positions with 64 CA rules and 999 time steps to generate CA images
2. Extract ASM parameters from GLCM analysis on the CA images generated in Step 1. Frontline visualization is given due importance in this step.
3. Calculate Median of ASM Parameters for each rule on all positions

   Mark the residue positions that has k fold or greater value of the median value for each rule, where $k \geq 15$. Marked residue positions are probable hot-spots (HS).

Prediction of Hot-Spot for a Benchmark Data Set

The algorithm for hot-spot detection is tested on the dataset used in the publication [7]. A total of 534 mutations from 53 different protein complexes are present in the dataset. Sample results derived out of CA-based model are reported in Tables 5.6, 5.7, 5.8 and 5.9 for 4 proteins complexes from this dataset. Each of the tables displays PDB ID, Chain ID, residue Position, Amino Acid of the residues tested, Known hot-spot and our prediction in Column 1–6, respectively. Column 7 reports —True Positive (TP), True Negative (TN), False Positive (FP) and False Negative (FN) instances. The last column computes the quality of the prediction. The prediction quality is calculated by the following equation:

$$\text{Accuracy} = \frac{(\text{True Positives (TP)} + \text{True Negatives (TN)})}{(\text{True Positives (TP)} + \text{True Negatives (TN)} + \text{False Positives (FP)} + \text{False Negatives (FN)})}$$

The overall prediction quality can be calculated by adding up true or false predictions of the 4 proteins reported here. Total number of residues tested is 36, while total true positive and true negative instances are 8 and 19. False positive and false negative instances are 5 and 4. So, the overall accuracy is 75%. We have reported results based on limited number of experiments. Large-scale experimentation is being implemented to improve the quality of prediction for diverse set of applications.

Next two subsections report application of our PCAM model for (i) hot-spot prediction for Monoclonal Antibodies (MAbs) and (ii) modelling mutational effect on MAb-PD-L1 interaction.

Table 5.6 Result for colicin E9 immunity protein 1BXI

PDB	Chain	Position	Amino acid	Hot-spot	Prediction	Inference	Quality
1BXI	A	23	CYS	NS	NS	TN	TP: 4
	A	24	ASN	NS	NS	TN	TN: 9
	A	27	THR	NS	NS	TN	FP: 2
	A	28	SER	NS	NS	TN	FN: 2
	A	29	SER	NS	NS	TN	Accuracy = (TP + TN)/(TP + TN + FP + FN) = 76.4%
	A	30	GLU	NS	HS	FP	
	A	33	LEU	HS	HS	TP	
	A	34	VAL	HS	HS	TP	
	A	37	VAL	NS	NS	TN	
	A	38	THR	NS	NS	TN	
	A	41	GLU	HS	NS	FN	
	A	48	SER	NS	NS	TN	
	A	49	GLY	NS	HS	FP	
	A	50	SER	HS	HS	TP	
	A	51	ASP	HS	HS	TP	
	A	55	TYR	HS	NS	FN	
	A	56	PRO	NS	NS	TN	

Table 5.7 Results of integrin protein

PDB	Chain	Position	Amino acid	Hot-spot	Prediction	Inference	Quality
1DZI	A	154	ASN	NS	NS	TN	TP: 2
	A	157	TYR	HS	HS	TP	TN: 5
	A	215	GLN	HS	NS	FN	FP: 0
	A	219	ASP	NS	NS	TN	FN: 1
	A	220	LEU	NS	HS	TN	Accuracy = (TP + TN)/(TP + TN + FP + FN) = 87.5%
	A	221	THR	HS	HS	TP	
	A	256	GLU	NS	NS	TN	
	A	258	HIS	NS	NS	TN	

Table 5.8 Result for Staphylococcal Enterotoxin C3

PDB	Chain	Position	Amino acid	Hot-spot	Prediction	Inference	Quality
1JCK	B	20	THR	NS	HS	FP	TP: 2
	B	26	TYR	NS	NS	TN	TN: 3
	B	60	ASN	NS	NS	TN	FP: 2
	B	90	TYR	HS	HS	TP	FN: 0
	B	91	VAL	HS	HS	TP	Accuracy = (TP + TN)/(TP + TN + FP + FN) = 71.4%
	B	103	LYS	NS	NS	TN	
	B	176	PHE	NS	HS	FP	

Table 5.9 Result for bone morphogenetic protein-2

PDB	Chain	Position	amino acid	Hot-spot	Prediction	Inference	Quality
1ES7	A	26	VAL	NS	HS	FP	TP: 0
	A	31	TRP	HS	NS	FN	TN: 2
	A	49	PHE	NS	NS	TN	FP: 1
	A	50	PRO	NS	NS	TN	FN: 1
							Accuracy = (TP + TN)/(TP + TN + FP + FN) = 50%

5.8   Study on Monoclonal Antibodies (MAbs)

Cancer immunotherapy utilizing antibodies to mask the inhibitory receptor have drawn considerable attention in the recent years [38–41]. Several Monoclonal Antibodies (MAbs), targeting human cell signal protein PD-L1, have been approved for clinical applications in cancer treatment. The PD-L1 protein is expressed widely in both lymphoid and non-lymphoid tissues [42] and it binds with another cell signalling protein PD-1, which suppresses the T-cell immuno response. Blockage of PD-L1 binding by MAbs is an attractive strategy for restoring tumour-specific T-cell immunity in patients with several forms of cancer [37].

It has been experimentally determined that many residues are involved in a protein binding interface between PD1 and PD-L1 (Fig. 5.3), but only a few of these residues make critical contributions towards formation of the PD-L1-PD1complex. These key residues or hot-spots (HS) are the general targets for rational drug design in blocking protein interactions.

Although MAb-based immunotherapy has achieved great successes in recent years, some basic questions still exist. What are the hot-spots in the PD-L1-MAb interaction surface for checkpoint blockade MAb targeting? Can we predict the possible mutational escapes on PD-L1 under the immune selective pressure of the MAbs during immune checkpoint blockade therapy?

To answer the above questions, which are essential for designing more specific MAbs, the accuracy of MAb hot-spot prediction is a necessity. The computational prediction of probable hot-spots on MAbs will decrease the time of actual wet lab trial. This will also help to predict the complex structure formed between MAb and PD-L1, which is necessary for crystallization of the complex structure.

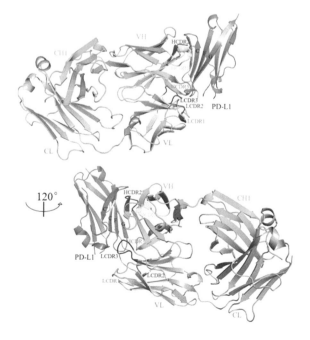

Fig. 5.3 3D structure of PD-L1/Atezolizumab complex—binding domain of PD-L1 is shown in grey and the heavy chain and light chain of Atezolizumab are shown in cyan and pink, respectively

Recent structural studies of MAbs reports the high-resolution complex structure of Avelumab [43], Atezolizumab [44] and Durvalumab [45] with PD-L1. These MAbs are approved by USA-FDA for cancer treatments [46]. Each MAb has a heavy chain and a light chain. Their sequences are available on DrugBank [47]. The aligned sequences of these MAbs are shown in Fig. 5.4.

(a)

```
DurvalumabEVQLVESGGGLVQPGGSLRLSCAASGFTFSRYWMSWVRQAPGKGLEWVANIKQDGSEKYY      60
AtezolizumabEVQLVESGGGLVQPGGSLRLSCAASGFTFSDSWIHWVRQAPGKGLEWVAWISPYGGSTYY    60
AvelumabEVQLLESGGGLVQPGGSLRLSCAASGFTFSSYIMMWVRQAPGKGLEWVSSIYPSGGITFY        60
        ****:******************** :  ************:  **. .:*

DurvalumabVDSVKGRFTISRDNAKNSLYLQMNSLRAEDTAVYYCAREGGWFGELAFDYWGQGTLVTVS      120
AtezolizumabADSVKGRFTISADTSKNTAYLQMNSLRAEDTAVYYCARR-HWP--GGFDYWGQGTLVTVS    117
AvelumabADTVKGRFTISRDNSKNTLYLQMNSLRAEDTAVYYCARI-KLGTVTTVDYWGQGTLVTVS        119
        .*:******** *.:**: *****************.***********

DurvalumabSASTKGPSVFPLAPSSKSTSGGTAALGCLVKDYFPEPVTVSWNSGALTSGVHTFPAVLQS      180
AtezolizumabSASTKGPSVFPLAPSSKSTSGGTAALGCLVKDYFPEPVTVSWNSGALTSGVHTFPAVLQS    177
AvelumabSASTKGPSVFPLAPSSKSTSGGTAALGCLVKDYFPEPVTVSWNSGALTSGVHTFPAVLQS        179
        ***********************************************************

DurvalumabSGLYSLSSVVTVPSSSLGTQTYICNVNHKPSNTKVDKRVEPKSCDKTHTCPPCPAPEFEG      240
AtezolizumabSGLYSLSSVVTVPSSSLGTQTYICNVNHKPSNTKVDKKVEPKSCDKTHTCPPCPAPELLG    237
AvelumabSGLYSLSSVVTVPSSSLGTQTYICNVNHKPSNTKVDKKVEPKSCDKTHTCPPCPAPELLG        239
        *************************************:****************  *

DurvalumabGPSVFLFPPPKPKDTLMISRTPEVTCVVVDVSHEDPEVKFNWYVDGVEVHNAKTKPREEQY      300
AtezolizumabGPSVFLFPPPKPKDTLMISRTPEVTCVVVDVSHEDPEVKFNWYVDGVEVHNAKTKPREEQY    297
AvelumabGPSVFLFPPPKPKDTLMISRTPEVTCVVVDVSHEDPEVKFNWYVDGVEVHNAKTKPREEQY        299
        ************************************************************

DurvalumabNSTYRVVSVLTVLHQDWLNGKEYKCKVSNKALPASIEKTISKAKGQPREPQVYTLPPSRE      360
AtezolizumabASTYRVVSVLTVLHQDWLNGKEYKCKVSNKALPAIEKTISKAKGQPREPQVYTLPPSRE     357
AvelumabNSTYRVVSVLTVLHQDWLNGKEYKCKVSNKALPAPIEKTISKAKGQPREPQVYTLPPSRD        359
        *****************************************************:

DurvalumabEMTKNQVSLTCLVKGFYPSDIAVEWESNGQPENNYKTTPPVLDSDGSFFLYSKLTVDKSR      420
AtezolizumabEMTKNQVSLTCLVKGFYPSDIAVEWESNGQPENNYKTTPPVLDSDGSFFLYSKLTVDKSR    417
AvelumabELTKNQVSLTCLVKGFYPSDIAVEWESNGQPENNYKTTPPVLDSDGSFFLYSKLTVDKSR        419
        *:**********************************************************

DurvalumabWQQGNVFSCSVMHEALHNHYTQKSLSLSPGK      451
AtezolizumabWQQGNVFSCSVMHEALHNHYTQKSLSLSPGK    448
AvelumabWQQGNVFSCSVMHEALHNHYTQKSLSLSPGK        450
        *******************************
```

(b)

```
DurvalumabEIVLTQSPGTLSLSPGERATLSCRASQR--VSSSYLAWYQQKPGQAPRLLIYDASSRATG      58
AtezolizumabDIQMTQSPSSLSASVGDRVTITCRASQD--VST-AVAWYQQKPGKAPKLLIYSASFLYSG    57
Avelumab-QSALTQPASVSGSPGQSITISCTGTSSDVGGYNYVSWYQQHPGKAPKLMIYDVSNRPSG        59
        .*.::* * *:*::* .::.::****:**:**:**,.*:*

DurvalumabIPDRFSGSGSGTDFTLTISRLEPEDFAVYYCQQYGSL-PWTFGQGTKVEIK-RTVAAPSV      116
AtezolizumabVPSRFSGSGSGTDFTLTISSLQPEDFATYYCQQYLYH-PATFGQGTKVEIK-RTVAAPSV    115
AvelumabVSNRFSGSKSGNTASLTISGLQAEDEADYYCCSSYTSSSTRVFGTGTKVTVLGQPKANPTV        119
        :  .***** **.:**** *: ** *.***..*.** **** ::* *:*

DurvalumabFIFPPSDEQLKSGTASVVCLLNNFYPREAKVQWKVDNALQSGNSQESVTEQDSKDSTYSL      176
AtezolizumabFIFPPSDEQLKSGTASVVCLLNNFYPREAKVQWKVDNALQSGNSQESVTEQDSKDSTYSL    175
AvelumabTLFPPSSEELQANKATLVCLISDFYPGAVTVAWKADGSPVKAGVE-TTKPSKQSNNKYAA        178
        :****.*:*:. .*:***:..* **.*:.... : ::. ......:*:

DurvalumabSSTLTLSKADYEKHKVYACEVTHQGLSSPVTKSFNRGEC-      215
AtezolizumabSSTLTLSKADYEKHKVYACEVTHQGLSSPVTKSFNRGEC-    214
AvelumabSSYLSLTPEQWKSHRSYSCQVTHEGST--VEKTVAPTECS        216
        ** *:*::::.*: *:*:***:* :* *:.**
```

Fig. 5.4 Aligned sequences **a** Heavy chain and **b** Light chain of 3MAbs—Durvalumab, Atezolizumab and Avelumab by CLUSTAL multiple sequence alignment program

Predicted Results Derived out of PCAM Model

The Python code developed for the algorithm predict Hot-Spots (HS) from a mixture of HS and NS (Non- Hot-Spots) reported in Sect. 5.7.2 is operated on each of the MAbs.

PCAMsfor each of the MAbs, for both heavy and light chain, are evolved for 999 time steps. Next we have introduced in silico ALA-mutagenesis in each of residue positions of these MAbs, and evolved the PCAM for 999 time steps. The CA evolution data is analysed by GLCM algorithm and Difference Score is calculated for each residue positions as discussed in Sect. 5.3.

The 'Difference Score' shows an interesting pattern among all the MAbs. In Rule 197 CA evolution, Ala-mutagenesis of each of the 20 amino acids across the chain shows few clusters of difference scores. The number of cluster varies from 1 to 3. Further, it is also noticed that Ala-mutagenesis in few of residue positions generate unique difference scores, which does not fit into any of the clusters of that amino acid. We mark these residue positions as the probable hot-spots. The experimental and structural data of PDL1-MAb binding confirms that the majority of hot-spot residues necessary for the binding are present in our predicted hot-spot list. The method is next explained for an illustration.

There are 19 Tyrosine (Y) amino acids present at Avelumab in positions: 32, 52, 60, 80, 94, 95, 109, 152, 183, 201, 281, 299, 303, 322, 352, 376, 394, 410 and 439. The difference scores with Rule 197 for ALA-mutagenesis at all the positions are plotted in Fig. 5.5. The difference score values of Tyrosine forms two clusters: (i) 12 residue positions show value same or close to Difference Score 7300, and (ii) 4 positions show Difference Score same or close to 2010; but, there are (iii) 3 residue positions 32, 52 and 60, which show unique difference score—4974, 4579 and 2788, respectively. This 3 Difference Scores do not fall in any clusters. Thus, from our analysis, we will mark this residue positions as probable hot-spot sites. Residue position 52 is a known hot-spot reported in the crystallographic structural data of Avelumab-PD-L1 binding [43]. This method is employed for the MAbs are results are reported next.

The detailed result of PD-L1 binding MAbs—Avelumab and Atezolizumab are reported in Tables 5.10 and 5.11. Each MAb has two chains—heavy chain and light chain (Fig. 5.3). For each chain, the known hot-spots positions with amino acid are shown in second column. Third column reports our prediction where HS means Hot-Spot and NS means Non-Hot-Spot. The prediction quality and accuracy are reported in the last two columns—for each of the cases the accuracy is above 90%.

The PCAM-based hot-spot identification method shows prediction accuracy of more than 90%. The scheme is able to predict most of the hot-spot residues with significantly smaller false positive prediction compared to the sequence length.

There is no benchmark data for MAbs hot-spot prediction. Due to its highly variable regions in amino acid sequence, any standard homology-based scheme is not suitable for MAb analysis. The PCAM-based modelling is a novel approach to analyse such proteins, where standard homology-based methods do not perform well due to their high dependency on training data.

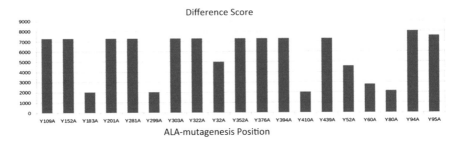

Fig. 5.5 Difference score of Ala-mutagenesis at 19 positions with TYR (Y is single letter code for TYR). Y109A means TYR at 109 position is substituted with ALA (A is the single letter code for ALA)

Table 5.10 Result table for Avelumab

Chain	Known hot-spot	Predicted hot-spot	Prediction quality	Accuracy
Heavy chain	F27	HS	Total residue: 450	Accuracy = (TP + TN)/ (TP + TN + FP + FN) = 96.6%
	T28	HS	TP: 8	
	S31	HS	TN: 427	
	I33	HS	FP: 11	
	Y52	NS	FN: 4	
	P53	HS		
	S54	HS		
	G55	HS		
	I57	NS		
	F59	HS		
	L101	NS		
	G102	NS		
Light chain	Y32	NS	Total residue: 216	Accuracy = (TP + TN)/ (TP + TN + FP + FN) = 91.2%
	Y34	NS	TP: 4	
	Y93	HS	TN: 193	
	S95	HS	FP: 17	
	S97	HS	FN: 2	
	R99	HS		

This is important to note that, our PCAM-based method does not use any physical domain data to train the model. It exclusively depends on the encoding of amino acids based on the molecular configuration and interaction of encoded amino acids by a 3NCA rule.

Table 5.11   Result table for Atezolizumab

Chain	Known hot-spot	Predicted hot-spot	Prediction quality	Accuracy
Heavy chain	G55	HS	Total residue: 448	Accuracy = (TP + TN)/ (TP + TN + FP + FN) = 93.9%
	S57	HS	TP: 6	
	T54	HS	TN: 412	
	T58	NS	FP: 28	
	D31	HS	FN: 2	
	R99	HS		
	W101	NS		
	S30	HS		
Light chain	Y93	HS	Total residue: 214 TP: 1 TN: 194 FP: 19 FN: 0	Accuracy = (TP +TN)/ (TP + TN + FP + FN) = 91.1%

5.9   Study of Mutational Effect on Binding

Study of MAb–protein interaction has received considerable attention in recent years. Further, the evaluation of binding affinity on insertion of mutations on PD-L1is also investigated. The aim of the study is to compare the binding affinity of wild and mutant PD-L1 with a MAb. We report the mutational effect based on the CA model framework reported in earlier sections. This section establishes the prediction capability of the proposed model on a case study recently reported [44]. Further, the CA rule to be employed for the study of any mutant of PD-L1 with a specific MAb gets identified from this study.

5.9.1   A Case Study—Effect of Mutation on PD-L1 on its Binding Affinity with MAbs

A recent study [44] reported the crystal structure of an anti-PD-L1 nanobody KN035 and compared it with the clinically used PD-L1 blocker Atezolizumab [44, 48]. It analyses the contribution of each PD-L1 residue of the interface in the binding affinity. The molecular biology method, used for HS detection, follows through Ala-mutagenesis and affinity measurement. The PD-L1 residues I54, Y56, E58, Q66 and R113 are identified as binding Hot-Spots (HS). It is experimentally established that the five HS residues are all involved in binding to all known PD-L1 targeting MAbs.

The PD-L1 mutants are tested for their binding affinity towards anti-PD-L1 nanobody KN035 and Atezolizumab. The results establish the fact that replacement

of ala mutation at HS residue E58 results to 18-fold decrease in the binding affinity
of PD-L1 towards Atezolizumab. Other HS residues play a relatively insignificant
role in binding. This signifies PD-L1's E58is critical for binding with Atezolizumab
(Fig. 5.3). Furthermore, studies of PD-L1 inhibitors reports PD-L1 inhibitions
involves two more HS residues I54 and Y56 [49, 50].

In Table 5.10, the K_d values for KN035 and Atezolizumab are shown for 10
mutants and wild (W) type of PD-L1 protein. K_d value is the measure of binding
affinity, which reflects the strength of the binding interaction between a single
biomolecule (e.g. PD-L1) to its ligand/binding partner (e.g. MAb). Binding affinity
is typically measured by the equilibrium dissociation constant (K_d), which is used to
evaluate and rank order strengths of bimolecular interactions.

For a general reaction:

$$A_x B_y \leftrightharpoons xA + yB$$

in which a complex $A_x B_y$ breaks down into xA subunits and yB subunits, the
dissociation constant is defined

$$K_d = \frac{[A]^x [B]^y}{[A_x B_y]}$$

where $[A], [B]$ and $[A_x B_y]$ are the concentrations of A, B and the complex $A_x B_y$,
respectively.

The smaller the K_d value, the greater the binding affinity of the ligand for its
target. The larger the K_d value, the more weakly the target molecule and ligand are
attracted to and bind to one another. There are many methods to measure binding
affinity and dissociation constants, such as ELISAs, gel-shift assays, pull-down
assays, equilibrium dialysis, analytical ultracentrifugation, SPR and spectroscopic
assays. Column 1 of Table 5.12 shows the ten mutants analysed under this case
study with wild type noted as W. K_d for binding of the wild/mutants on two MAbs
are reported on columns 2, 3 (for KN035) and columns 5, 6 (for Atezolizumab).

5.9.2 Results Derived Out of PCAM Evolution

The PD-L1 protein has been extensively studied using the PCAM. The 10 mutations reported in column 1 of Table 5.12 are analysed through GLCM analysis. The
1st column shows for PD-L1 variants, where W signifies Wild type. The second and
fifth columns report the experimentally determined K_d values—the equilibrium
dissociation constant between the PD-L1 and MAbs—KN035 and Atezolizumab,
respectively. The third and sixth columns show the ratio values of variants K_d and
Wild-type K_d. Higher ratio value signifies lesser binding affinity. The fourth and
seventh columns (grey coloured) report the difference score calculated from the
GLCM analysis of PCAM model from Rule 201 (Fig. 5.6a) and Rule 99 (Fig. 5.6b)
respectively. Computation of Difference Score has been detailed in Sect. 5.6.1.

Table 5.12 The known K_d values of PD-L1 Variants for KN035 and Atezolizumab and Difference Scores calculated from PCAM models with rule 99 and rule 201. The significantly higher changes of K_d values and Difference Scores are marked in bold

1	2	3	4	5	6	7
PD-L1 variants	K_d of KN035 (M)	K_d variant/ K_d wild (W)	Difference Score Rule 201	K_d of Atezolizumab (M)	K_d variant/ K_d wild (W)	Difference Score Rule 99
Wild	3.0E−09	1	0	9.96E−09	1	0
I54A	2.42E−07	80.7	0	3.23E−08	3.2	1.42
Y56A	1.24E−06	**413.3**	**2441.32**	2.68E−08	2.7	0.57
E58A	1.49E−07	49.7	1	1.81E−07	**18.2**	**334.43**
D61A	1.99E−08	6.6	0	9.99E−09	1.0	0.35
N63A	2.30E−08	7.7	1	1.73E−08	1.7	12.55
Q66A	4.88E−07	**162.7**	**998**	2.46E−09	0.25	**335.32**
R113A	5.34E−07	**178**	0	8.52E−08	8.6	0.18
M115A	5.51E−08	18.4	1	4.57E−08	4.6	0.36
Y123A	4.24E−08	14.1	1	4.66E−08	4.7	**374.78**
R125A	2.97E−08	9.9	**998**	5.89E−08	6.0	0.51

Fig. 5.6 Difference Score graph PDL1 mutants compared to Wild (W) type for **a** CA Rule 99 and **b** CA Rule 201

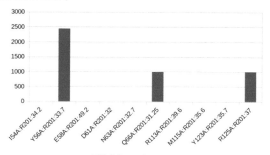

(a)
Difference Score

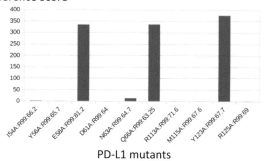

(b)
Difference Score

5.9.3 Comparison of PCAM Model Results with the Wet Lab Experimental Results—Identification of CA Rule

This section reports a comparison of CA model results with the wet lab experimental results. Based on the comparative study with different PCAMs we identify the 3NCA rules relevant for study of interaction of any other mutant on PD-L1 with the specific MAb.

MAb KN035: For MAb KN035, the PD-L1 variant Y56A shows more than 400 fold change in K_d value. While Q66A and R113A show K_d value changes more than 150 fold. Rest of the residue positions show a significantly smaller change in K_d values. The difference scores estimated by our study with Rule 201, reported in the fourth column of Table 5.12, models the behaviour of PD-L1 mutants and KN035 binding. It generates high Difference Score of 2441.32, and 998 for mutants Y56A and Q66A, both having high K_d value changes (Fig. 5.6a). It misses mutant R113A and falsely predict R125A. For any other mutational study on KN035-PD-L1, the interaction the target rule to be used is Rule 201.

MAb Atezolizumab: On the other hand, for MAb Atezolizumab, the K_d values do not vary as much as that for KN035. The only mutant E58A shows a significantly higher change in K_d value (18 folds). The difference score, estimated by our method using Rule 99, which models the behaviour of Atezolizumab—PD-L1 binding, reports a high value for the mutant E58A (7th column in Table 5.12; Fig. 5.6b). It also generates high Difference Scores for mutants Q66A and Y123A, which are false positive. The prediction quality can be improved by the removal of false positive predictions using other sequence or structural analysis. For any other mutational study on MAb Atezolizumab–PD-L1 interaction the target rule to be used is Rule 99.

5.10   The Scope of Future Study

Proteomics is a wide domain of study. In this chapter, we have studied the protein–protein interaction with respect to predict the binding hot-spots and the effect of mutation on such binding. We have selected the interaction of Monoclonal Antibodies (MAbs) and cell signalling protein PD-L1 to illustrate our method. MAb-based immunotherapy is currently the most advanced and well-researched method of cancer treatment. The PCAM model can be extended to be applied in structural proteomics, protein stability, drug–protein interaction, and other areas of proteomics.

Structural proteomics is the process of the high-throughput characterization of the three-dimensional structures of biological macromolecules. Recently, the process for protein structure determination has become highly automated and structural proteomics platforms have been established around the globe, utilizing X-ray crystallography as a tool. Although protein structures often provide clues about the biological function of a target, once the three-dimensional structures have been determined, bioinformatics and proteomics-driven strategies can be employed to derive their biological activities and physiological roles. PCAM-based model can be used to study the effect of amino acid mutation due to Single Nucleotide Polymorphism (SNP) or Post-Translational Mutation (PTM) on the 3D structure of protein. Such studies can help to identify the important residue positions which can destabilize protein structure on mutation, which leads to the high-throughput protein stability analysis. PCAM-based analysis can be further extended to predict binding affinity of drug–protein complexes. The effect mutation on binding affinity and mutation escapes of the drug can be also be predicted from PCAM analysis.

5.11   Summary

Advances in Genomics have enriched the discipline of Proteomics. This has resulted in an explosion in the number of uncharacterized protein sequences. Wet lab characterization of protein interaction with other biomolecules demands high cost and time. This chapter addresses the problem of predicting protein interactions employing the in silico experiment based on CA model. Bottom-up design of the protein model is based on the first principle of analysing the information content of its building block (amino acid) along with the design of three neighbourhood CA (3NCA) rules for its backbone and side chain. Protein Modelling CA Machine (PCAM) is designed with CA rules. For a candidate protein chain to be analysed for its interaction with other biomolecules, 64 PCAMs are designed. Such a large number of PCAMs for one protein chain is a necessity to represent the wide variation of protein interaction with different biomolecules. For top-down validation of the model, these PCAMs are employed for consistent mapping of patterns/ parameters of one PCAM to the features extracted from a wet lab experiment.

The relevant pattern/parameters of a PCAM are extracted by generating binary image out of the evolution of a PCAM for 999 generations. For texture analysis of binary image, we have employed GLCM (Gray-Level Co-occurrence Matrix). The parameters extracted from GLCM matrices lay the foundation for top-down validation of the model.

A generic scheme to model protein–protein interaction has been proposed to predict binding Hot-Spot (HS) residues of a protein chain. Predicted results are reported for four protein chains by executing the program code developed for the algorithm designed for the scheme.

The applications of the generic scheme are next reported to address the burning problem of current interest in the field of cancer immunotherapy. The first application deals with the prediction of hot-spots on Monoclonal Antibodies (MAbs), while the second one reports the analysis of the mutational effect on MAb—PD-L1 interaction. Programmed Death-Ligand 1 (PD-L1) is a protein found in cancer cells which help them to evade immune attack. Consequently, the design of MAbs as PD-L1 inhibitor has received considerable attention in recent years. In this context, we report our prediction results on the mutational effect of MAb–PD-L1 interaction for the MAbAtezolizumab.

Questions

1. What are the underlying reasons for keeping a provision of 64 different PCAM designs for a protein chain? How the specificity of a protein chain is incorporated in the PCAM design?
2. Note the sequential steps implemented for design of CA rules for PCAM and design of seed for the CA.
3. Hydroxylysine is an uncommon amino acid found in collagen tissues. Design the 8-bit binary side chain encoding for it using its molecular structure.
4. Write a brief survey on the schemes proposed by different publications on the application of CA for modelling a protein chain. Please add current publications.
5. Write a Python code for CA Seed generation and 3NCA evolution as discussed in Sect. 5.4.2.
6. Define the terms texture analysis and GLCM in the context of analysis of image derived out of PCAM evolution. Write a survey on the application GLCM in various biological or clinical studies.
7. Write a Python code using Scikit-learn and skimage library for calculating GLCM parameters for a binary CA evolution matrix.
8. Study the evolution patterns of 64 PCAMs for Human Insulin (UniProt ID P01308) and its mutated versions with ALA-mutagenesis at each residue positions. Summarize your conclusion from the study of the PCAM evolution for wild and mutants using GLCM parameters.
9. 'In order to execute biological functions, single protein rarely acts in isolation'—True or False? If 'True' justify your answer from the study of published literature on the importance of protein–protein interaction network for biological functions.
10. How do you define 'Hot-Spots' in the context of protein–protein interaction? From the study of published literature, write short notes—(a) computational methods, and (b) experimental techniques employed for prediction of 'Hot-Spots'.
11. Write short notes on—antibody, monoclonal antibody and cancer immunotherapy.
12. Describe the sequential steps normally observed for binding of a drug molecule on a protein structure.
13. (a) 'A kinase is an enzyme that catalyzes the transfer of phosphate groups from high-energy, phosphate-donating molecules to specific substrates'—True or False? If 'True' justify your answer with an illustration (b) what is the possible effect of the mutation on an enzyme gene?
14. From the study of recent publications note the sequential steps implemented for protein structure prediction from its primary chain.

15. Write a short note on important functions performed by different proteins with suitable illustrations.
16. What are PD-1 and PD-L1 molecules? How cancer cells implement immune escape for their growth?
17. Why hot-spot prediction on MAbs is a major focus to fight deadly cancer disease? Note the sequential steps implemented in the PCAM model for MAb Hot-Spot prediction.
18. Study the paper 'Shuguang Tan et al. An unexpected N-terminal loop in PD-1 dominates binding by nivolumab, Nature Communications volume 8, Article number: 14,369 (2017)'. Generate 64 PCAMs for PD-1 binding MAb Nivolumab and detect which 3NCA Rule can identify most of the binding hot-spots reported in the paper. (Download the supplementary materials of paper for hot-spots positions).
19. From the study of recent publications, write a short note on the autoimmune disease with illustration of a common disease rheumatoid arthritis.

References

1. Y, Ofran, Rost, B.: Protein–protein interaction hot-spots carved into sequences. PLoS Comput. Biol. 3(7), 119 (2007)
2. Kundrotas, P., Belkin, S., Vakser, I.: Structure-function relationships in protein complexes. Biophys. J. 114(3), 46a (2018)
3. Fersht, A.: Structure and mechanism in protein science: a guide to enzyme catalysis and protein folding. World Sci. 9 (2017)
4. Webb, B., Sali, A.: Protein structure modeling with MODELLER, pp. 1–15. Humana Press, New York, NY (2014)
5. Lee, J., Freddolino, P.L., Zhang, Y.: Ab initio protein structure prediction. From protein structure to function with bioinformatics, pp. 3–35. Springer, Dordrecht (2017)
6. Moult, J., et al.: Critical assessment of methods of protein structure prediction (CASP)—round XII. Proteins Struct. Funct. Bioinf. 86, 7–15 (2017)
7. Moreira, I.S., et al.: SpotOn: high accuracy identification of protein-protein interface hot-spots. Sci. Rep. 7(1), 8007 (2017)
8. Burks, C., Farmer, D.: Towards modeling DNA sequences as automata. Physica D: nonlinear phenomena 10(1–2), 157–167 (1984)
9. Sirakoulis, G., Karafyllidis, I., Mizas, C., Mardiris, V., Thanailakis, A., Tsalides, P.: A cellular automaton model for the study of dna sequence evolution. Comput. Biol. Med. 33(5), 439–453 (2003)
10. Mizas, C., Sirakoulis, G., Mardiris, V., Karafyllidis, I., Glykos, N., Sandaltzopoulos, R.: Reconstruction of dna sequences using genetic algorithms and cellular automata: towards mutation prediction? Biosystems 92(1), 61–68 (2008)
11. de Sales, J.A., Martins, M.L., Stariolo, D.A.: Cellular automata model for gene networks. Phys. Rev. E 55, 3262–3270 (1997)
12. Xiao, X., Shao, S., Ding, Y., Chen, X.: Digital coding for amino acid based on cellular automata. In: 2004 IEEE international conference on systems, man and cybernetics, vol. 5, pp. 4593–4598. Oct 2004

13. Xiao, X., Shao, S., Ding, Y., Huang, Z., Chou, K.-C.: Using cellular automata images and pseudo amino acid composition to predict protein subcellular location. Amino Acids 30(1), 49–54 (2006)
14. Xiao, X., Wang, P., Chou, K.-C.: Gpcr-ca: a cellular automaton image approach for predicting g-protein-coupled receptor functional classes. J. Comput. Chem. 30(9), 1414–1423 (2008)
15. Xiao, X., Ling, W.: Using cellular automata images to predict protein structural classes. In: The 1st International Conference on Bioinformatics and Biomedical Engineering, pp. 346–349. ICBBE 2007, July 2007
16. Xiao, X., Wang, P., Chou, K.-C.: Predicting protein structural classes with pseudo amino acid composition: An approach using geometric moments of cellular automaton image. J. Theor. Biol. 254(3), 691–696 (2008)
17. Chou, K.-C.: Prediction of protein cellular attributes using pseudo amino acid composition. Proteins Struct. Funct. Genet. 43, 246–255 (2001)
18. Chou, K.-C.: Some remarks on protein attribute prediction and pseudo amino acid composition. J. Theor. Biol. 273(1), 236–247 (2011)
19. Xiao, X., Wang, P., Chou, K.-C.: Cellular automata and its applications in protein bioinformatics. Curr. Protein Pept. Sci. 12(6), 508–519 (2011)
20. Santos, J., Villot, P., Dieguez, M.: Cellular automata for modeling protein folding using the HP model. In: 2013 IEEE Congress on Evolutionary Computation (CEC), pp. 1586–1593. June 2013
21. Santos, J., Villot, P., Dieguez, M.: Emergent protein folding modeled ´ with evolved neural cellular automata using the 3D HP model. J. Comput. Biol. 21(11), 823–845 (2014)
22. Chopra, P., Bender, A.: Evolved cellular automata for protein secondary structure prediction imitate the determinants for folding observed in nature. Silico Biology 7(7), 87–93 (2006)
23. Cristea P.: Independent component analysis for genetic signals. In: SPIE Conference BIOS 2001-International Biomedical Optics Symposium, pp. 20–26. San Jose, USA, January 2001
24. Pan, Y.X., et al.: Application of pseudo amino acid composition for predicting protein subcellular location: stochastic signal processing approach. J. Protein Chem. 22(4), 395–402 (2003)
25. Haralick, R.M., Shanmugam, K.: Textural features for image classification. IEEE Trans. Syst. Man Cybernet. 6, 610–621 (1973)
26. De Maesschalck, R., DelphineJouan, R., Massart, D.L.: The mahalanobis distance. Chemometr. Intell. Lab. Syst. 50(1), 1–18 (2000)
27. Petta, I., Lievens, S., Libert, C., Tavernier, J., De Bosscher, K.: Modulation of protein–protein interactions for the development of novel therapeutics. Mol. Ter. 24, 707–718 (2016). https://doi.org/10.1038/mt.2015.214
28. Clackson, T., Wells, J.A.: A hot-spot of binding energy in a hormone-receptor interface. Science 267, 383–386 (1995)
29. Te Moreira, I.S.: Role of water occlusion for the definition of a protein binding hot-spot. Curr. Top. Med. Chem. 15, 2068–2079 (2015)
30. Moreira, I.S., Fernandes, P.A., Ramos, M.J.: Hot-spots—a review of the protein-protein interface determinant amino-acid residues. Proteins 68, 803–812 (2007). https://doi.org/10.1002/prot.21396
31. Ramos, R.M., Moreira, I.S.: Computational Alanine scanning mutagenesis—an improved methodological approach for protein DNA complexes. J. Chem. Theory Comput. 9, 4243–4256 (2013). https://doi.org/10.1021/ct400387r
32. Brender, J.R., Zhang, Y.: Predicting the effect of mutations on protein-protein binding interactions through structure-based interface profiles. PLoS Comput. Biol. 11, e1004494 (2015). https://doi.org/10.1371/journal.pcbi.1004494
33. Xue, L.C., Dobbs, D., Bonvin, A.M.J.J., Honavar, V.: Computational prediction of protein interfaces: a review of data driven methods. FEBS Lett. 589, 3516–3526 (2015). https://doi.org/10.1016/j.febslet.2015.10.003
34. Melo, R., et al.: A machine learning approach for hot-spot detection at protein-protein interfaces. Int. J. Molec. Sci. 17, 1215 (2016). https://doi.org/10.3390/ijms17081215

35. Chou, K.C.: Some remarks on protein attribute prediction and pseudo amino acid composition. J. Theor. Biol. **273**, 236–247 (2011). https://doi.org/10.1016/j.jtbi.2010.12. 024

36. Chen, W., Feng, P., Ding, H., Lin, H.: PAI: predicting adenosine to inosine editing sites by using pseudo nucleotide compositions. Sci. Rep. **6**, 35123 (2016). https://doi.org/10.1038/srep35123

37. Herbst, R.S., Soria, J.C., Kowanetz, M., Fine, G.D., Hamid, O., Gordon, M.S., Sosman, J.A., McDermott, D.F., Powderly, J.D., Gettinger, S.N., Kohrt, H.E., Horn, L., Lawrence, D.P., et al.: Predictive correlates of response to the anti-PD-L1 antibody MPDL3280A in cancer patients. Nature **515**, 563–567 (2014)

38. Chen, L., Flies, D.B.: Molecular mechanisms of T cell co-stimulation and co-inhibition. Nat. Rev. Immunol. **13**, 227–242 (2013)

39. Greenwald, R.J., Freeman, G.J., Sharpe, A.H.: The B7 family revisited. Ann. Rev. Immunol. **23**, 515–548 (2005)

40. Lenschow, D.J., Walunas, T.L., Bluestone, J.A.: CD28/B7 system of T cell costimulation. Ann. Rev. Immunol. **14**, 233–258 (1996)

41. Carreno, B.M., Collins, M.: The B7 family of ligands and its receptors: new pathways for costimulation and inhibition of immune responses. Ann. Rev. Immunol. **20**, 29–53 (2002)

42. Dong, H., Zhu, G., Tamada, K., Chen, L.: B7-H1, a third member of the B7 family, co-stimulates T-cell proliferation and interleukin-10 secretion. Nat. Med. **5**, 1365–1369 (1999)

43. Tan, S., Zhang, C.W., Gao, G.F.: Seeing is believing: anti-PD-1/PD-L1 monoclonal antibodies in action for checkpoint blockade tumor immunotherapy. Signal Trans. Target. Therap. **1**, 16029 (2016)

44. Zhang, F., et al.: Structural basis of the therapeutic anti-PD-L1 antibody atezolizumab. Oncotarget **8**(52), 90215–90224 (2017)

45. Tan, S., et al.: Distinct PD-L1 binding characteristics of therapeutic monoclonal antibody durvalumab. Protein Cell **9**(1), 135–139 (2018)

46. Gay, C.L., et al.: Clinical trial of the anti-PD-L1 antibody BMS-936559 in HIV-1 infected participants on suppressive antiretroviral therapy. J. Infect. Dis. **215**(11), 1725–1733 (2017)

47. Wishart, D.S., et al.: DrugBank 5.0: a major update to the DrugBank database for 2018. Nucleic Acids Res. **46**(D1), D1074–D1082 (2017)

48. Zhang, F., Wei, H., Wang, X., Bai, Y., Wang, P., Wu, J., Jiang, X., Wang, Y., Cai, H., Xu, T., Zhou, A.: Structural basis of a novel PD-L1 nanobody for immune checkpoint blockade. Cell Discov. **3**, 17004 (2017)

49. Zak, K.M., Grudnik, P., Guzik, K., Zieba, B.J., Musielak, B., Dömling, A., Dubin, G., Holak, T.A.: Structural basis for small molecule targeting of the programmed death ligand 1 (PD-L1). Oncotarget **7**, 30323–30335 (2016)

50. Guzik, K., Zak, K.M., Grudnik, P., Magiera, K., Musielak, B., Törner, R., Skalniak, L., Dömling, A., Dubin, G., Holak, T.A.: Small-molecule inhibitors of the programmed cell death-1/programmed death-ligand 1 (PD-1/PD-L1) interaction via transiently induced protein states and dimerization of PD-L1. J. Med. Chem. **60**, 5857–5867 (2017)

Chapter 6
Current and Next Phases
of Development of cLife (CA-Based Life)

> *"Focusing on artificial life frees us to explore novel chemical systems, but what we learn from these systems helps us to understand possible pathways leading to the origin of life".*
> —Jack W. Szostak

6.1 Introduction

'A New kind of Computational Biology' has been proposed in the earlier five chapters of this book. This framework lays the foundation of the current design of cLife—Cellular Automata (CA)-based model of current life form supported by three major biomolecules—DNA, RNA and protein. This chapter summarizes the model we presented in the earlier chapters. In Sect. 6.4, we have presented a comparative analysis of the in silico framework of cLife model and other two popular approaches—Molecular Dynamic (MD) simulation and Artificial Intelligence (AI)-based tools.

As far as our present state of understanding of the evolution of the universe and life on earth goes, both living organisms and nonliving objects/entities coexisted on earth for millions of years. But even with the immense advances in science and technology, we are yet to decipher how the combination of the most common elements came up to form the primitive life forms. So, we thought it was worthwhile to undertake a brief journey through the scientific discoveries of the nested constituent blocks of molecule, atom, electron, proton, etc., common for both living and nonliving objects. Subsequent to discussions on the mathematical modelling and analysis of non-living objects and fundamentals of biomolecules, we introduced Cellular Automata (CA) preliminaries in the last part of Chap. 1. We proposed a model of artificial life mimicking the current life form. We have termed the model as cLife (CA-based life). The underlying motivation of the model is to establish the foundation of a new kind of in silico platform—a framework for a 'new kind of computational biology'.

The long-term goal of cLife is to travel in time in the reverse direction—from current to primitive life form. In the present model of cLife, in Chap. 3, we have

© Springer Nature Singapore Pte Ltd. 2018
P. P. Chaudhuri et al., *A New Kind of Computational Biology*,
https://doi.org/10.1007/978-981-13-1639-5_6

tried to establish a link between primitive life form, as we visualize, to the current life form. In the next phase of development, we plan to strengthen the foundation of primitive life form.

6.2   Short-Term Goal of Current Phase of Development

Our short-term goal for cLife is to extend support to the drug design and biomolecule discovery community for developing new class of drugs to fight various deadly diseases. Outbreak of such deadly diseases around the globe results in loss of human life and high cost to society. Degrading environment, climate change and current lifestyle of modern living are associated with various known and new kinds of diseases. Different classes of virus and bacteria have been also detected which are found to be immune to the current drugs (e.g. antibiotic resistance of bacteria). There is also a need for personalized and targeted drugs for the most effective treatment of patients. However, the current drug design methodology is an enormously costly affair with over a billion dollars spent on any successful drug and a decade to take a new drug to market.

The current effort is focused on using the cLife models to accelerate drug and biomolecule discovery with ultra-fast capability to visualize, synthesize, test, interact and experiment in silico—with a wide variety of biomolecules. The sole purpose is to narrow down the number of candidates to be tested through wet lab experiment that involves high cost and time.

We have designed cLife to model the current life form evolved around three basic macromolecules—RNA, DNA and protein. The CA models for these three macromolecules are presented in Chaps. 3, 4 and 5. A number of case studies are covered in each of these chapters. These case studies are collected out of the recently published wet lab results reported for different biomolecules. We presented top-down validation of the results derived out of cLife through these case studies.

Design of CAMs (Cellular Automata Machines) proceeds from the analysis of the building blocks of the macromolecules. The rule design reported in Chap. 2 is guided by the information retrieved from the constituent micromolecules—(i) codon of mRNA string, (ii) molecular structure of sugar–phosphate and nucleotide bases of nucleic acid strand, amino acids of protein. This bottom-up model development process enables us to design simple one dimensional three or five neighbourhood CA representing structure and behaviour of the macromolecules in the next three chapters—RNA in Chap. 3, DNA in Chap. 4, and protein in Chap. 5. Each of these chapters reports how we have moved forward for top-down validation of the proposed models designed from the bottom-up approach of analysing the information content of the six micromolecules.

Through this model building approach, we have shown:

(i) How complex molecular phenomenon in Genomics and Proteomics can be addressed computationally as state machine evolution.

(ii) How the nonlinear dynamics of CA evolution can model the behavioural characteristics of biomolecules.

The in silico experimental set up for prediction of results out of CA model demands validation against the known wet lab experimental results. Fortunately, one of the largest databases of current time has grown from the study of biomolecules. We have used the experimental results reported in such databases to validate our prediction results. The major hurdles we have encountered for top-down validation are summarized in Sect. 6.6. The solution of these problems have been proposed in the current phase of development and will continue to be addressed in the next phase of development of cLife.

6.3 CA Model for RNA, DNA, Protein

The discovery of different classes and types of RNAs has revolutionized the study of life sciences. For example, the discovery of ribo enzyme in 1982 established the fact that RNA could function like any other protein enzyme. Ribo switches discovered in 2002 emphasized the role of RNA for gene expression. A ribo switch is a regulatory segment of a messenger RNA molecule that binds a small molecule, resulting in a change in production of the proteins encoded by the mRNA. Subsequent identification of different classes of small RNA molecules enhanced the importance of RNA as the controlling mechanism for protein production. Such developments encouraged the authors of this book to initiate a detailed study of RNA molecules.

In the above background, in Chap. 3 we initiated discussions on 'RNA First' hypothesis. We next introduced the CA model based on the analysis of information embedded in nucleotide base triplets referred to as 'codon' observed in all forms of life. Design of 64 CA rules modelling 64 codons is reported along with the presentation of CA rule equivalency representing codon degeneracy. One of the major contributions of this chapter is the CA model for co-evolutional folding of nascent peptide chain during ribosomal translation. In subsequent sections of Chap.3, we reported CA model to study different types of RNA molecules—RNA transcript, mRNA precursor, matured mRNA and special classes of non-coding RNA—tRNA, precursor RNA, small length RNAs (e.g. siRNA, sgRNA, etc.). Different classes of CAMs (CA Machines) are designed out of RNA base sequences. Signal graphs derived out of evolution of these CAMs lays the foundation for analysis of structure and function of these biomolecules. The signal graph analytics methodology has been designed based on the study of wet lab (or clinical study) results. Predicted results derived out of CA models are tabulated on executing Python codes developed for the algorithms reported in the chapter.

Next Chap. 4 presents the analysis of DNA molecule. Different DCAMs (DNA modelling CA Machines) are designed with CA rules designed for sugar–phosphate backbone and sequence of nucleotide bases A, T, C, G. We have shown that the regularity of spatio-temporal characteristics of BBDCAM (BackBone Modelling DNA CA Machine) signal graph represent the regularity of DNA helical structure. Rather than study of DNA molecule from DNA strand, we have proposed the study of RNA transcript derived out of the DNA strand. The methodology presented in Chap. 3 for RNA analysis is utilized for the analysis of RNA transcript.

One of the major focuses of Chap. 4 is the study of alternative splicing of introns leading to derivation of different protein products out of the same gene. This differential splicing, as projected in recent publications, is the end result of the prevailing physical conditions of living cells which demand production of different proteins under different physical conditions along with suppression or enhancement of protein production. The CA model for prediction of exon–intron boundary is presented along with the provision of skipping an exon in matured mRNA if a 'skip' tag is set on. This tag is assumed to represent physical domain conditions that demand exclusion of a specific exon in the protein product.

The next landmark of Chap. 4 is the analysis SNP (Single-Nucleotide Polymorphism) observed in DNA molecules. While Chap. 3 reported SNP on codon string of DNA coding region, Chap. 4 concentrated on the study of SNP on both coding and non-coding region of DNA. We have mapped the SNP of DNA to its transcript. Next we presented the CA model for SNP highlighting the deviation of CA model parameters for a mutant from that of its wild version. Based on the signal graph analytics of RCAM (RNA CA Machine) output for transcript, we have co-related—(i) the deviation of CA model parameters of mutant from that of wild to (ii) the deviation of physical domain features of mutant from its wild version. A few case studies are presented based on wet lab analysis (or clinical study) results reported in published literature or available in various databases. These case studies validate the results derived out of CA model of cLife with respect to the qualitative deviation of mutant features compared to that of the wild version. Subsequent to finalizing the experimental set up for SNP study, we have reported results on random insertion of mutations on DNA strand.

The last section of Chap. 4 presented the CA model for gene editing technology of CRISPR/Cas9. While for earlier applications we analysed RNA transcript and RCAMs (RNA CA Machines), for gene editing we analyse evolution of both DCAM (DNA CA Machine) and RCAMs. We have analysed the RCAM output of sgRNA (single guide RNA) molecule and DCAM output of its target gene region where a sgRNA binds for its editing. The outcome of this analysis is the derivation of CA model parameters for prediction of on-target and off-target binding of sgRNA on the target gene. The emphasis of this study is to add an additional binding criteria derived out of CA model to the existing list of criteria identified by different wet lab or in silico analysis. We expect this will add value for reduction of off-target binding of a sgRNA designed for its target gene.

Following the presentation of CA model for nucleic acids, the third macro-molecule in our study list is protein—the workhorse of all living cells across the species. While DNA mainly stores the biological information, RNA transcript and its derivatives are associated with the essential functions of a living cell including synthesis of protein products. Different classes of proteins (or protein complexes) execute each of the essential functions necessary to sustain life. In the process, a protein interacts with wide varieties of micro and macro biomolecules. For example, small non-coding RNAs interact with different proteins to form protein complexes, which bind on DNA and RNA to implement transcription and translation functions. Chapter 5 presents the CA model for a protein referred to as PCAM (protein modelling CA machine).

The building blocks of a protein chain are amino acid residues. CA rule design for micromolecules of nucleic acids, presented in Chap. 2, is based on the analysis of the information content of the atomic structure of nucleotide bases bonded to the sugar–phosphate backbone. A similar procedure is followed in Chap. 2 for CA rule design for amino acids of the peptide chain of a protein. We analysed the information content of each of the residues in respect of their common backbone and the associated side chain, which differ for each amino acid. However, major differences exist between these two biomolecules—nucleic acid and protein, as elaborated next.

There is a limited variation of four bases in nucleic acids, while variation in a protein chain is much wider with twenty widely diverse amino acids. Further, sugar–phosphate backbone of nucleic acid is much stronger compared to that of the peptide backbone chain of a protein. Divergence of such structural characteristics lead to much wider variation of structure and function of a protein compared to that of nucleic acids. Such variations get reflected as a protein interacts with different biomolecules. Consequently, the design of CA rules for amino acids and PCAM (Protein modelling CA Machine), as reported in Chap. 2, significantly deviate from that of nucleic acid molecule.

Section 5.3 reports a brief survey on CA applications in the field of Genomics and Proteomics. The underlying methodology reported in Sect. 5.4 of Chap. 5 for PCAM design is next summarized.

The inputs for the biological process of amino acid synthesis are the 64 codons. The design of three neighbourhood CA (3NCA) rules for 64 codons has been reported in Chap. 2. Detailed analysis of the 64 rules is covered in Sect. 3.2 of Chap. 3 along with the representation of codon degeneracy in the CA model with CA rule equivalence. In this background, the information content of amino acid residue backbone is modelled with 8-bit 3NCA balanced rules. We have employed the 64 3NCA rules designed for codons to represent the backbone of a protein chain. Thus, a polypeptide chain of a protein is modelled with a uniform three neigbourhood CA, each residue of the chain is represented with eight CA cells. Consequently, a protein chain with n residues is represented by a uniform 3NCA with 8n number of cells. In the next step of PCAM design, side chain of amino acids are encoded as 8-bit strings based on the information retrieved from the molecular structure of side chains. The 8-bit pattern assigned to a side chain is used as the initial state (seed) of the eight cells modelling the residue backbone. Provision of 64 different uniform 3NCAs has been kept so that based on the analysis of wet lab results, a particular 3NCA can be selected to model interaction of a protein with a specific biomolecule.

For an input protein chain, we have 64 different PCAMs. Based on the similarity of CA machine evolution patterns, the 64 PCAMs for a protein are grouped into (say) k number of classes. Each of these k classes represents different features of protein interaction with a specific biomolecule. For a protein chain, provision of 64 PCAMs is provided for modelling its interaction with different biomolecules. This interaction modelling of a protein is the major focus of Chap. 5.

Based on the foundation of PCAM design in Sect. 5.4, next two Sects. 5.5 and 5.6 cover the methodology for extracting CA parameters out of PCAM evolution. PCAM for a protein chain evolves for 1000 generation steps. The features are next extracted from the texture analysis of binary image of PCAM evolution. A statistical method of examining texture that considers the spatial relationship of pixels is the Grey-Level Co-Occurrence Matrix (GLCM). The GLCM characterizes the texture of an image by calculating how often pairs of pixel with specific values and in a specified spatial relationship occur in an image.

Section 5.7 of Chap. 5 covers a generic model for protein–protein interaction. The focus of this study is to predict protein 'hotspots' defined as the residues which impede the protein interaction on mutation. To identify the interaction hotspots, we have introduced in silico alanine mutagenesis on each of the residues. The mutagenesis changes the behaviour of PCAM output of the mutant compared to its wild version. We analyse the GLCM parameters reported in Sects. 5.5 and 5.6. Hotspots are predicted from the analysis of the parameters derived out of GLCM analysis of wild and mutant. Section 5.6 reports the application of hotspot prediction methodology for Monoclonal Antibodies (MAbs). Predicted results are validated against the known wet lab experimental results reported in databases and publications.

The last Sect. 5.9 of Chap. 5 reports mutational study on a protein. The target application is to predict the effect of mutation on a protein molecule referred to as PD-L1 (Programmed Death Ligand 1). The PD-L1 is a trans-membrane protein that suppresses immune system. Cancer immunotherapy and the PD-1/PD-L1 checkpoint pathway have attracted major attention due to large increase of cancer patients around the globe. The effect of mutation on PD-L1 for its binding towards two MAbs (KN035 and Atezolizumab) is reported. The predicted results are validated against the wet lab experimental results recently published.

In addition to validation of the results, each of the cases studies identifies which specific PCAM (out of the available 64) is appropriate for modelling the interaction —the major focus of this chapter.

6.4 Major Advantages of cLife

We have presented a comparative analysis of different in silico approaches in Sect. 1.4.9 of Chap. 1 and also in Sect. 5.2 of Chap. 5.

Molecular Dynamic (MD) simulation has been the de facto bottom-up mathematical framework to model interacting forces at atomic/molecular level. MD simulation handles a problem of exponential complexity typically associated with many-body problem. The cost and time associated with MD simulation is high and hence limited to local interactions rather than running comprehensive system level simulations.

The difference between MD simulation and the approach in this book can be summarized with an interesting analogy: atoms and molecules that compose the micromolecules and macromolecules which in turn compose the biological system are comparable to letters and words in a sentence we write to convey our feelings. Individually, a letter or word does not carry the idea contained in the sentence. The letters and words arranged in a specific sequence convey the meaning/feeling expressed in the sentence. They could be arranged and re-arranged in many ways to convey different meanings. The bottom-up approach of MD simulation is equivalent to trying to predict the feeling expressed in a sentence from the semantics of the letters and words used to frame the sentence.

Another relatively new approach is the Artificial Intelligence (AI)/Machine Learning (ML) tools developed in recent years with the availability of high computing power and large scale wet lab and clinical data. The top-down AI/ML approach is in sharp contrast to the historic bottom-up MD approach. Following the same analogy, it may be concluded that the AI/ML driven approach tries to extract semantic meaning of a sentence purely from the sentence itself without any semantic knowledge of letters and words used to frame the sentence. At best, such semantic knowledge is retrieved from the experience of drug designers. Further, to identify a consistent pattern, voluminous data is required and biological data is still atypical for AI/ML, too sparse and incomplete, too biased and too noisy.

The approach 'new kind of computational biology' proposed in the book is designed based on the information (that is, semantics) retrieved from the information content both at the bottom level of micromolecules, and also at the behavioural level information of macromolecules retrieved from wet lab experiments. So, rather than concentrating on study of individual words in a sentence, we start exploring the meaning of a sentence from the semantic knowledge of how the words are normally arranged in a sentence as per a given set of grammatical rules. Subsequent top-down algorithmic modelling maps CA evolution pattern to physical domain features of macromolecules retrieved from wet lab experimental data. Fortunately unlike natural language, the grammatical rules to model the behaviour of micromolecules is limited. It is true that the context under which a biomolecule is analysed can vary significantly. In building the in silico model, we have assumed availability of wet lab experimental data/information on considering such contextual variables.

To summarize, our approach is both bottom-up and top-down, leveraging both their respective advantages and reducing their disadvantages if considered inisolation. It uses the information of the building blocks of the biomolecules, but abstracts the information in a CA model. It is fast, runs on a normal desktop computer and can model a more complex system. For example, interaction of a protein with a macromolecule can be designed with the CA based in silico framework, as illustrated in Chap. 5. We take advantage of the wealth of wet lab experimental data available for top-down validation. But the need of such data is much more restricted and constrained compared to relying only on AI/ML. It just needs enough information to map it to the underlying model based on information retrieved from physical domain.

The underlying basic principle of cLife model design is summarized next.

Based on the information retrieved from the building blocks of the biomolecules, we design the CA rules and CA machines. For top-down validation, we search for wet lab experimental results from databases and research publications. Next, based on the information retrieved from these experimental results followed by signal graph analytics, we design our algorithm, so that CA model parameters get mapped to the physical domain features extracted from experimental results. This cycle of validation may run through a few iterations till the algorithmic steps get finalized. Once the algorithm design is complete, we proceed to develop program code for the algorithm for large-scale experimentation followed by prediction of results for any input. The conventional approaches of access to voluminous data, availability of an extensive training set and subsequent testing set to design in silico experimental setup are not necessary in cLife model.

6.5 Major Hurdles and Future Roadmap

The major hurdles we encountered and the roadmap to address these hurdles in the next phase of development are summarized below.

1. The very first step of top-down validation of results derived out of CA model is initiated with the front-line visualization of CA evolution parameters followed by mapping of such parameters to physical domain features extracted from wet lab experimental data available in databases and published literature. It is essential to reduce this time of manual visualization by taking help of Artificial Intelligence (AI) and other soft computing tools. The initial front-line visualization, no doubt, will enable the designer of the CA model to select the AI technique and soft computing paradigm appropriate for the specific application. Nevertheless, reduction of time for front-line visualization is a necessity and will be undertaken in the future development of cLife.

2. Currently, we have mostly used the wet lab experimental data for top-down validation. In rare cases where the data we are looking for are not available in databases, we have used the molecular dynamic simulation results. Automation of the search process for relevant information from such large databases for top-down validation is going to be a major focus for future development.

3. There are some incompleteness and uncertainty in some of the physical data models. Further, there are instances where contradictory wet lab results are reported in published literature. Similar uncertainty may show up in the predictive models of cLife. Design of a well-defined methodology to iron out uncertainties and contradiction of results derived out of CA model will be addressed in ongoing development.

4. The CA evolution displaying specific pattern/parameters provide an alternative representation of behaviour of the biomolecule under the specific environment of the wet lab experimental setup. This is done in a time space format of signal graph patterns derived out of CAM evolution. Development of this behavioural model of cLife demands extensive experiment and multiple iteration cycles to fine tune cLife to represent life as we see in the current form. Semi-automation of this step will be a major focus for future development of cLife.

5. The results reported in different chapters are derived for normal physical condition of pH value and temperature. However, depending on the environment and prevailing physical condition, variation of these parameter values in a living cell have been reported. The CA models for future cLife will cover methodologies with due consideration to variation of these two physical domain parameters.

6. Chapter 5 presents CA model for protein–protein interaction. The model for interaction of protein with other classes of biomolecules is not covered in the current version of Chap. 5. The cLife model will be extended in future to model —(i) RNA—Protein binding, (ii) DNA—Protein binding and (iii) the model for protein complex associated with binding of small RNA molecule. Development of CA model for binding of a protein complex on functionally important region is a challenging task to be addressed in the future development of cLife.

7. The current version of Chap. 5 does not cover the model for binding of small size drug molecules on a protein. Future extension of cLife will cover this model.

8. A mathematical formalism for analysis of evolution patterns and parameters derived out of different classes of cellular automata machine will be a major focus of the next phase of cLife development.

6.6 Future Extension

In addition to addressing the above issues, we have planned the future development of cLife model as an 'Information' processing platform to address the problems of Life Sciences. Future extension will cover the study of different biomolecules with emphasis on different classes of RNA molecules, human neural system and primitive life form in order to throw some light on the 'origin of life'.

Printed in the United States
By Bookmasters